DATE DUE

NOBEL LECTURES IN CHEMISTRY
1991-1995

NOBEL LECTURES

Including Presentation Speeches
And Laureates' Biographies

PHYSICS

CHEMISTRY

PHYSIOLOGY OR MEDICINE

LITERATURE

PEACE

ECONOMIC SCIENCES

NOBEL LECTURES

INCLUDING PRESENTATION SPEECHES
AND LAUREATES' BIOGRAPHIES

CHEMISTRY

1991–1995

EDITOR

Bo G Malmström

Department of Biochemistry and Biophysics
Göteborg University

World Scientific
Singapore • New Jersey • London • Hong Kong

, NJ 07661

UK office: 57 Shelton Street, Covent Garden, London WC2H 9HE

NOBEL LECTURES IN CHEMISTRY (1991–1995)

ISBN 981-02-2679-9
ISBN 981-02-2680-2 (pbk)

Printed in Singapore by Uto-Print

FOREWORD

Since 1901 the Nobel Foundation has published annually "Les Prix Nobel" with reports from the Nobel award ceremonies in Stockholm and Oslo as well as the biographies and Nobel lectures of the laureates. In order to make the lectures available for people with special interests in the different prize fields the Foundation gave Elsevier Publishing Company the right to publish in English the lectures for 1901–1970, which were published in 1964–1972 through the following volumes:

Physics 1901–1970	4 vols.
Chemistry 1901–1970	4 vols.
Physiology or Medicine 1901–1970	4 vols.
Literature 1901–1967	1 vol.
Peace 1901–1970	3 vols.

Thereafter, and onwards the Nobel Foundation has given World Scientific Publishing Company the right to bring the series up to date and also publish the Prize lectures in Economics from the year 1969. The Nobel Foundation is very pleased that the intellectual and spiritual message to the world laid down in the laureates' lectures, thanks to the efforts of World Scientific, will reach new readers all over the world.

Bengt Samuelsson
Chairman of the Board

Michael Sohlman
Executive Director

Stockholm, October 1996

PREFACE

The Nobel Lectures in all fields, from Physics to Peace, together with the biographical sketches, portraits of the laureates and the presentation speeches, are published collectively each year in *Les Prix Nobel*. The Nobel Foundation has, however, felt a need instead to collect the lectures in each separate field, and the present volume is the third one reproducing the Nobel Lectures in Chemistry, together with the auxiliary material, this time covering the period 1991–1995. It is the hope of the Foundation and the Editor that these volumes will serve as a source of education and inspiration for chemists and other persons interested in chemistry, particularly for the young.

As I mentioned in the preface to an earlier volume, The Royal Swedish Academy of Sciences, and its Nobel Committee for Chemistry, takes a broad view of Chemistry. I also pointed out that the inclusion of Biochemistry, sometimes criticized by representatives of more classical branches of chemistry, has a long tradition, starting with the prize awarded by Eduard Buchner in 1907. This book illustrates beautifully that this tradition has continued in the period 1991–1995, when one prize was explicitly biochemical but others had strong biochemical implications. The simple reason for this is, of course, that Life is essentially a chemical phenomenon.

The award to **R R Ernst** was specifically for his methodological development of the NMR technique, but perhaps the most important applications of Ernst's methods are in the determination of the solution structures of biological macromolecules. Electron-transfer reactions, the theory of which formed the basis of the prize in 1992 to **R A Marcus**, are important in all branches of chemistry and, not least, in the energy metabolism of living organisms, an area which is one of the main interests of Marcus himself. In 1993 the award went to **K B Mullis** and **M Smith** for work on DNA-based chemistry, and their methods have revolutionized the study of structure-function relationships in proteins. The prize to **G A Olah** in 1994 was for fundamental contributions in classical organic chemistry, and the mechanisms of organic reactions are very fundamental to our understanding of life processes. Finally, the award in 1995 to **P Crutzen**, **M J Molina** and **F S Rowland** was for work in atmospheric chemistry, particularly concerning the ozone layer, which is of fundamental importance for life on our planet.

The text of the lectures is taken directly from *Les Prix Nobel*, but some errors have been corrected in consultation with the laureates. In addition, the laureates have been given the opportunity to update their biographical sketches, but not all of them have availed themselves of this possibility.

Bo G. Malmström
Editor

CONTENTS

Chemistry 1991

RICHARD R. ERNST

for his contributions to the development of the methodology of high resolution nuclear magnetic resonance (NMR) spectroscopy

THE NOBEL PRIZE IN CHEMISTRY

Speech by Professor Sture Forsén of the Royal Swedish Academy of Sciences.
Translation from the Swedish text.

Your Majesties, Your Royal Highnesses, Ladies and Gentlemen,

The Nobel Prize in Chemistry for 1991 is being awarded for methodological developments in an important spectroscopic field — nuclear magnetic resonance spectroscopy. Scientists usually refer to this method by its acronym, "NMR," and few would dispute that it is the single most important spectroscopic tool in modern chemistry. NMR spectroscopy allows detailed studies of the structure of small and large molecules in solution and provides unique information about the way molecules move and interact with each other.

The expression "methodological developments" — that is, the emergence of new theoretical or experimental tools or substantial improvements of old ones — merits further comments. A brief review of historical developments in chemistry, or for that matter in natural science as a whole, will convincingly show that methodological developments often have had an enormous and sometimes quite dramatic impact on the progress of science. Consider for example the invention of the microscope. The step from the primitive magnifying glass to devices in which two or more glass lenses were combined, made it possible to observe, in unprecedented detail, a whole new world of small dimensions. The detailed and astonishingly delicate anatomy of tiny insects was revealed. The compartmental, "cellular" structure of all living organisms was discovered. Eventually, microorganisms, yeast cells, bacteria, etc. were discovered — opening up a whole new branch of science, microbiology, and in the end providing a rational explanation for the causes of disease as well as how to treat or prevent them. This example alone will suffice to illustrate the importance of methodological developments in science.

One particular methodological development in which this year's Laureate was a leading figure early in his scientific career was the introduction of Fourier transformation and pulse techniques in NMR spectroscopy, thereby improving the sensitivity of the technique tenfold or even hundredfold. Now I am sure that most of you will shake your head, Fourier transformation and pulse techniques; what is that? Let me try to illustrate it through an analogy. Remember first that spectroscopy is very much concerned with the detection of signals from a sample containing some compound. Assume that you are interested in finding out how well tuned a piano is. The traditional, "old-fashioned", way of doing this would of course be to hit each key in succession and record the frequencies — the signals from our sample if you wish. Now a modern piano usually has 88 keys and it would take some time to go through them one by one, let us say 10 minutes, i.e. 600 seconds. Now there is a much faster way of getting the same results: stretch out both your arms and hit all keys at once, like this [sound effect]. You have now performed a pulse experiment. The result sounds awkward, but remember that all the tones are there in the response. But how could you possibly extract the individual tones from this cacaphony? That you can do by a mathematical analysis called — well you may have guessed it — Fourier transformation. A fast modern computer would perform this analysis in less than a second and the output from your computer would be the individual notes [new sound effect — a scale]. So the new way — the FT way — of checking the tuning of a piano would perhaps take six seconds instead of 600 seconds, a substantial improvement in time. This may sound senseless, why this hurry, even if this new method would allow you to tune 100 pianos in the time it took to tune one in the "old-fashioned" way? But, savings in time can be used in another way, to increase sensitivity. To continue our analogy, assume that you had encountered a piano with "signals" from the strings barely audible above the background noise in the room. Now you could improve the detection of these weak signals by hitting the keys of this same piano 100 times every sixth second and *adding* the result. This would improve the signal-to-noise ratio tenfold, in the jargon of scientists.

When Fourier transform NMR was introduced around 1970 it had a tremendous impact on the applicability of the NMR technique to chemistry. It now became feasible to study very weak signals from small amounts of material or from important elements with magnetic nuclei that are rare in nature, for example ^{13}C and ^{15}N. The Achilles heel of the NMR technique had hitherto been its poor sensitivity, but now *this* obstactle was largely removed.

A later revolutionary development in NMR, in which this year's Chemistry Laureate played a leading role, was the introduction of more than one frequency dimension, 2, 3 or higher. In 2D NMR, the chemical "piano" is hit with pulses of varying lengths and intervals. This allows chemists to extract many parameters of interest from the NMR spectra with great ease. Hopelessly muddled and hard-to-interpret spectra can be spread out and thereby simplify the analysis — much as a two-dimensional map of a landscape would be superior to a mere silhouette. 2D NMR makes it possible to find out which specific atoms in a molecule are closely linked by chemical bonds or which atoms are near each other in space, or which atoms take part in chemical exchange reactions, and much more. A whole new range of experiments have become possible and multidimensional NMR has substantially increased the range of applications to chemistry. The new method has been a prerequisite for the very important applications of NMR in structural biology that have taken place during the past decade.

Professor Ernst,
You have played a leading role in several of the most significant method-ological developments that have taken place in the field of NMR spectros-copy over the past two decades; developments that have had a lasting impact on the way modern chemistry is conducted. You have, in an admirable way, combined excellent experimental know-how with extraordinary theoretical insight. In recognition of your services to chemistry, and to natural science as a whole, the Royal Swedish Academy of Sciences has decided to confer upon you this year's Nobel Prize for Chemistry.

Professor Ernst, I have been granted the privilege of conveying to you the warmest congratulations of the Academy, and I now invite you to receive your Prize from the hands of His Majesty the King.

Richard H. B.

RICHARD R. ERNST

I was born 1933 in Winterthur, Switzerland, where our ancestors resided at least since the 15th century. We lived in a home built in 1898 by my grandfather, a merchant. My father, Robert Ernst, was teaching as an architect at the technical high school of our city. I had the great luck to grow up, together with two sisters, in a town that combined in a unique way artistic and industrious activities. Invaluable art collections and a small but first rank symphony orchestra carry the fame of Winterthur far across the borders of Switzerland. On the other hand, industries producing heavy machinery, like Diesel motors and railway engines, provided the commercial basis of prosperity.

I soon became interested in both sides. Playing the violoncello brought me into numerous chamber and church music ensembles, and stimulated my interest in musical composition that I tried extensively while in high school. At the age of 13, I found in the attic a case filled with chemicals, remainders of an uncle who died in 1923 and was, as a metallurgical engineer, interested in chemistry and photography. I became almost imme-diately fascinated by the possibilities of trying out all conceivable reactions with them, some leading to explosions, others to unbearable poisoning of the air in our house, frightening my parents. However, I survived and started to read all chemistry books that I could get a hand on, first some 19th century books from our home library that did not provide much reliable information, and then I emptied the rather extensive city library. Soon, I knew that I would become a chemist, rather than a composer. I wanted to understand the secrets behind my chemical experiments and behind the processes in nature.

Thus, after finishing high school, I started with high expectations and enthusiasm to study chemistry at the famous Swiss Federal Institute of Technology in Zürich (ETH-Z). I was rapidly disappointed by the state of chemistry in the early fifties as it was taught at ETH-Z; we students had to memorize incountable facts that even the professors did not understand. A good memory not impeccable logic was on demand. The physical chemistry lectures did not reveal much insight either, they were limited just to classical thermodynamics. Thus, I had to continue, similar as in high school, to gain some decent chemical knowledge by reading. A book from which I learned a lot at that time was "Theoretical Chemistry" by S. Glasstone. It revealed to me the fundamentals of quantum mechanics, spectroscopy, statistical me-chanics, and statistical thermodynamics, subjects that were never even mentioned in lectures, except in a voluntary and very excellent lecture course given by the young enthusiastic Professor Hans H.Günthard who had studied chemistry and physics in parallel.

It was clear to me, after my diploma as a "Diplomierter Ingenieur-Chemiker" and some extensive military service, I had to start a PhD thesis in the laboratory of Professor Günthard. Fortunately, he accepted me and associated me with a young most brilliant scientist Hans Primas, who never went through any formal studies but nevertheless acquired rapidly whatever he needed for his work that was then concerned with high resolution nuclear magnetic resonance (NMR), a field in its infancy at that time. Much of his and also my time was spent on designing and building advanced electronic equipment for improved NMR spectrometers. In parallel, we developed the theoretical background for the experiments we had in mind as well as for the optimum performance of the instruments. Signal-to-noise ratio calculations and optimizations were daily routine as NMR suffers from a disappointingly low sensitivity that severely limits its applications. Hans Primas developed and analyzed field modulation techniques, constructed a field frequency lock system, and contributed a new design of shaped pole caps for the electromagnet that was supposed to deliver an extremely homogeneous magnetic field. These developments led to two types of spectrometers that were adopted by Trüb-Täuber, a Swiss electronics company, and sold all over Europe. Later in 1965, Trüb-Täuber was dissolved, and the NMR spectroscopy section led to the foundation of Spectrospin AG that is, together with Bruker Analytische Messtechnik, nowadays the world leading producer of NMR spectrometers.

My own work dealt with the construction of high sensitivity radio frequency preamplifiers and in particular high sensitivity probe assemblies, initially for a 25 MHz, later for a 75 MHz proton resonance spectrometer. On the theoretical side, I was concerned with stochastic resonance. The goal set by Hans Primas was the usage of random noise for the excitation of nuclear magnetic resonance, following the famous concepts of Norbert Wiener for the stochastic testing of non-linear systems. The theoretical treatment was based on a Volterra functional expansion using orthogonal stochastic polynomials. I tried in particular to design a scheme of homonuclear broadband decoupling to simplify proton resonance spectra. By applying a stochastic sequence with a shaped power spectral density that has a hole at the observation frequency, all extraneous protons should be decoupled without perturbing the observed proton spin. The theoretical difficulties were mainly concerned with the computation of the response to non-white noise. Experiments were not attempted at that time, we did not believe in the usefulness of the concept anyway, and I finished my thesis in 1962 with a feeling like an artist balancing on a high rope without any interested spectators.

I thus decided to leave the university forever and tried to find an industrial job in the United States. Among numerous offers, I decided for Varian Associates in Palo Alto where famous scientists, like Weston A.Anderson, Ray Freeman, Jim Hyde, Martin Packard, and Harry Weaver, were working along similar lines as we in Zürich but with a clear commercial goal in mind. This attracted my interest, hoping to find some motivation for my own work. And indeed, I was extremely lucky. Weston Anderson was on his way to invent Fourier transform spectroscopy in order to improve the sensitivity of NMR by parallel data acquisition. After his involvement in the development of a cute mechanical device, the "wheel of fortune", to generate and detect several frequencies in parallel, he proposed to me in 1964 to try a pulse excitation experiment that indeed led to Fourier transform (FT) NMR as we know it today. The first successful experiments were done in summer 1964 while Weston Anderson was abroad on an extensive business trip. In this work I could take advantage in an optimum way of my knowledge in system theory gained during my studies with Primas and Günthard. The response to our invention was however meager. The paper that described our achievements was rejected twice by the Journal of Chemical Physics to be finally accepted and published in the Review of Scientific Instruments. Varian also resisted to build a spectrometer that incorporated the novel Fourier transform concept. It took many years before in the competitive company Bruker Analytische Messtechnik Tony Keller and his coworkers demonstrated in 1969 for the first time a commercial FT NMR spectrometer to the great amazement of Varian that had the patent rights on the invention.

Still at Varian, I was further extending my earlier work on stochastic resonance with the introduction of heteronuclear broadband decoupling by noise irradiation, the "noise decoupling" that led to a rapid development in carbon-13 spectroscopy. It has been replaced later by the much more efficient multiple pulse schemes of Malcolm H. Levitt and Ray Freeman using composite pulses.

Of major importance for the success of more advanced experiments and measurement techniques in NMR was the availability of small laboratory computers that could be hooked up directly to the spectrometer. During my last years at Varian (1966—68), we developed numerous computer applications in spectroscopy for automated experiments and improved data processing.

In 1968 I returned, after an extensive trip through Asia, to Switzerland. A brief visit to Nepal started my insatiable love for Asian art. My main interest is directed towards Tibetan scroll paintings, the so-called thangkas, a unique and most exciting form of religious art with its own strict rules and nevertheless incorporating an incredible exuberance of creativity.

Back in Switzerland, I had a chance to take over the lead of the NMR research group at the Laboratorium für Physikalische Chemie of ETH-Z after Professor Primas turned his interests more towards theoretical chemistry. Despite an initial lack of suitable instrumentation, I continued to work on methodological improvements of time-domain NMR with repetitive pulse experiments and Fourier double resonance. In addition, we performed the first pulsed time-domain chemically-induced dynamic nuclear polarization (CIDNP) experiments. We developed at that time also stochastic resonance as an alternative to pulse FT spectroscopy employing binary pseudo-random noise sequences for broadband excitation, correlating input and output noise. Similar work was done simultaneously by Prof. Reinhold Kaiser at the University of New Brunswick.

The next fortunate event occurred in 1971 when my first graduate student, Thomas Baumann, visited the Ampère Summer School in Basko Polje, Yugoslavia, where Professor Jean Jeener proposed a simple two-pulse sequence that produces, after two-dimensional Fourier transformation, a two-dimensional (2D) spectrum. In the course of time, we recognized the importance and universality of his proposal. In my group, Enrico Bartholdi performed at first some analytical calculations to explore the features of 2D experiments. Finally in the summer of 1974, we tried our first experiments in desperate need of results to be presented at the VIth International Conference on Magnetic Resonance in Biological Systems, Kandersteg, 1974.

At the same time, it occurred to me that the 2D spectroscopy principle could also be applied to NMR imaging, previously proposed by Paul Lauterbur. This led then to the invention of Fourier imaging on which the at present most frequently used spin-warp imaging technique relies. First experiments were done by Anil Kumar and Dieter Welti.

From then on, the development of multi-dimensional spectroscopy went very fast, inside and outside of our research group. Prof. John S. Waugh extended it for applications to solid state resonance, and the research group of Prof. Ray Freeman, particularly Geoffrey Bodenhausen, contributed some of the first heteronuclear experiments. We started 1976 an intense collaboration, lasting for 10 years, with Professor Kurt Wüthrich of ETH-Z to develop applications of 2D spectroscopy in molecular biology. He and his research group have been responsible for most essential innovations that enabled the determination of the three-dimensional structure of biomolecules in solution.

During the following years, a large number of ingenious coworkers, in particular Geoffrey Bodenhausen, Lukas Braunschweiler, Christian Griesinger, Anil Kumar, Malcolm H. Levitt, Slobodan Macura, Luciano Müller, Ole W. Sørensen, and Alexander Wokaun, contributed numerous modifications of the basic 2D spectroscopy concept, such as relay-type coherence transfer, multiple quantum filtering, multiple quantum spectroscopy, total correlation spectroscopy, exclusive correlation spectroscopy, accordion spectroscopy, spy experiments, three-dimensional spectroscopy, and many more. In parallel, numerous other research groups contributed an even larger number of innovative methods.

Besides these activities in high resolution NMR, we always had a research program in solid state NMR going aiming at methodological developments, such as improved 2D spectroscopy techniques and spin diffusion, and applications to particular systems such as one-dimensional organic conductors, polymer blends, and dynamics in hydrogen-bonded carboxylic acids in collaboration with Thomas Baumann, Pablo Caravatti, Federico Graf, Max

Linder, Beat H.Meier, Rolf Meyer, Thierry Schaffhauser, Armin Stöckli, and Dieter Suter.

More recently, I had also the pleasure to closely collaborate with Prof. Arthur Schweiger, an extremely innovative EPR spectroscopist, in the development of pulsed EPR and ENDOR techniques. This turned out to be a specially challenging field due to the inherent experimental difficulties and the many ways to overcome the problems.

In recent years, more and more of my time has become absorbed by administrative work for the research council of ETH-Z of which I am presently the president. I recognized that teaching and research institutions vitally depend on the involvement of active scientists also in management functions.

Looking back, I realize that I have been favored extraordinarily by external circumstances, the proper place at the proper time in terms of my PhD thesis, my first employment in the USA, hearing about Jean Jeener's idea, and in particular having had incredibly brilliant coworkers. At last, I am extremely grateful for the encouragement and for the occasional read-justment of my standards of value by my wife Magdalena who stayed with me so far for more than 28 years despite all the problems of being married to a selfish work-addict with an unpredictable temper. Magdalena has, without much input from my side, educated our three children: Anna Magdalena (kindergarden teacher), Katharina Elisabeth (elementary school teacher), and Hans-Martin Walter (still in high school). I am not surprised that they show no intention to follow in my footsteps, although if I had a second chance myself, I would certainly try to repeat my present career.

NUCLEAR MAGNETIC RESONANCE FOURIER TRANSFORM SPECTROSCOPY

Nobel Lecture, December 9, 1992

by

Richard R. Ernst

Laboratorium für Physikalische Chemie, Eidgenössische Technische Hochschule, ETH-Zentrum 8092 Zürich, Switzerland

The world of the nuclear spins is a true paradise for theoretical and experimental physicists. It supplies, for example, most simple test systems for demonstrating the basic concepts of quantum mechanics and quantum statistics, and numerous textbook-like examples have emerged. On the other hand, the ease of handling nuclear spin systems predestinates them for testing novel experimental concepts. Indeed, the universal procedures of coherent spectroscopy have been developed predominantly within nuclear magnetic resonance (NMR) and have found widespread application in a variety of other fields.

Several key experiments of magnetic resonance have already been honored by physics Nobel prizes, starting with the famous molecular beam experiments by Isidor I. Rabi (1—3) acknowledged in 1944, followed by the classical NMR experiments by Edward M. Purcell (4) and Felix Bloch (5,6), honored with the 1952 prize, and the optical detection schemes by Alfred Kastler (7), leading to a prize in 1966. Some further physics Nobel prize winners have been associated in various ways with magnetic resonance: John H.Van Vleck developed the theory of dia- and paramagnetism and introduced the moment method into NMR, Nicolaas Bloembergen had a major impact on early relaxation theory and measurements; Karl Alex Müller has contributed significantly to electron paramagnetic resonance; Norman F.Ramsey is responsible for the basic theory of chemical shifts and J couplings; and Hans G. Dehmelt has developed pure nuclear quadrupole resonance.

But not only for physicists is nuclear magnetic resonance of great fascination. More and more chemists, biologists, and medical doctors discover NMR, not so much for its conceptual beauty but for its extraordinary usefulness. In this context, a great number of magnetic resonance tools have been invented to enhance the power of NMR in view of a variety of applications (8—15). This Nobel lecture provides a glimpse behind the scene of an NMR toolmaker's workshop.

Nuclear spin systems possess unique properties that predestinate them for molecular studies:

(i)   The nuclear sensors provided by nature are extremely well localized, with a diameter of a few femtometers, and can report on local affairs in their immediate vicinity. It is thus possible to explore molecules and matter in great detail.

(ii)   The interaction energy of the sensors with the environment is extremely small, with less than 0.2 J/mol, corresponding to the thermal energy at 30 mK, and the monitoring of molecular properties is virtually perturbation-free. Nevertheless, the interaction is highly sensitive to the local environment.

(iii)   Geometrical information can be obtained from nuclear pair interactions. Magnetic dipolar interactions provide distance information, while scalar J-coupling interactions allow one to determine dihedral bond angles.

On first sight, it may be astonishing that it is possible to accurately determine internuclear distances by radio frequencies with wavelengths $\lambda \simeq 1$ m, that seemingly violate the quantum mechanical uncertainty relation, $\sigma_q \cdot \sigma_p \geq \hbar/2$, with the linear momentum $p = 2\pi \, \hbar/\lambda$, as it applies to scattering experiments or to a microscope. It is important that in magnetic resonance the geometric information is encoded in the spin Hamiltonian, $\mathcal{H} = \mathcal{H}\,(\mathbf{q}_1, \ldots, \mathbf{q}_k)$, where $\mathbf{q}_k$ are the nuclear coordinates. An accurate geometric measurement, therefore, boils down to an accurate energy measurement that can be made as precise as desired, provided that the observation time t is extended according to $\sigma_E \cdot t \geq \hbar/2$. An upper limit of t is in practice given by the finite lifetime of the energy eigenstates due to relaxation processes. Thus, the accuracy of NMR measurements is not restricted by the wavelength but rather by relaxation-limited lifetimes.

The information content of a nuclear spin Hamiltonian and the associated relaxation superoperator of a large molecule, e.g. a protein, is immense. It is possible to determine the chemical shift frequencies of hundreds of spins in a molecule to an accuracy of 16—18 bits. Internuclear distances for thousands of proton pairs can be measured to about 0.1 Å. Several hundred dihedral angles in a molecule can be determined with an uncertainty of less than 10°.

The weakness of the nuclear spin interactions, so far described as an advantage, leads on the other hand to severe detection problems. Large numbers of spins are required to discriminate the weak signals from noise. Under optimum conditions with modern high field NMR spectrometers, 10^{14}—10^{15} spins of one kind are needed to detect a signal within a performance time of one hour. The low signal-to-noise ratio is the most limiting handicap of NMR. Any increase by technical means will significantly extend the possible range of NMR applications.

This clearly defines the two goals that had to be achieved during the past three decades to promote NMR as a practical tool for molecular structure determination:

(i)   Optimization of the signal-to-noise ratio.

(ii)   Development of procedures to cope with the enormous amount of inherent molecular information.

ONE-DIMENSIONAL FOURIER TRANSFORM SPECTROSCOPY

A major improvement in the signal-to-noise ratio of NMR has been achieved in 1964 by the conception of Fourier transform spectroscopy. The basic principle, parallel data acquisition, leading to the multiplex advantage, was applied already by Michelson in 1891 for optical spectroscopy (16) and explicitly formulated by Fellgett in 1951 (17). However, the approach used in optics, spatial interferometry, is unsuited for NMR where an interferometer with the necessary resolution would require a path length of at least $3 \cdot 10^8$m.

Weston A. Anderson at Varian Associates, Palo Alto, was experimenting in the early sixties with a mechanical multiple frequency generator, the "wheel of fortune" that was conceived to simultaneously excite the spin system with N frequencies in order to shorten the performance time of an experiment by a factor N, recording the response of N spectral elements in parallel (18). It was soon recognized that more elegant solutions were needed for a commercial success.

Numerous possibilities are conceivable for the generation of a broad band frequency source that allows the simultaneous irradiation of an entire spectrum. We mention four schemes: (i) Radio frequency pulse excitation, (ii) stochastic random noise excitation, (iii) rapid scan excitation, (iv) excitation by a computer-synthesized multiple-frequency waveform. For each scheme, a different type of data processing is required to derive the desired NMR spectrum.

The application of radio frequency (rf) pulse excitation was suggested by Weston A. Anderson to the author for a detailed experimental study in 1964 (19—21). The experiment is explained in Fig. 1. To the sample that is polarized in a static magnetic field along the z-axis, an rf pulse is applied along the y-axis. It rotates the magnetization vectors $\mathbf{M}_k$ of all spins I_k by $\pi/2$ into an orientation perpendicular to the static field:

$$M_{kz} \xrightarrow{(\pi/2)_y} M_{kx},   \quad\quad [1]$$

using a convenient arrow notation (23) with the acting operator, here a $(\pi/2)_y$ rotation, on the top of the arrow. The following free induction decay (FID) consists of the superposition of all eigenmodes of the system. An observable operator D is used to detect the signal that is Fourier-transformed for separating the different spectral contributions. Figure 1 contains an early example of Fourier transform spectroscopy using the sample 7-ethoxy-4-methyl-coumarin of which 500 FID's were co-added and Fourier-transformed to produce the Fourier transform spectrum (FT) shown (22). A slow passage continuous wave spectrum (cw), recorded in the same total time of 500 s, is shown also in Fig. 1 for comparison of the signal-to-noise ratios.

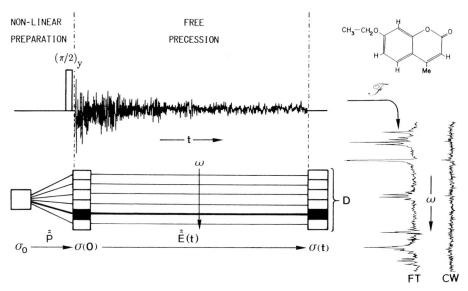

Figure 1. Schematic representation of pulse Fourier transform spectroscopy by the example of 60MHz proton resonance of 7-ethoxy-4-methyl-coumarin (22). An initial $(\pi/2)_y$ rf pulse, represented by the rotation superoperator $\hat{P}$, excites from the equilibrium state σ_0 transverse magnetization $\sigma(0)$. Free precession of all coherences in parallel under the evolution superoperator $\hat{E}(t)$ leads to the final state $\sigma(t)$. Detection with the detection operator D produces the shown FID (sum of 500 scans) which, after Fourier transformation, produces the spectrum FT. For comparison, a continuous wave spectrum CW is shown that has been recorded in the same total time of 500 s under identical conditions.

To please the more mathematically inclined reader, the experiment can also be expressed by the evolution of the density operator $\sigma(t)$ under the preparation superoperator $\hat{P} = \exp\{-i\hat{F}_y\pi/2\}$ and the evolution superoperator $\hat{E}(t) = \exp\{-i\hat{\mathcal{H}} - \hat{\Gamma}t\}$. The superoperator $\hat{F}_y$ is defined by $\hat{F}_y A = [F_y, A]$ with $F_y = \Sigma_k I_{ky}$ where I_{ky} is a component angular momentum operator of spink. $\hat{\mathcal{H}}$ is the Hamiltonian commutator superoperator, $\hat{\mathcal{H}}A = [\mathcal{H}, A]$, and $\hat{\Gamma}$ is the relaxation superoperator. The expectation value $<D>(t)$ of the observable operator D is then given by

$$<D>(t) = \text{Tr}\{D\hat{E}(t)\hat{P}\sigma_0\} \qquad [2]$$

where σ_0 represents the thermal equilibrium density operator of the spin system.

The reduction in performance time is determined by the number of spectral elements N, i.e. the number of significant points in the spectrum, roughly given by $N = F/\Delta f$ where F is the total spectral width and Δf a typical signal line width. A corresponding increase in the signal-to-noise ratio of $\sqrt{N}$ per unit time can be obtained by co-adding an appropriate number of FID signals originating from a repetitive pulse experiment. The signal-to-noise gain can be appreciated from Fig. 1.

It has been known since a long time that the frequency response function (spectrum) of a linear system is the Fourier transform of the impulse response (free induction decay). This was already implicitly evident in the work of Jean Baptiste Joseph Fourier who investigated in 1822 the heat conduction in solid bodies (24). Lowe and Norberg have proved in 1957 this relation also to hold for spin systems despite their strongly nonlinear response characteristics (25).

Stochastic testing of unknown systems by white random noise has been proposed in the 1940s by Norbert Wiener (26). So to say, the color of the output noise carries the spectral information on the investigated system. The first applications of random noise excitation in NMR have been proposed independently by Russel H. Varian (27) and by Hans Primas (28, 29), for broadband testing and for broadband decoupling, respectively. The first successful experiments using random noise irradiation led to heteronuclear "noise decoupling" (30, 31), a method that proved to be essential for the practical success of carbon-13 resonance in chemical applications.

In 1971, Reinhold Kaiser (32) and the author (33) independently demonstrated stochastic resonance as a means to improve the signal-to-noise ratio of NMR by broadband irradiation. Here, the computed cross-correlation function

$$c_1(\tau) = \overline{n_o(t)\, n_i\, (t-\tau)} \tag{3}$$

of the input noise $n_i(t)$ and the output noise $n_o(t)$ is equivalent to the FID of pulse Fourier transform spectroscopy. This is illustrated in Fig.2 for flu-

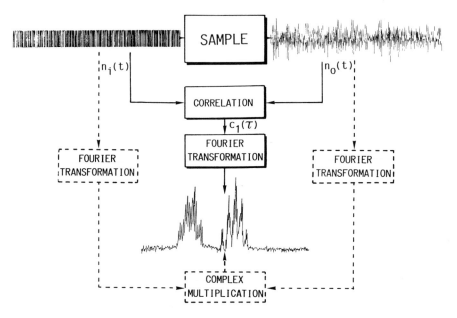

Figure 2. Schematic representation of stochastic resonance by the example of 56.4 MHz fluorine resonance of 2,4-difluorotoluene (33). Excitation with a binary pseudo-random sequence $n_i(t)$ of length 1023 generates the response $n_o(t)$. Cross-correlation of the two signals produces $c_1(\tau)$ which, after Fourier transformation, delivers the shown spectrum. An alternative procedure, that has actually been used in this case, computes the individual Fourier transforms of $n_i(t)$ and $n_o(t)$ and multiplies the complex conjugate $\mathcal{F}\{n_i(t)\}^*$ with $\mathcal{F}\{n_o(t)\}$ to obtain the same spectrum.

orine resonance of 2,4-difluorotoluene. A binary pseudo-random sequence of length 1023 with a maximal white spectrum is used for excitation. Its advantages are the predictable spectral properties and the constant rf power. The low peak power puts less stringent requirements on the electronic equipment. Disadvantages concern the simultaneous irradiation and detection that can lead to line broadening effects which are absent in pulse Fourier transform spectroscopy where perturbation and detection are separated in time. A further disadvantage, when using real random noise, is the probabilistic nature of the response that requires extensive averaging to obtain a stable mean value. Higher order correlation functions, such as

$$c_3(\tau_1, \tau_2, \tau_3) = \overline{n_o(t)\, n_i(t-\tau_1)\, n_i(t-\tau_2)\, n_i(t-\tau_3)}, \qquad [4]$$

allow also the characterization of nonlinear transfer properties of the investigated system (26). This has been exploited extensively by Blümich and Ziessow for NMR measurements (34, 35).

A third approach, rapid scan spectroscopy, initially proposed by Dadok and Sprecher (36), achieves a virtually simultaneous excitation of all spins by a rapid frequency sweep through the spectrum (37, 38). The resulting spectrum is strongly distorted, but can be corrected mathematically because of the deterministic nature of the distortions. Correction amounts to convolution with the signal of a single resonance measured under identical conditions or simulated on a computer. An example is given in Fig.3. It is interesting to note the similarity of a rapid scan spectrum with a FID except for the successively increasing oscillation frequency.

Finally, by computer synthesis, it is possible to compute an excitation function with a virtually arbitrary excitation profile. This has originally been utilized for decoupling purposes by Tomlinson and Hill (39) but is also the

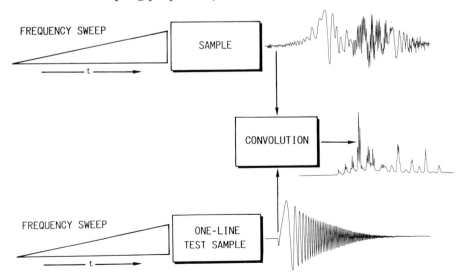

Figure 3. Schematic representation of rapid scan spectroscopy. The strongly distorted sample spectrum obtained by a rapid frequency sweep can be corrected by convolution with the equally sweep-distorted spectrum of a one-line test sample.

basis for composite pulse excitation schemes that have proved to be very powerful (40,41).

Among the broadband excitation techniques, pulse excitation is the only one that allows for a rigorous analytical treatment irrespective of the complexity of the spin system. It does not lead to any method-inflicted line broadening as in stochastic resonance nor to correction-resistant signal distortions as in rapid scan spectroscopy of coupled spin systems (38). Pulse Fourier transform spectroscopy is conceptually and experimentally simple and, last but not least, it can easily be expanded and adapted to virtually all conceivable experimental situations. Relaxation measurements, for example, require just a modified relaxation-sensitive preparation sequence, such as a π—$\pi/2$ pulse pair for T_1 measurements (42) and a $\pi/2$—π pulse pair for T_2 measurements (43). Also the extension to chemical exchange studies using the saturation transfer experiment of Forsén and Hoffman (44) is easily possible.

It should be mentioned at this point that pulsed NMR experiments were suggested already by Felix Bloch in his famous 1946 paper (6), and the first time-domain magnetic resonance experiments have been performed 1949 by H.C. Torrey (45) and, in particular, by Erwin L.Hahn (46—48) who may be regarded as the true father of pulse spectroscopy. He invented the spin echo experiment (46) and devised extremely important and conceptually beautiful solid state experiments (49, 50).

Pulse Fourier transform spectroscopy has not only revolutionized high resolution liquid state NMR spectroscopy, but it has unified NMR methodology across all fields, from solid state resonance, through relaxation measurements, to high resolution NMR, with numerous spill-overs also into other fields such as ion cyclotron resonance (51), microwave spectroscopy (52), and electron paramagnetic resonance (53). In the present context, it provided also the germ for the development of multidimensional NMR spectroscopy.

TWO-DIMENSIONAL FOURIER TRANSFORM SPECTROSCOPY

As long as purely spectroscopic measurements are made, determining the eigenfrequencies or normal modes of a system, one-dimensional spectroscopy is fully adequate. In NMR, this applies to the measurement of the chemical shifts that characterize the local chemical environment of the different nuclei. However, no information can be obtained in this manner on the spatial and topological relations between the observed nuclei.

There are two important pair interactions in nuclear spin systems, the scalar through-bond electron-mediated spin-spin interaction, the so-called J coupling, and the through-space magnetic dipolar interaction. They are illustrated in Fig. 4. The J coupling is represented by the scalar term $\mathscr{H}_{kl} = 2\pi J_{kl} \, \mathbf{I}_k \, \mathbf{I}_l$ in the spin Hamiltonian. It is responsible for the multiplet splittings in high resolution liquid-state spectra. Under suitable conditions, it can lead to an oscillatory transfer of spin order between the two spins I_k

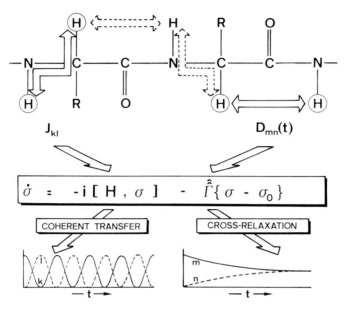

Figure 4. The two pair interactions relevant in NMR. The through-bond scalar J_{kl} coupling contributes to the Hamiltonian and leads to a coherent transfer of spin order between spins I_k and I_l. The time-modulated through-space dipolar interaction $D_{mn}(t)$ causes multiexponential cross relaxation between spins I_m and I_n. The two interactions allow a sequential assignment of the resonances of neighboring spins in the shown peptide fragment and the determination of geometric parameters. The three-bond J coupling is a measure for the dihedral angle about the central bond. The dipolar interaction allows the measurement of internuclear distances.

and I_l. The magnetic dipolar interaction D_{mn}, on the other hand, is represented by a traceless tensor of second rank. Its average in isotropic solution is zero and it can lead to a line splitting only in anisotropic media. However, its time modulation causes also in isotropic solution relaxation processes that are responsible for a multiexponential recovery towards thermal equilibrium among spins after a perturbation. Knowledge of these interactions allows one to deduce geometric information on molecular structure in solution (54, 55) and atomic arrangements in solids. In the optimum case, a complete 3-dimensional structure of a molecule can be deduced (56).

Although these interactions affect 1D spectra, special techniques are needed for their measurement. In the linear response approximation, it is, by first principle, impossible to distinguish between two independent resonances and a doublet caused by a spin-spin interaction. Experiments to explore the nonlinear response properties of nuclear spin systems have been known since the fifties. Saturation studies using strong rf fields yield multiple quantum transitions that contain connectivity information through the simultaneous excitation of several spins belonging to the same coupled spin system (57). Particularly fruitful were double and triple resonance experiments where two or three rf fields are applied simultaneously, resulting in decoupling and spin tickling effects (58—60).

The early multiple resonance experiments have in the meantime been replaced by multi-dimensional experiments. Pair interactions among spins are most conveniently represented in terms of a correlation diagram as shown in Fig. 5. This suggests the recording of a "two-dimensional spectrum" that establishes such a correlation map of the corresponding spectral features. The most straight forward approach may be a systematic double-resonance experiment whose result can be represented as an amplitude $S(\omega_1, \omega_2)$ depending on the frequencies ω_1 and ω_2 of two applied rf fields (8, 58).

A new approach to measure two-dimensional (2D) spectra has been proposed by Jean Jeener at an Ampere Summer School in Basko Polje, Yugoslavia, 1971 (61). He suggested a 2D Fourier transform experiment consisting of two $\pi/2$ pulses with a variable time t_1 between the pulses and the time variable t_2 measuring the time elapsed after the second pulse as shown in Fig. 6 that expands the principles of Fig. 1. Measuring the response $s(t_1, t_2)$ of the two-pulse sequence and Fourier-transformation with respect to both time variables produces a two-dimensional spectrum $S(\omega_1, \omega_2)$ of the desired form (62, 63).

This two-pulse experiment by Jean Jeener is the forefather of a whole class of 2D experiments (8,63) that can also easily be expanded to multi-dimensional spectroscopy. Each 2D experiment, as shown in Figs. 6 and 7, starts with a preparation pulse sequence $\hat{P}$ which excites coherences, i.e. coherent superpositions represented by the density operator $\sigma(0)$, that are allowed to precess for an evolution time t_1 under the evolution superoperator $\hat{E}(t_1)$. During this period, the coherences are so-to-say frequency-labelled. A following mixing sequence $\hat{R}$ performs a controlled transfer of coherence to different nuclear spin transitions that evolve then during the detection period as a function of t_2 under the evolution superoperator $\hat{E}(t_2)$.

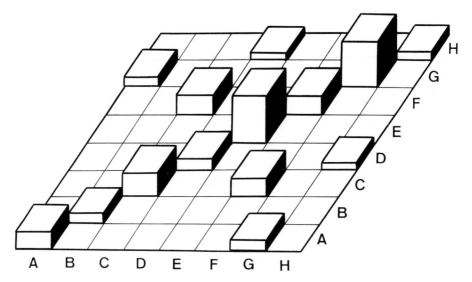

Figure 5. Schematic correlation diagram for the representation of internuclear pair interactions.

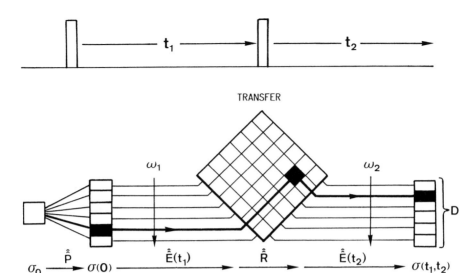

Figure 6. Schematic representation of a 2D experiment, here with a simple two-pulse sequence. The first pulse excites coherences that precess during t_1 and are transfered by the second pulse to different transitions where the coherences continue to precess with a new frequency. The 2D spectrum obtained by a 2D Fourier transformation of $<D>(t_1, t_2)$ provides a visual representation of the transfer matrix R.

Detection is performed with the detection operator D in analogy to Fig.1, leading to the expression

$$<D>(t_1, t_2) = \text{Tr}\left\{ D\hat{\hat{E}}(t_2)\hat{\hat{R}}\hat{\hat{E}}(t_1)\hat{\hat{P}}\sigma_0 \right\}. \tag{5}$$

It is not sufficient to perform a single two-pulse experiment. To obtain the necessary data $<D>(t_1, t_2)$ to compute a 2D spectrum $S(\omega_1, \omega_2)$, it is required to systematically vary t_1 in a series of experiments and to assemble a 2D data matrix that is then Fourier transformed in two dimensions as is indicated schematically in Fig. 7. The resulting 2D spectrum correlates the precession frequencies during the evolution period with the precession frequencies during the detection period. It represents a vivid and easily interpretable representation of the mixing process. Diagonal and cross peaks are measures for the elements of the transfer matrix of the mixing pulse sequence in Fig. 6.

Among the numerous transfer processes that can be represented in this manner, the most important ones are (8)

(i)   the scalar J coupling, leading to "2D correlation spectroscopy", abbreviated COSY,

(ii)   internuclear cross relaxation, leading to "2D nuclear Overhauser effect spectroscopy", abbreviated NOESY, and

(iii)   chemical exchange, leading to "2D exchange spectroscopy", abbreviated EXSY.

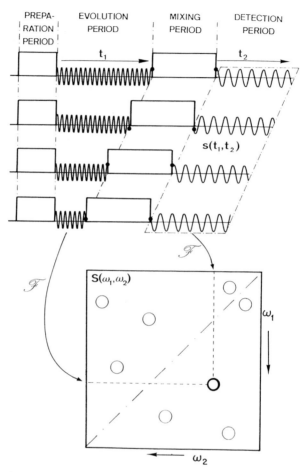

Figure 7. General 2D experiment consisting of a preparation, an evolution, a mixing, and a detection period. The duration t_1 of the evolution period is varied systematically from experiment to experiment. The resulting signal $s(t_1, t_2)_\alpha <D>(t_1, t_2)$ is Fourier-transformed in two dimensions to produce the 2D spectrum $S(\omega_1,\omega_2)$.

The COSY transfer (i), which proceeds through J coupling, is truly a quantum mechanical effect that does not find a satisfactory classical explanation. By means of a single $(\pi/2)_x$ rf mixing pulse, as in Fig.6, it is possible to transfer coherence of spin k, anti-phase with respect to spin l and represented in the density operator by the operator term $2I_{ky}I_{lz}$ into coherence of spin l, anti-phase with respect to spin k, represented by $-2I_{kz}I_{ly}$:

$$2I_{ky}I_{lz} \xrightarrow{\;(\pi/2)_x\;} -2I_{kz}\,I_{ly} \qquad\qquad [6]$$

whereby each factor of the above product spin operator can be considered to be rotated by $\pi/2$ about the x axis. Anti-phase coherence of the type $2I_{ky}\,I_{lz}$ is only formed during the evolution period when there is a direct spin-spin coupling between the spins I_k and I_l:

$$I_{kx} \xrightarrow{\;2\pi J_{kl}I_{kz}I_{lz}t_1\;} I_{kx}\cos(\pi J_{kl}t_1)+2I_{ky}\,I_{lz}\sin(\pi J_{kl}\,t_1). \qquad [7]$$

This implies that in a two-dimensional correlation spectrum there are cross peaks only between directly coupled spins (as long as the weak coupling approximation holds). It is also obvious from Eq.[7] that there is no net coherence transfer, e.g. $I_{kx} \rightarrow I_{lx}$, and the cross-peak integral must disappear, in other words, there is an equal number of cross-peak multiplet lines with positive and negative intensity.

A COSY spectrum, such as the one shown in Fig. 8 for the cyclic decapeptide antamanide (I)

Pro⁸   Phe⁹   Phe¹⁰   Val¹   Pro²

(I)

can be used to find pairs of spins belonging to the same coupling network of an amino acid residue in the molecule. All strong cross peaks arise from two-bond and three-bond couplings that allow, first of all, the assignment of the pairs of NH and CαH backbone protons, as indicated by C in Fig.9 for the six amino acid residues with NH protons. In addition, it is also possible to assign the side-chain protons.

The transfers (ii) and (iii) of NOESY and EXSY experiments involve incoherent, dissipative processes that drive the system back to equilibrium after an initial perturbation in an exponential or multiexponential manner. They require an extended mixing time during which the random processes are given a chance to occur. Both processes can be investigated with the same three-pulse scheme shown in Fig. 10b (8, 64—67). The mixing period is bracketed by two $\pi/2$ pulses that transform coherence into static spin order and back into coherence. The exchange processes transfer the spin order between different spins or between different chemical species, respectively. This type of transfer can be understood based on classical kinetic models. The resulting 2D spectrum represents a kinetic matrix with cross-peak intensities proportional to the exchange-rate constants of pseudo first order reactions.

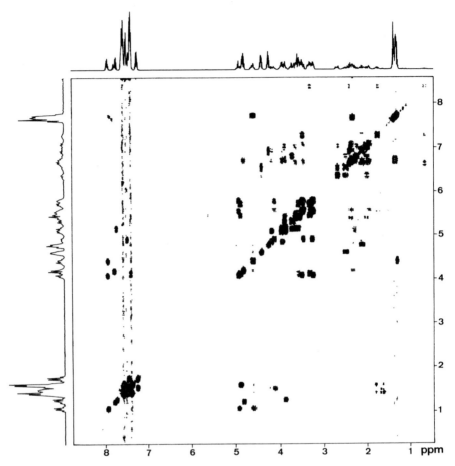

Figure 8. Phase-sensitive 400 MHz proton resonance COSY spectrum of antamanide in chloro-form solution (at 250 K) in a contour-line representation. Positive and negative contours are not distinguished. The spectrum has been recorded by Dr. Martin Blackledge.

For the NOESY transfer, the exchange-rate constants are given by the cross-relaxation rate constants, that are due to magnetic dipolar interaction, and are proportional to $1/r^6_{kl}$ for nuclear pairs I_k and I_l, in addition to a dependence on the correlation time τ_c of the molecular tumbling in solution. The distance dependence can be used to measure relative or, if τ_c is known, absolute distances in molecules. In the course of the assignment process, the NOESY cross peaks allow an identification of spatially neighboring protons in a molecule, important for protons that belong to adjacent amino acid residues in peptides. A NOESY spectrum of antamanide is given in Fig. 11. The cross peaks between sequential backbone protons of adjacent amino acid residues, contained in Fig. 11, are marked in Fig. 9 by N. It is seen that, together with the J-cross peaks from the COSY spectrum of Fig. 8, two unbroken chains of connectivities are found that can be used for the identification of the backbone protons. The two chains are disjoint due to the absence of NH protons in the four proline residues. The general assignment procedure of proton resonance frequencies based on COSY

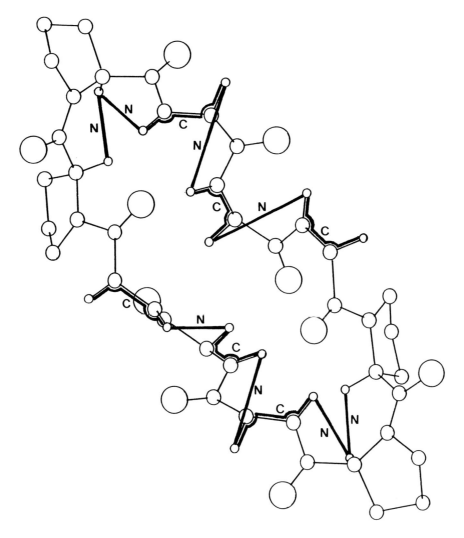

Figure 9. Assignment of backbone protons in antamanide by the combination of COSY (C) and NOESY (N) cross peaks. The missing NH protons in the four proline residues break the chain of sequential C,N connectivities.

and NOESY spectra has been established by Wüthrich and his research group (56).

Based on a complete or partial set of assigned resonances, it is then possible to deduce molecular structural information. Each NOESY cross-peak intensity delivers an internuclear distance that can be used in a manual or computerized process to construct a molecular model that is compatible with the experimental data. In this process it is also possible to employ scalar coupling constants extracted from COSY-type spectra (most conveniently from E.COSY spectra, as mentioned later). According to the Karplus-relations (54), there is an accurate relation between vicinal coupling constants and dihedral angles. Ingenious computer procedures to determine molecular structures based on NMR data have, for the first time, been

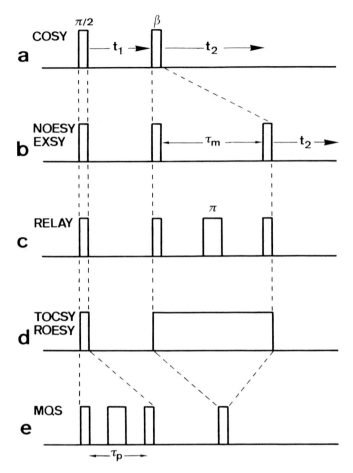

Figure 10. Some of the most useful homonuclear 2D experiments: (a) COSY, (b) NOESY or EXSY, (c) relayed COSY, (d) TOCSY or ROESY in the rotating coordinate system, (e) multiple quantum spectroscopy.

developed by Kurt Wüthrich and his research team and tested on a large number of small to medium-size proteins (56, 68—71). At present, mainly two computer algorithms for the structure determination are in use, the distance geometry algorithm (72, 73) and modifications of it and the re-strained molecular dynamics algorithm (74, 75), again with many variations. The structural problem in antamanide will be discussed later, as it involves intramolecular dynamic processes that complicate the situation.

Cross peaks in a NOESY-type exchange spectrum can also originate from process (iii), i.e. from chemical exchange, and the three-pulse experiment of Fig. 10b is indeed very suited to investigate chemical exchange networks (64, 65, 76). A distinction of the two types of peaks is not possible by inspection of a single 2D spectrum. However, variable temperature studies are often conclusive. At sufficiently low temperature where chemical ex-change becomes slow, only NOESY cross peaks should remain. Another way of distinction are rotating frame experiments as mentioned in the next section.

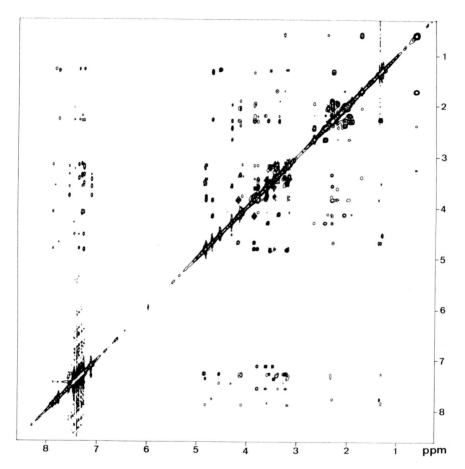

Figure 11. 400 MHz proton resonance NOESY spectrum of antamanide in chloroform (at 250K) in a contour-line representation. The spectrum has been recorded by Dr. Martin Blackledge.

A typical ^{13}C chemical exchange spectrum of a mixture of cis-decalin and trans-decalin is given in Fig. 12. The spectrum demonstrates the well known conformational stability of trans-decalin whereas four pairs of carbon spins of cis-decalin are involved in a conformational exchange process, giving rise to two pairs of cross peaks (76).

MODIFIED TWO-DIMENSIONAL FOURIER EXPERIMENTS

Starting from the two prototype 2D Fourier experiments, an enormous number of modified, expanded, and improved experiments has been suggested. Many of them have found a place in the routine arsenal of the NMR spectroscopist. A first class of experiments, as visualized in Fig. 13, causes extended correlation through two or more transfer steps: Relayed correlation experiments involve two-step correlation and total correlation spectroscopy (TOCSY) achieves multiple step correlation. The latter experiment leads to the important class of rotating frame experiments, including rotating frame Overhauser effect spectroscopy (ROESY) an alternative to

NOESY. Finally, also multiple quantum spectroscopy allows one to investigate connectivity in spin systems. A second class of experiments attempts the simplification of spectra by exclusive correlation (E.COSY), multiple quantum filtering, and spin-topology filtration.

Relayed Correlation

In a standard COSY experiment, coherence is transfered exclusively between two directly coupled spins by means of a single mixing pulse. By a

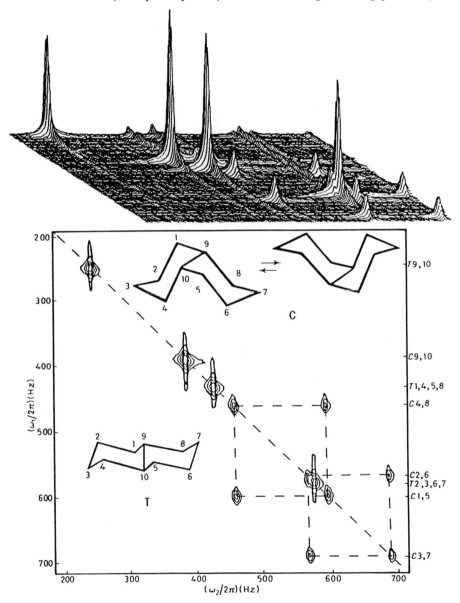

Figure 12. 2D ^{13}C chemical exchange spectrum (EXSY) of a mixture of cis- and trans-decalin recorded at 22.5 MHz and 241 K (76). A stacked plot and a contour representation are given with the assignment of the peaks.

sequence of two $\pi/2$ pulses, as in Fig. 10c, it is possible to effect a transfer across two sequential couplings from spin I_k to spin I_l through the relay spin I_r (77, 78)

$$I_{kz} \xrightarrow{(\pi/2)I_{ky}} I_{kx} \xrightarrow{2\pi J_{kr}I_{kz}I_{rz}t_1} 2I_{ky}I_{rz} \xrightarrow{(\pi/2)(I_{kx}+I_{rx})}$$

$$-2I_{kz}I_{ry} \xrightarrow{2\pi J_{kr}I_{kz}I_{rz}\tau_m + 2\pi J_{rl}I_{rz}I_{lz}\tau_m} 2I_{ry}I_{lz} \xrightarrow{(\pi/2)(I_{rx}+I_{lx})} -2I_{rz}I_{ly} \quad [8]$$

(assuming $J_{kr}t_1 = J_{kr}\tau_m = J_{rl}\tau_m = 1/2$). During the extended mixing period τ_m, it is thus necessary to refocus the anti-phase character of the I_r spin coherences with respect to spin I_k and create anti-phase character with respect to spin I_l to allow for a second transfer by the second mixing pulse. Relayed correlation is useful whenever the resonance of the relay spin I_r cannot unambiguously be identified. With a relay experiment it is then nevertheless possible to assign spins I_k and I_l to the same coupling network (e.g. belonging to the same amino acid residue in a polypeptide chain). It is

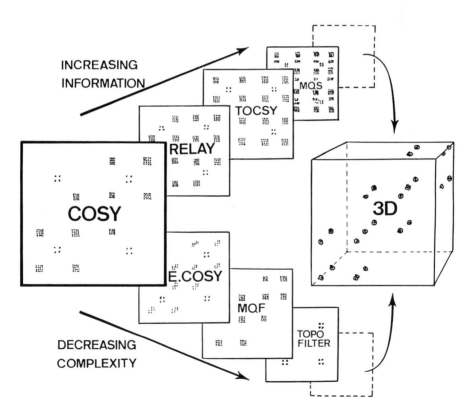

Figure 13. Extensions of the standard COSY experiment. Relayed correlation, total correlation spectroscopy (TOCSY), and multiple quantum spectroscopy (MQS) increase the information content, while exclusive correlation (E.COSY), multiple quantum filtering (MQF), and spin topology filtration reduce the complexity. Both avenues can lead to three-dimensional spectroscopy.

usually of advantage to refocus the effects of the chemical shift precession during the mixing period by incorporating a central π pulse as in Fig.

Relayed coherence transfer is demonstrated by 300 MHz proton resonance spectra of the linear nonapeptide buserilin, pyro-Glu-His-Trp-Ser-Tyr-D-Ser-Leu-Arg-Pro-NHCH$_2$CH$_3$. Figure 14a shows a (double-quantum filtered) COSY spectrum and Fig. 14b the corresponding relayed COSY spectrum (79). In both spectra, the resonance connectivities for the residue leucine are marked. It is evident that in the COSY spectrum only nearest neighbor protons are connected by cross peaks: NH—C$_\alpha$H, C$_\alpha$H—C$_\beta$H^{1,2}, C$_\beta$H^{1,2}—C$_\gamma$H, and C$_\gamma$H—(C$_\delta$H$_3$)^{1,2}. On the other hand in the relayed COSY spectrum, also the next nearest neighbors NH—C$_\beta$H^{1,2} and C$_\beta$H^{1,2}—(C$_\delta$H$_3$)^{1,2} are connected. The third pair of relayed cross peaks C$_\alpha$H—C$_\gamma$H is weak due to the high multiplicity of the C$_\gamma$H resonance and not visible in the contour representation of Fig. 14b. Similar relayed cross peaks can be found for the other amino acid residues.

Rotating Frame Experiments

By means of an extended mixing pulse sequence, transfer of coherence over an arbitrary number of steps is in principle possible. In particular, continuous wave irradiation leads to the mixing of all eigenmodes of a spin system and correspondingly to transfers of coherence between all of them. This is exploited in total correlation spectroscopy (TOCSY) with the sequence of Fig. 10d. All spins belonging to the same J-coupling network can be identified with TOCSY (80,81). The accurate matching of the precession frequencies of the various spins in the presence of a radio frequency field is crucial to enable an efficient transfer of coherence. Either very strong radio frequency fields or specially designed pulse sequences are needed for this purpose (81). Coherence transfer is possible when the effective average magnetic field strengths B_k^{eff} in the rotating frame are equal within a J-coupling constant, $|\gamma(B_k^{eff} - B_l^{eff})| < |2 \pi J_{kl}|$, corresponding to a strong coupling case in the rotating frame.

The TOCSY experiment is of interest for assigning proton resonances to individual amino acid residues in a protein. Of particular value is that its transfer rate is enhanced by a factor 2 in comparison to COSY or relayed transfer experiments in the laboratory frame (80). Another property is that, due to the presence of a radio-frequency field, in-phase coherence transfer of the type

$$I_{kx} \xrightarrow{\quad 2\pi J_{kl} \, I_k \, I_l \tau_m \quad} I_{lx} \qquad [9]$$

is possible, leading to in-phase cross-peak multiplet structures.

A TOCSY spectrum of buserilin is included in Fig. 14c for comparison with the relayed and standard COSY spectra given. Here also three-step transfers C$_\alpha$H—(C$_\delta$H$_3$)^{1,2} and even four-step transfers NH—(C$_\delta$H$_3$)^{1,2} are visible. Again, some expected cross peaks involving C$_\gamma$H are missing because of the extensive multiplet structure of C$_\gamma$H.

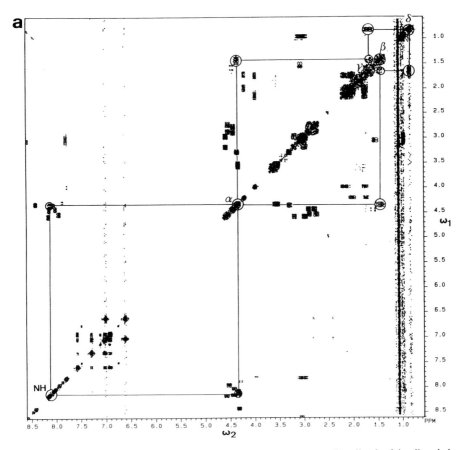

Figure 14. 300 MHz correlation spectra of the nonapeptide buserilin dissolved in dimethyl sulfoxide. Phase-sensitive plots with equal representation of positive and negative contours are shown. The resonance connectivities are indicated for leucine (79). (a) Double quantum-filtered COSY spectrum using the sequence of Fig. 18. (b) Relayed COSY spectrum using the sequence of Fig. 10c with τ_m = 25 ms. (c) TOCSY spectrum using the sequence of Fig. 10d with τ_m = 112 ms and an MLEV-17 pulse sequence applied during τ_m.

The elimination of the chemical shift precession by the rf irradiation leads, in addition to the coherent transfer through the J-coupling network, also to an incoherent transfer of spin order through transverse cross relaxation. The transverse cross-relaxation terms are, in principle, always present. However, strong differential chemical shift precession of spin pairs causes normally a quenching of the transfer in the sense of first order perturbation theory. In the presence of a strong rf field, this quenching is no longer operative and transverse cross relaxation occurs. This is the transfer mechanism of the rotating frame Overhauser effect spectroscopy (ROESY) experiment (82).

ROESY has similar properties as NOESY but differs in the dependence of the cross-relaxation rate constant Γ_{kl} on the correlation time τ_c of the molecular rotational motion that modulates the internuclear dipolar interaction, responsible for cross relaxation:

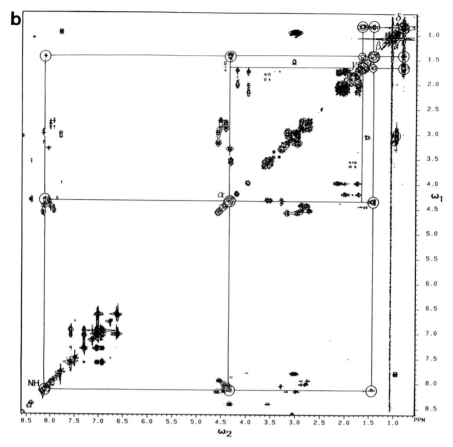

Figure 14 b.

$$\Gamma_{k\iota}^{\text{NOE}} = \frac{\gamma^4 \, \hbar^2}{10 \, r_{k\iota}^6} \left(\frac{\mu_0}{4\pi} \right)^2 \quad [- \frac{1}{2} J(0) + 3J(2\omega_0)], \tag{10}$$

$$\Gamma_{k\iota}^{\text{ROE}} = \frac{\gamma^4 \, \hbar^2}{10 \, r_{k\iota}^6} \left(\frac{\mu_0}{4\pi} \right)^2 \quad [J(0) + \frac{3}{2} J(\omega_0)], \tag{11}$$

with the spectral density

$$J(\omega) = 2\tau_c / (1 + (\omega\tau_c)^2), \tag{12}$$

ω_0 is as usual the Larmor frequency of the two nuclei with the internuclear distance $r_{k\iota}$. This implies that $\Gamma_{k\iota}^{\text{NOE}}$ changes sign for an intermediate correlation time $\tau_c = (5/4)^{1/2}/\omega_0$, whereby the cross-relaxation rate constant becomes small in the neighborhood of this condition. Depending on the viscosity of the solvent and the resonance frequency ω_0, this occurs for globular molecules within a range of molecular mass of 500—2000 Dalton. Γ_{kl}^{ROE}, on the other hand, is less sensitive to τ_c and remains positive for any molecular mass. The ROESY experiment is therefore of advantage for molecules of intermediate size.

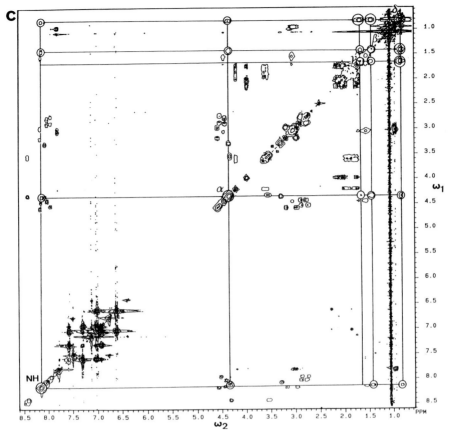

Figure 14 c.

The different sensitivity of NOE and ROE on τ_c allows one, in addition, to deduce information on intramolecular mobility by comparison of the two measurements (83). An advantage of ROESY in comparison to NOESY is the negative cross-peak amplitude for ROESY while the simultaneously occurring chemical exchange cross peaks are positive and allow for an easy distinction unless they overlap.

It should be recognized that in the rotating frame coherence transfer through J couplings and cross relaxation occur simultaneously, TOCSY cross peaks being positive while ROESY cross peaks appear with negative amplitude. This complicates the 2D spectra and calls for separation procedures. The suppression of the coherent transfer through J couplings (TOCSY) is easy as it is just necessary to mismatch the condition $|\gamma(B_k^{\text{eff}} - B_l^{\text{eff}})| < |2\pi J_{kl}|$, for example by a slight frequency offset in the presence of not too strong rf fields. The cross-relaxation rates are much less sensitive to such a mismatch such that a clean ROESY spectrum results.

To obtain a clean TOCSY spectrum is more demanding as relaxation cannot easily be manipulated. A "clean TOCSY" technique has been proposed by C. Griesinger (84). It relies on a combination of Eqs. [10] and [11] to cause a vanishing average cross-relaxation rate constant:

$$\bar{\Gamma}_{kl} = p \, \Gamma^{NOE}_{kl} + (1-p) \, \Gamma^{ROE}_{kl} \stackrel{!}{=} 0. \qquad [13]$$

A suitable weight p can be found whenever $\Gamma^{NOE}_{kl} < 0$, i.e. for sufficiently large molecules with $\tau_c > (5/4)^{1/2}/\omega_0$. This requires the magnetization to move on a trajectory that spends a fractional time p along the z axis and a fraction $(1-p)$ in the transverse plane. For $\tau_c \to \infty$, one finds $p = 2/3$ for $\bar{\Gamma}_{kl} = 0$. A suitable pulse sequence, modifying an MLEV-17 spin locking sequence, has been proposed in Ref. 84. Another optimized sequence, called 'clean CITY', has been developed by J. Briand (85). A clean TOCSY spectrum of basic pancreatic trypsin inhibitor (BPTI) using the clean CITY sequence is compared in Fig. 15 with a conventional TOCSY spectrum to demonstrate the efficient suppression of the (negative) ROESY peaks.

Multiple Quantum Spectroscopy
In spectroscopy, in general, only those transitions are directly observable for which the observable operator has matrix elements different from zero, leading to the so-called allowed transitions. In high field magnetic resonance with weak continuous wave perturbation or observing the free induction decay in the absence of rf, the transverse magnetization observable operator $F_x = \sum_k I_{kx}$ has matrix elements only between eigenstates of the Hamiltonian differing in the magnetic quantum number M by ± 1. Thus, single quantum transitions are the allowed transitions, multiple quantum transitions with $|\Delta M| > 1$ being forbidden. Multiple quantum transitions can however be induced by strong continuous wave rf fields that cause a mixing of states (8, 57) or by a sequence of at least two rf pulses (8, 63, 86, 87). Observation is possible again in the presence of a strong rf field (8, 57) or after a further detection pulse (8, 63, 86, 87).

For spin $I = 1/2$ systems, multiple quantum transitions invariably involve several spins, and multiple quantum spectra contain information on the connectivity of spins within the J-coupling network in analogy to 2D correlation spectra. In particular, the highest order transition allows one to determine the number of coupled spins. Relaxation rate constants of multiple quantum coherences are dependent on the correlation of the random perturbations affecting the involved spins and deliver information on motional processes (88).

A simple instructive example of a 2D double quantum spectrum is given in Fig. 16 to demonstrate the use of multiple quantum transitions for the assignment of resonances (89). Along ω_1, double quantum transitions and along ω_2 single quantum transitions are displayed for the six-spin system of 3-aminopropanol-d$_3$ (DOCH$_2$CH$_2$CH$_2$ND$_2$). In general, there are three different categories of double quantum transitions:

(I)   Double quantum transitions involving two directly coupled spins. They lead to pairs of cross peaks displaced symmetrically from the double quantum diagonal ($\omega_1 = 2\omega_2$) with ω_2 coordinates corresponding to the Larmor frequencies of the two spins (e.g. $\omega_1 = \Omega_A + \Omega_M$, $\Omega_M + \Omega_X$).

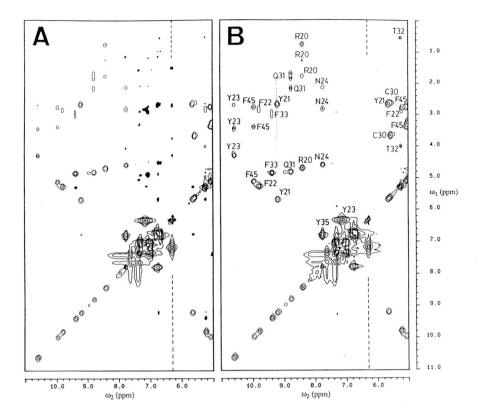

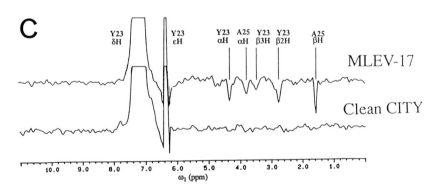

Figure 15. Phase-sensitive 300 MHz ^{1}H TOCSY spectra of 15 mM bovine pancreatic trypsin inhibitor in D_2O recorded with a 69ms mixing time (85). (a) Mixing process with MLEV-17 pulse sequence. Negative peaks are shown by contours filled in black. (b) Mixing process with clean CITY pulse sequence. (c) Cross sections along ω_1 through the Tyr^{23} εH diagonal peak at 6.33ppm in the two spectra (a) and (b) (see broken lines).

(II) Double quantum transition involving two magnetically equivalent spins. They lead to one or more cross peaks at an ω_1 frequency that intersects the double quantum diagonal at the ω_2 frequency corresponding to the common Larmor frequency of the two spins (e.g. $\omega_1 = 2\,\Omega_A$, $2\,\Omega_M$, $2\,\Omega_x$, although the spins are only within experimental accuracy magnetically equivalent).

(III) Double quantum transitions involving two remotely coupled spins. They lead to single cross peaks at an ω_1 frequency that intersects the double quantum diagonal at ω_2 equal to the mean of the two Larmor frequencies (e.g. $\omega_1 = \Omega_A + \Omega_x$). These cross peaks carry information identical to that in relayed correlation spectra.

For the practical application it is essential that a multiple quantum spectrum never contains a strong diagonal peak array. It should be mentioned that a

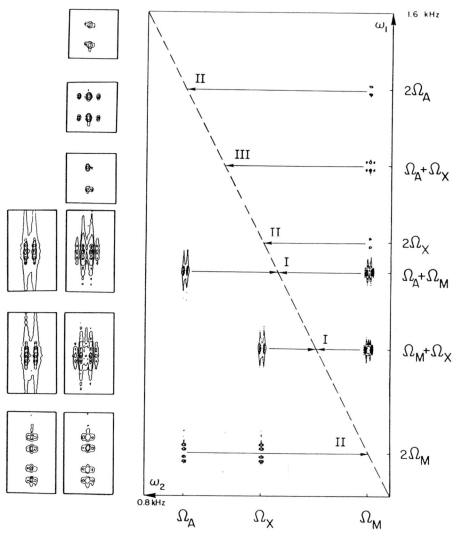

Figure 16. 90 MHz 2D correlation spectrum of 3-amino-propanol-d$_3$ with double quantum transitions along ω_1 and single quantum transitions along ω_2. The three categories I, II, and III of double quantum transitions mentioned in the text are indicated. Blow-ups of all cross peaks are shown on the left. The spectrum is shown in an absolute value representation (from Ref.89).

beautiful and useful form of a double quantum experiment is 2D INAD-
EQUATE spectroscopy proposed by Bax, Freeman and Kempsell (90,91).
There, only type I peaks can occur.

The methods mentioned so far produce additional cross peaks that provide information not accessible with the standard COSY and NOESY experiments. In the following, techniques are discussed that lead to simplified
spectra which may facilitate their interpretation.

Multiple Quantum Filtering

A selective filtering effect can be achieved by exciting intermediately multiple quantum coherence, selecting a particular quantum order, and recon-
verting the selected order into observable magnetization. Depending on the
selected order, this leads to multiple-quantum filtering of various orders.
The spin-system-selective effect relies on coherence transfer selection rules
that limit the allowed transfers for weakly coupled spins (8,92):

(i)   It is impossible to excite p-quantum coherence in spin systems with
less than p coupled spins $I = 1/2$.

(ii)   For the appearance of a diagonal peak of spin I_k in a p-quantum
filtered COSY spectrum, the spin I_k must be directly coupled to at
least $p-1$ further spins.

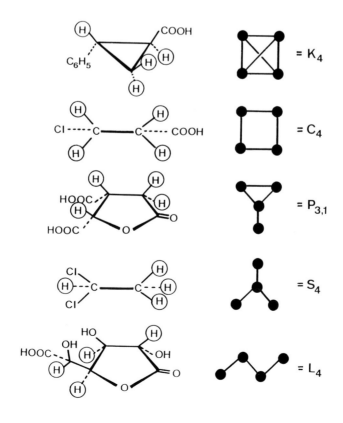

(iii)   For the appearance of the cross peaks between spins I_k and I_l in a p-quantum filtered COSY spectrum, both spins must simultaneously be coupled to at least p-2 further spins.

Violations of these coherence transfer selection rules occur for strong coupling and for certain special relaxation situations (93).

In Fig. 17, the effect of 4-quantum filtering on various four-spin systems is demonstrated. The sample consists of the five molecules trans-phenyl-cyclopropanecarboxylic acid (K_4), DL-isocitric acid-lactone ($P_{3,1}$), 1,1-dichloroethane (S_4), 2-chloropropionic acid (C_4), and D-saccharic acid-1,4-lactone (L_4) with the coupling topologies shown on the previous page (94).

Figure 17a gives a conventional (double-quantum-filtered) COSY spectrum of the mixture while in Fig. 17b the corresponding 4-quantum filtered spectrum is reproduced. The filtering effect can easily be understood based on the given rules and the shown coupling topologies. The interpretation is left to the reader. Only cross peaks of the molecule with K_4 topology and diagonal peaks of molecules with $P_{3,1}$, S_4, and K_4 topology remain.

Technically, multiple quantum filtering exploits the characteristic dependence of a multiple quantum coherence transfer on the rf phase of the acting pulse sequence (8,92,95,96). Let us assume a transfer of coherence $c_{p1}(t)$ by a unitary transformation U(0), representing a particular pulse sequence, to coherence $c_{p2}(t)$, where p_1 and p_2 are the orders of coherence,

$$c_{p1}(t) \xrightarrow{U(0)} c_{p2}(t). \qquad [14]$$

All rf pulses in the sequence shall now be phase-shifted by Φ, leading to $U(\Phi)$. Then it can be shown that the resulting coherence $c_{p2}(t)$ is phase-shifted by $(p_2 - p_1)\,\Phi$

$$c_{p1}(t) \xrightarrow{U(\Phi)} c_{p2}(t)\,\exp\{i(p_2 - p_1)\,\Phi\}. \qquad [15]$$

The phase shift is therefore proportional to the change in coherence order $\Delta p = p_2 - p_1$. After performing a series of experiments with the phase Φ incremented in regular intervals $2\pi/N$ from 0 to $2\pi(N-1)/N$, it is possible to select for a particular Δp by computing the corresponding Fourier coefficient of Δp: Let $s(t,\Phi)$ be the recorded signal of an experiment with phase shift Φ, then we obtain the filtered signal

$$s(t, \Delta p) = \sum_{k=0}^{N-1} s(t, 2\pi k/N)\,\exp(-i2\pi k\Delta p/N). \qquad [16]$$

The required number of increments N of the phase Φ depends on the number of Δp values that have to be discriminated (96). It is obvious that unless the initial order of coherence p_1 is known, no particular order of coherence p_2 can be filtered out in this manner. Most conveniently the initial state is selected to be in thermal equilibrium with $p_1 = 0$. Then, the entire pulse sequence preceding the point at which a coherence order

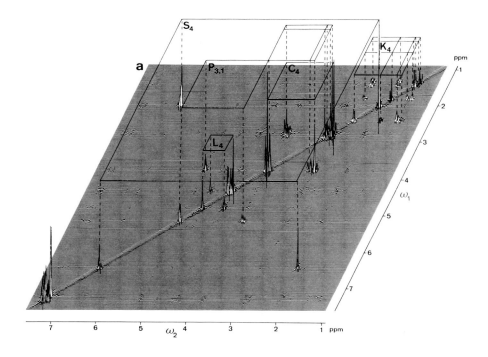

Figure 17. Multiple quantum-filtered and spin-topology-filtered 300 MHz COSY spectra of a mixture of the four-spin systems trans-phenyl cyclopropanecarboxylic acid (K$_4$), DL-isocitric acid-lactone (P$_{3,1}$), 1,1-dichloroethane (S$_4$), 2-chloro-propionic acid (C$_4$), and D-saccharic acid-1,4-lactone (L$_4$). (a) Double-quantum-filtered spectrum using the pulse sequence of Fig. 18. (b) 4-quantum-filtered spectrum using the pulse sequence of Fig. 18. (c) C$_4$ spin-topology-filtered spectrum using the pulse sequence of Fig. 19 (from Ref. 94).

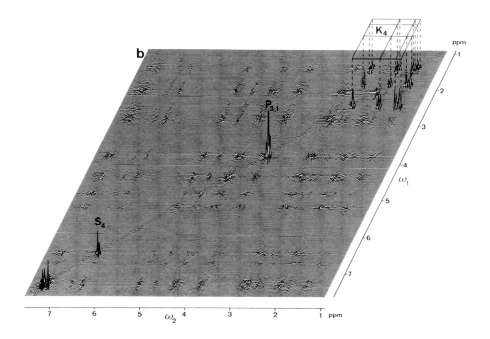

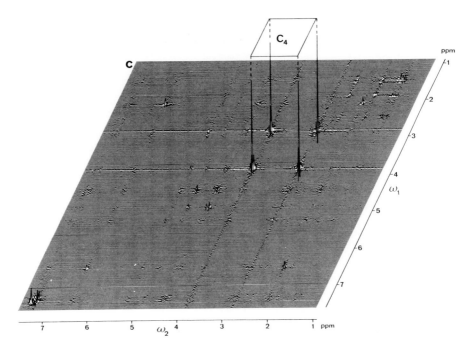

should be selected must be phase-cycled. For multiple-quantum filtered COSY, this leads to the pulse sequence shown in Fig. 18.

Obviously, multiple quantum filtering and phase-cycling require N-times more experiments to be performed. However, no information is lost as in each term of Eq. [16] just the phase factor is compensated and identical signals are co-added for the relevant pathways. Thus the longer perform-ance time is refunded in terms of an increased signal-to-noise ratio.

Spin Topology Filtration

It may be desirable to enhance the filtering effect demonstrated in Fig. 17 and to select individual spin coupling topologies. Indeed it is possible to design extended pulse sequences, in combination with multiple quantum filtration, that are tailor-made for specific spin coupling topologies (94, 97, 98). A pulse sequence, built into a 2D COSY experiment, that is selective for cyclic C_4 spin coupling topologies is shown in Fig. 19. Applied to the previous mixture of four-spin systems, the 2D spectrum of Fig. 17c is obtained. It shows efficient suppression of all other spin systems. It should however be noted that the situation is here rather ideal. Often, these filters do not perform as well because their design relies on the equality of all non-zero spin couplings. In reality, there are weak and strong couplings that cannot be characterized by topological considerations alone. Often also the signals decay during the extended pulse sequences due to relaxation. This limits the practical usefulness of these designs.

Exclusive Correlation Spectroscopy

Multiple quantum filtering suppresses not only diagonal and cross peaks in 2D spectra but also changes the sign pattern in the cross-peak multiplet structure. By appropriate combination of differently multiple-quantum filtered 2D spectra, it is possible to simplify the multiplet structure by reducing the number of multiplet components. The recipe of exclusive correlation spectroscopy (E.COSY), proposed by O.W. Sørensen, eliminates all multiplet components from a COSY spectrum except for those belonging to pairs of transitions with an energy level in common (99—101). In practice, it is not necessary to literally combine multiple-quantum-filtered spectra but it is possible to directly co-add the experimental results from a phase cycle with the appropriate weight factors.

Figure 20 shows schematically the combination of cross-peak multiplets connecting spins I_1 and I_2 in a three-spin system after 2- and 3-quantum filtering. The remaining pattern consists of two basic squares with side lengths equal to the active coupling constant J_{12} responsible for the coherence transfer. The displacement vector between the two squares is given by the two passive couplings J_{13} and J_{23} to the third (passive) spin. It should be mentioned that this multiplet structure is identical to the one obtained by a COSY experiment with an extremely small flip angle of the mixing pulse (102).

E.COSY is of practical use whenever the cross-peak multiplet structure has to be analyzed in order to determine J-coupling constants. This can be done conveniently by hand by measuring the displacement of peripheral multiplet components (101) or by an automatic recursive contraction procedure (103).

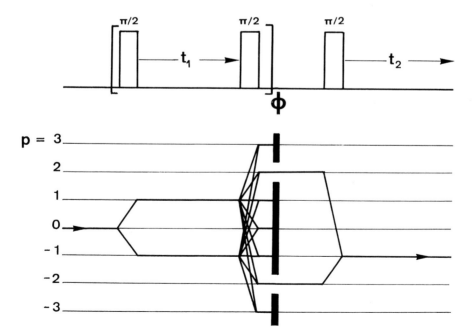

Figure 18. Pulse sequence for multiple-quantum-filtered COSY with the coherence transfer diagram for double-quantum filtering. The phase Φ is incremented systematically in a set of N experiments and the resulting experimental results combined according to Eq. [16].

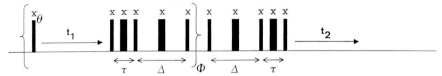

Figure 19. Pulse sequence for C_4 spin topology filtration consisting of $\pi/2$ and π pulses. The delays are adjusted to $\tau = 1/(8J)$ and $\Delta = 1/(2J)$ where J is the uniform J-coupling constant. Φ is phase-cycled for four-quantum selection and Θ for the suppression of axial peaks (94).

Heteronuclear 2D Experiments

In addition to the homonuclear 2D experiments discussed so far, at least an equal number of heteronuclear experiments has been proposed and intro-duced in the routine spectroscopy laboratory. Of greatest practical impor-tance are heteronuclear shift correlation spectra that correlate the chemical shifts of directly bonded or remotely connected heteronuclei (104, 105). In this context, so-called inverse detection experiments where proton I-spin coherence is observed in t_2 while low-abundance, low sensitivity S-spin coherence is evolving in t_1, are of particular interest (104). The most efficient schemes create heteronuclear two-spin coherence that evolves in t_1 and that acquires the frequency information of the S-spin resonance (106). Also in the heteronuclear environment, relayed coherence transfer is of importance (78) as well as experiments in the rotating frame (107). Spin filtering is used in the form of multiplicity selectivity, distinguishing S spins coupled to one, two, or three I spins (108), and in the form of J filtering for the distinction of one-bond and multiple-bond couplings (109). This enu-meration of heteronuclear experiments is by no means exhaustive.

THREE-DIMENSIONAL FOURIER SPECTROSCOPY

No new principles are required to develop 3D spectroscopy that is just a logical extension of 2D spectroscopy. Instead of a single mixing process which relates two frequency variables, two sequential mixing processes relate three frequencies: the origin frequency ω_1, the relay frequency ω_2, and the detection frequency ω_3, as shown in Fig.21. In this sense, a 3D experiment can be considered as the combination of two 2D experiments. Obviously, a very large number of possible 3D experiments can be con-ceived. However, only few of them have so far proved to be indispensible (110—118).

Two applications of the 3D spectroscopy concept have emerged: (i) 3D correlation and (ii) 3D dispersion (see also Fig.13). Three-dimensional correlation is of importance in homonuclear experiments. It has been mentioned that the assignment procedure in biomolecules requires a COSY-type and a NOESY-type 2D spectrum. The two 2D experiments could be contracted into one 3D experiment, combining a J-coupling-mediated and a cross-relaxation transfer. A 3D COSY-NOESY spectrum possesses the advantage that the entire assignment process can be carried

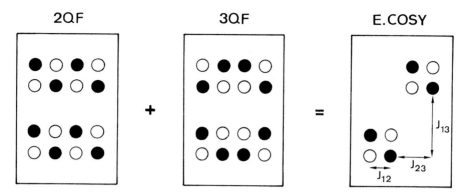

Figure 20. E.COSY experiment to simplify the multiplet structure of cross peaks. The double-quantum- and the triple-quantum-filtered cross peak between spins I_1 and I_2 of a three-spin system are combined to produce an E.COSY pattern. Positive and negative multiplet components are distinguished by empty and filled circles.

out with a single homogeneous data set (115, 116). It incorporates also redundancies that allow cross checks of the assignments. For obtaining quantitative information, however, 3D spectra are less suited as all peak intensities are products of two transfer coefficients that are some times difficult to separate.

A 3D ROESY-TOCSY spectrum of the linear nonapeptide buserilin is shown in Fig.22 (116). A ROESY instead of a NOESY step is required for buserilin, being a molecule of intermediate size where the NOE intensities are small. The TOCSY step has the advantage that chains of multiple-step cross peaks extending into the side chains are obtained that facilitate the identification of the amino acid residues.

It should be recognized that recording a 3D spectrum is considerably more time-consuming than two 2D spectra as two time parameters t_1 and t_2 have to be incremented independently, leading to a 2D array of experiments. Here the question arises; when is it worth the effort to record a 3D spectrum? This question has been discussed before (116, 119, 120).

Let us consider a particular cross peak in a 3D spectrum that correlates the coherences $\{tu\}$ in ω_1, $\{rs\}$ in ω_2, and $\{pq\}$ in ω_3 dimensions. Its intensity is determined by the following product of matrix elements (in the eigenbasis of the unperturbed Hamiltonian $\mathscr{H}_0$) (116):

$$Z_{\{pq\}\,\{rs\}\,\{tu\}} = D_{qp}\, R^{(2)}_{\{pq\}\,\{rs\}}\, R^{(1)}_{\{rs\}\,\{tu\}}\, (\hat{\hat{P}}\sigma_0)_{tu}. \qquad [17]$$

A non-vanishing intensity establishes a two-step correlation $\{tu\}$—$\{rs\}$—$\{pq\}$.

The 3D experiment can be compared with two 2D experiments that employ the mixing processes $\hat{\hat{R}}^{(1)}$ and $\hat{\hat{R}}^{(2)}$, respectively. The corresponding intensities would be

$$Z^{(1)}_{\{rs\}\,\{tu\}} = D^{(1)}_{sr} R^{(1)}_{\{rs\}\,\{tu\}}\, (\hat{\hat{P}}\sigma_0)_{tu}. \qquad [18]$$

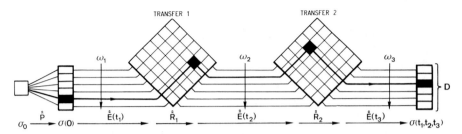

Figure 21. Schematic representation of a 3D experiment, extending Figs. 1 and 6. Three evolution periods with the time variables t_1, t_2, and t_3 are separated by two transfer or mixing processes. A 3D experiment can be conceived as the contraction of two 2D experiments.

$$\text{and } Z^{(2)}_{\{pq\} \{rs\}} = D_{qp} \, R^{(2)}_{\{pq\} \{rs\}} \, (\hat{P}^{(2)}\sigma_0)_{\, rs}. \tag{19}$$

When in the 2D spectra the two relevant peaks with intensities $Z^{(1)}_{\{rs\} \{tu\}}$ and $Z^{(2)}_{\{pq\} \{rs\}}$ can be identified, possibly in crowded regions, the two-step correlation, represented by a 3D peak, could also be established based on the two 2D spectra $\{tu\}$—$\{rs\}$ and $\{rs\}$—$\{pq\}$. Provided that $Z_{\{pq\} \{rs\} \{tu\}} \neq 0$, the intensities $Z^{(1)}_{\{rs\} \{tu\}}$ and $Z^{(2)}_{\{pq\} \{rs\}}$ are different from zero when in addition $D^{(1)}_{sr} \neq 0$ and $(\hat{P}^{(2)}\sigma_0)_{\, rs} \neq 0$. This implies that the "relay-transition" $\{rs\}$ must be excited in the preparation state $P^{(2)}$ and must be detectable by the observable $D^{(1)}$. For "allowed" one-spin single-quantum coherences, this condition is fulfilled for single pulse excitation and direct detection. On the other hand, "forbidden" multiple-spin single-quantum coherences (combination lines) and multiple-quantum coherences can neither be excited by a single non-selective pulse nor directly detected. Such coherences regularly occur in the ω_2-dimension of a 3D spectrum. The excitation and indirect detection of these coherences in 2D experiments requires special excitation and detection pulse sequences.

In conclusion, the two constituent 2D experiments deliver the same information on the spin system as the 3D spectrum, provided that (i) the relevant frequencies in the ω_2-dimension of the 3D spectrum can be excited and detected in the 2D experiments, and (ii) the cross-peaks are not hidden by spectral overlap and can be identified in the 2D spectra. The first condition is normally not severe as the 2D experiments can be modified for excitation and detection of forbidden transitions whenever required. On the other hand, the limited resolving power of 2D spectra is the most important motivation for justifying 3D (and possibly higher dimensional) spectroscopy.

Because the gain in resolution justifies 3D spectroscopy, it may be worthwhile to introduce a third frequency axis just for resolution purposes, rather than combining two processes relevant for the assignment requiring high resolution in all three dimensions. It is then possible to arbitrarily choose the extent of 3D resolution and to optimize the performance time of the 3D experiment. For the 3D spreading of a 2D spectrum, homonuclear or heteronuclear transfers can be used. Heteronuclear one-bond transfers

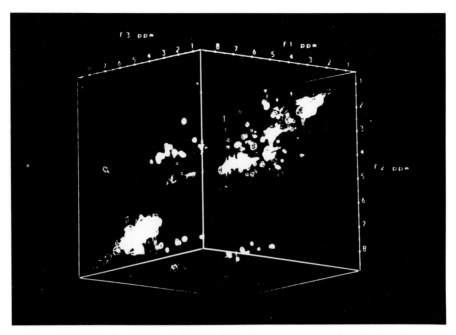

Figure 22. 3D view of a 300 MHz 3D homonuclear ROESY-TOCSY spectrum of buserilin in DMSO-d$_6$ photographed from a picture system (116).

are however far more efficient because the strong heteronuclear one-bond couplings prevent leakage to further spins. This allows an efficient transfer, virtually without loss of magnetization. In addition, nuclei like ^{13}C and ^{15}N exhibit large chemical shift ranges with high resolving power. The principle of spreading is indicated in Fig. 23.

A 3D ^{15}N-spreaded TOCSY spectrum of ribonuclease A is shown in Fig.24. The use of heteronuclear spreading requires usually isotopic labelling of the molecule. In this case, ribonuclease A has been grown in a ^{15}N-labelled nutrients-containing E.Coli medium (courtesy of Prof. Steven Benner). The spectrum has been obtained with the pulse sequence of Fig.25. Initially, proton coherence is excited and precesses during t$_1$ under ^{15}N refocusing by the applied π pulse. During the mixing time τ_m, coherence transfer from other protons to the NH protons is effected in the rotating frame by the application of a TOCSY multiple-pulse sequence. The NH coherence is then converted into ^{15}NH heteronuclear multiple-quantum coherence (HMQC) which precesses during t$_2$ and acquires ^{15}N resonance information (under proton refocusing). After reconversion into NH proton coherence, detection follows during t$_3$ under ^{15}N decoupling. For a complete assignment of the proton resonances, in addition a ^{15}N-spreaded NOESY spectrum is required.

The step to 4D spectroscopy (121) is a small and logical one: in 2D experiments, spins are pairwise correlated, e.g. NH and C$_\alpha$H protons. Three-dimensional dispersion uses either ^{15}N or ^{13}C$_\alpha$ resonance for spreading the resonances of NH or C_αH, respectively. In a 4D experiment, both spreading processes are applied simultaneously:

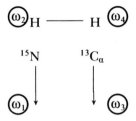

The order of the frequencies in the actual experiment is a matter of convenience. Normally, the detection frequency ω_4 refers to proton spins for sensitivity reasons. In most cases, the two spreading coordinates are rather coarsely digitized to limit the performance time, just enough to achieve separation of peaks overlapping in the 2D spectrum. Often 8 to 32 points in each of the two dimensions are sufficient.

MOLECULAR DYNAMICS INVESTIGATED BY NMR

The molecular structures, determined by NMR in solution, by X-ray diffraction in single crystals, or by other means, are invariably motionally averaged structures, whereby the averaging process is strongly dependent on the measurement technique. To interpret experimental "structures", some knowledge of the motional properties of the molecule is in fact indispensible. Molecular dynamics is also relevant for its own sake, in particular for understanding reactivity and interaction with other molecules. In many cases, active sites in a molecular pocket are only accessible due to the flexibility of the molecule itself.

The characterization of the motional properties of a molecule is by orders of magnitude more difficult than the description of an averaged molecular structure. While 3N—6 coordinates are sufficient to fix a structure containing N atoms, the characterization of molecular dynamics requires 3N—6 variances of the intramolecular coordinates, (3N—6)(3N—5)/2 covariances, and the same number of auto- and cross-correlation functions, respectively. In addition, also higher order correlation functions are needed for a more refined description of dynamics. In practice, a sufficient number of observables is never available for a full description of dynamics. In this sense, the study of dynamics is an open-ended problem. Numerous techniques are available to obtain data on dynamics: Debye-Waller factors in X-ray diffraction give hints on the variances of the nuclear coordinates, however without a measure for the time scale. Inelastic and quasi-elastic neutron scattering deliver correlation functions, but without a reference to the structure. Fluorescence depolarization allows one to determine the motional correlation function of fluorescent groups, such as tyrosine residues in proteins. Ultrasonic absorption gives an indication of the dominant motional mode frequencies, again without a structural reference.

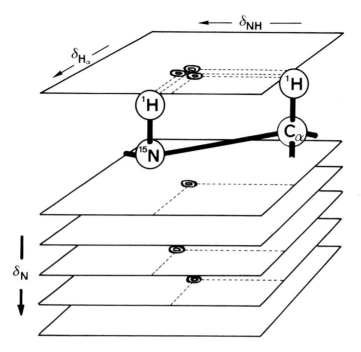

Figure 23. 3D resolution of a 2D proton-resonance spectrum by ^{15}N resonance spreading. The NH-C$_\alpha$H cross peaks are displaced in a third dimension by the corresponding ^{15}N chemical shifts.

NMR is more universally applicable to motional studies than most of the other techniques. The range of correlation times τ_c that can be covered by various NMR methods is enormous, from picoseconds to seconds and more:

$1 \text{ s} < \tau_c$:	Real time monitoring after initial perturbation
$10 \text{ ms} < \tau_c < 10 \text{ s}$:	2D exchange spectroscopy (EXSY)
$100 \text{ μs} < \tau_c < 1 \text{ s}$:	Line shape effects, exchange broadening and exchange narrowing
$1 \text{ μs} < \tau_c < 10 \text{ ms}$:	Rotating frame $T_{1\rho}$ relaxation measurements
$30 \text{ ps} < \tau_c < 1 \text{ μs}$:	Laboratory frame T_1 relaxation
$\tau_c < 100 \text{ ps}$:	Averaged parameter values .

Except for slow motions on a millisecond or slower time scale where lineshape, saturation transfer, and 2D exchange studies can be performed, many dynamics studies by NMR rely on relaxation measurements. The various relaxation parameters, such as the longitudinal relaxation time T_1, the transverse relaxation time T_2, the rotating frame relaxation time $T_{1\rho}$, and cross-relaxation rate constants Γ_{kl}, depend on the correlation time τ_c of the underlying random process.

The discussion shall be restricted to a recent study of the intramolecular dynamics in antamanide (I) (83, 122, 123) (see Figs. 8, 9, 11). Antamanide is an antidote for toxic components of the mushroom *Amanita phalloides*.

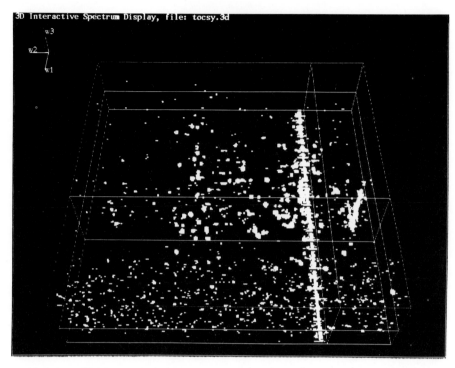

Figure 24. 3D ^{15}N-spreaded 600 MHz proton resonance TOCSY spectrum of ^{15}N-labelled ribonuclease A in H_2O solution. The 3D spectrum shows the ^{15}N resonances along the ω_2 axis. The spectrum has been recorded by C. Griesinger, using the pulse sequence of Fig. 25, and processed by S. Boentges. The sample was provided by Prof. S. Benner of ETH Zürich.

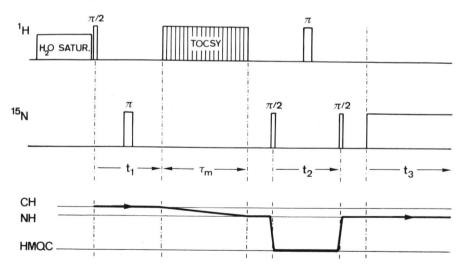

Figure 25. Pulse sequence for recording a 3D ^{15}N-spreaded TOCSY spectrum. After presaturation of the water resonance, proton resonance is excited and precesses during t_1. After the homonuclear TOCSY transfer from CH to NH protons, the coherence is converted into heteronuclear multiple quantum coherence (HMQC) that evolves during t_2 and acquires ^{15}N shift information. After reconversion to proton coherence, NH resonance is detected during t_3 under ^{15}N decoupling.

Astonishingly, the antidote occurs as a component of the same mushroom. Indications have been found in early ultrasonic absorption studies (124) that the peptide ring seems to undergo a conformational exchange process with a frequency of about 1 MHz. In the course of extensive investigations of antamanide by the research group of Professor Horst Kessler (125), it has also been noticed that the distance constraints obtained from NMR measurements could not be fitted by a single conformation. Martin Blackledge has performed in our laboratory rotating frame relaxation measurements and localized a hydrogen-bond exchange process with an activation energy of 25 kJ/mol and a lifetime of 25 μs at room temperature (unpublished results, see also Ref. 126). With a new dynamic structure determination procedure, called MEDUSA (123), the conformational space of antamanide has been investigated more systematically than ever before. 1176 feasible low-energy structures have been found. They have been combined in dynamically interchanging pairs in an attempt to fulfill all experimental constraints that consist of NOE distance constraints, J-coupling angular constraints and specific information on hydrogen bond dynamics. A large set of feasible structural pairs has been constructed. Many pairs are within experimental accuracy compatible with the experimental data. For a more restrictive description of the dynamic system of antamanide, additional and more accurate experimental data are required. Figure 26 shows, as an example, the dynamic pair of structures that so far fits the experimental data best. The two interconverting structures differ primarily in the hydrogen bonds Val^1NH—Phe^9O and Phe^6NH—Ala^4O, that exist only in one of the two conformations, and in the torsional angles φ_5 and φ_{10}.

A second study concentrated on the ring puckering dynamics of the four proline residues in antamanide (122). The conformation of the five-ring systems can be determined from the dihedral bond angles χ_1, χ_2, and χ_3 that in turn can be deduced from the vicinal proton-proton J-coupling constants using the Karplus relations (54). The relevant coupling constants (21 per residue) have been determined from E.COSY spectra. Based on these measurements, a model was constructed for each of the proline residues by least squares fitting. It was found that for Pro^3 and Pro^8 a good fit can be obtained with a single rigid conformation, while for Pro^2 and Pro^7 two rapidly exchanging conformations were required to reduce the fitting error into an acceptable range. At the same time, ^{13}C relaxation-time measurements confirmed that Pro^3 and Pro^8 appear to be rigid while Pro^2 and Pro^7 show dynamics with correlation times between 30 and 40 ps. This implies that the peptide ring dynamics and the proline ring dynamics are not correlated and proceed on entirely different time scales. The two exchanging conformations that have been found for Pro^2 are shown in Fig. 27. It is seen that the motion involves an envelope-type process where the 'flap of the envelope' (Cγ) is moving up and down.

MAGNETIC RESONANCE FOURIER IMAGING

Magnetic resonance imaging (MRI) has had an enormous impact on medical diagnosis and became rapidly a powerful routine tool. The basic procedure for recording a 2D or 3D NMR image of an object is due to Paul Lauterbur (127). A magnetic field gradient, applied along different directions in space in a sequence of experiments, produces projections of the nuclear spin density of the object onto the direction of the gradient. From a sufficiently large set of such projections it is possible to reconstruct an image of the object, for example by filtered backprojection in analogy to X-ray tomography.

A different approach is directly related to 2D and 3D Fourier transform spectroscopy. Frequency encoding of the three spatial dimensions is achieved by a linear magnetic field gradient applied successively along three orthogonal directions for the durations t_1, t_2, and t_3, respectively, in a pulse Fourier transform experiment (128). In full analogy to 3D spectroscopy, the time parameters t_1 and t_2 are incremented in regular intervals from experiment to experiment. The recorded signal $s(t_1,t_2,t_3)$ is Fourier-transformed in three dimensions to produce a function $S(\omega_1,\omega_2,\omega_3)$ that is equivalent to a 3D spatial image when the spatial information is decoded using the relations $x = \omega_1/g_x$, $y = \omega_2/g_y$, and $z = \omega_3/g_z$ with the three field gradients g_x, g_y, and g_z. The procedure is illustrated in Fig. 28 for two dimensions.

In a further refinement, proposed by Edelstein et al. (129), the time variables t_1 and t_2 are replaced by variable field gradient strengths g_x and g_y applied during a constant evolution time. With regard to the accumulated phase,

$$\gamma = x\, g_x t_1 + y\, g_y t_2 + z\, g_z t_3, \qquad [20]$$

it is immaterial whether the evolution time or the field gradients are varied. However, keeping the time t_k constant eliminates undesired relaxation effects.

In medical imaging, 3D experiments have a natural justification, although it is sometimes simpler to apply selective excitation techniques to select a 2D slice through the object to be imaged (130). Even extensions to higher dimensions are quite realistic. In a fourth dimension, for example, chemical shift information can be accommodated (131). Also 2D spectroscopic information could be combined with three spatial dimensions, leading to a 5D experiment. No limitations seem to exist for the human imagination. However, the practical limits will soon be reached when the required performance times is also taken into consideration.

CONCLUSION

I am not aware of any other field of science outside of magnetic resonance that offers so much freedom and opportunities for a creative mind to invent and explore new experimental schemes that can be fruitfully applied in a

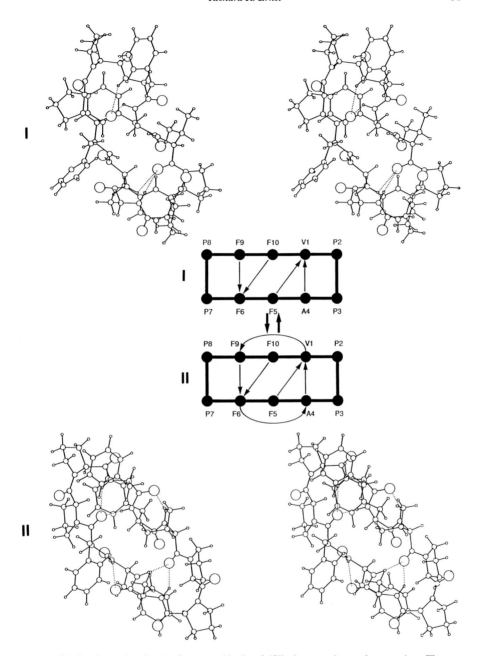

Figure 26. Conformational pair of antamanide that fulfills the experimental constraints. The two pairs are shown in stereographic as well as in abstract form. In the former, hydrogen bonds are indicated by broken lines, in the latter by arrows pointing towards the hydrogen-bonded oxygen (from Ref. 123).

Pro^2(1)   Pro^2(2)

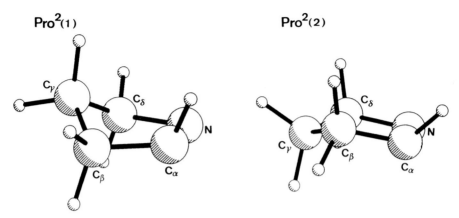

Figure 27. The two experimentally determined conformations of proline-2 in antamanide (see Ref. 122).

variety of disciplines. NMR is intellectually attractive because the observed phenomena can be understood based on a sound theory, and almost all conceits can also be tested by easy experiments. At the same time, the practical importance of NMR is enormous and can justify many of the playful activities of an addicted spectroscopist.

ACKNOWLEDGMENTS

Most of the credit for the inspiration and execution of the work described should go to my teachers Hans Primas and Hans H. Günthard, to my supervisor Weston A. Anderson, to the inspirator Jean Jeener, to my coworkers (in more or less chronological order): Thomas Baumann, Enrico Bartholdi, Robert Morgan, Stefan Schäublin, Anil Kumar, Dieter Welti, Luciano Müller, Alexander Wokaun, Walter P. Aue, Jiri Karhan, Peter Bachmann, Geoffrey Bodenhausen, Peter Brunner, Alfred Höhener, Andrew A. Maudsley, Kuniaki Nagayama, Max Linder, Michael Reinhold, Ronald Haberkorn, Thierry Schaffhauser, Douglas Burum, Federico Graf, Yongren Huang, Slobodan Macura, Beat H. Meier, Dieter Suter, Pablo Caravatti, Ole W. Sørensen, Lukas Braunschweiler, Malcolm H. Levitt, Rolf Meyer, Mark Rance, Arthur Schweiger, Michael H. Frey, Beat U. Meier, Marcel Müri, Christopher Councell, Herbert Kogler, Roland Kreis, Norbert Müller, Annalisa Pastore, Christian Schönenberger, Walter Studer, Christian Radloff, Albert Thomas, Rafael Brüschweiler, Herman Cho, Claudius Gemperle, Christian Griesinger, Zoltan L. Mádi, Peter Meier, Serge Boentges, Marc McCoy, Armin Stöckli, Gabriele Aebli, Martin Blackledge, Jacques Briand, Matthias Ernst, Tilo Levante, Pierre Robyr, Thomas Schulte-Herbrüggen, Jürgen Schmidt and Scott Smith; to my technical staff, Hansruedi Hager, Alexandra Frei, Janos A. Deli, Jean-Pierre Michot, Robert Ritz, Thomas Schneider, Markus Hintermann, Gerhard Gucher, Josef Eisenegger, Walter Lämmler, and Martin Neukomm; to my secretary Irène Müller; and to several research groups with which I had the pleasure to collaborate, first of all the research group of Kurt Wüthrich, and the

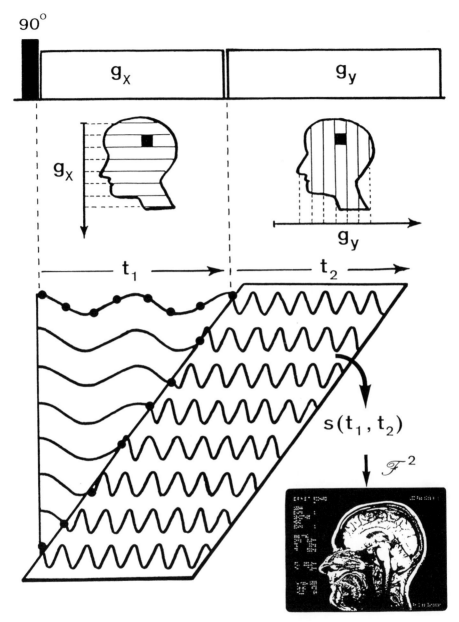

Figure 28. Schematic representation of Fourier NMR imaging, here shown in two dimensions. Two orthogonal gradients are applied during the t_1 and t_2 periods of a 2D experiment. A 2D Fourier transformation of the data set $s(t_1, t_2)$ produces a 2D image of the investigated subject (R.R.E.).

group of Horst Kessler. I also owe much gratitude for support in the early days to Varian Associates and more recently to the Swiss Federal Institute of Technology, the Swiss National Science Foundation, the Kommission zur Förderung der Wissenschaftlichen Forschung, and last but not least to Spectrospin AG.

REFERENCES

1. I.I. Rabi, Phys. Rev. **51**, 652 (1937).
2. I.I. Rabi, J.R. Zacharias, S. Millman, and P. Kusch, Phys. Rev. **53**, 318 (1938); I.I. Rabi, S. Millman, P. Kusch, J.R. Zacharias, Phys. Rev. **55**, 526 (1939).
3. J.M.B. Kellogg, I.I. Rabi, N.F. Ramsey, and J.R. Zacharias, Phys. Rev. **55**, 318 (1939); **56**, 728 (1939); **57**, 677 (1940).
4. E.M. Purcell, H.G. Torrey, and R.V. Pound, Phys. Rev. **69**, 37 (1946).
5. F. Bloch, W. Hansen, and M.E. Packard, Phys. Rev. **69**, 127 (1946).
6. F. Bloch, Phys. Rev. **70**, 460 (1946).
7. J. Brossel and A. Kastler, C. R. Acad. Sci. **229**, 1213 (1949); A. Kastler, J. Physique **11**, 255 (1950).
8. R.R. Ernst, G. Bodenhausen, and A. Wokaun, *Principles of NMR in One and Two dimensions*, Clarendon Press, Oxford, 1987.
9. A. Bax, *Two-dimensional NMR in Liquids*, Delft University Press, D.Reidel Publ. Comp., Dortrecht, 1982.
10. Attur-ur Rahman, *Nuclear Magnetic Resonance, Basic Principles*, Springer, New York, 1986.
11. N. Chandrakumar and S. Subramanian, *Modern Techniques in High-resolution FT-NMR*, Springer, New York, 1987.
12. H. Friebolin, *Ein- und zwei-dimensionale NMR-Spektroskopie*, VCH-Verlag, Weinheim, 1988.
13. G.E. Martin and A.S. Zektzer, *Two-dimensional NMR Methods for Establishing Molecular Connectivity*, VCH Verlagsgesellschaft, Weinheim, 1988.
14. J. Schraml and J.M. Bellama, *Two-dimensional NMR Spectroscopy*, Wiley Interscience, New York, 1988.
15. W.S. Brey, ed., 'Pulse methods in 1D and 2D liquid-phase NMR', Academic Press, New York, 1988.
16. A.A. Michelson, Phil. Mag. Ser.5, **31**, 256 (1891); A.A. Michelson, *Light Waves and their Uses*, University of Chicago Press, Chicago, 1902.
17. P. Fellgett, Thesis, Cambridge University, 1951; P. Fellgett, J. Phys. Radium, **19**, 187 (1958).
18. Varian Associates Magazine, **24**, No.7, 11 (Aug. 1979); IEEE Center for the History of Electrical Engineering Newsletter No.24, 2 (1990).
19. R.R. Ernst and W.A. Anderson, Rev. Sci. Instrum. **37**, 93 (1966).
20. R.R. Ernst, Adv. Magn. Reson. **2**, 1 (1966).
21. W.A. Anderson and R.R. Ernst, US Patent No. 3.475.680 'Impulse resonance spectrometer including a time averaging computer and a Fourier analyzer', filed May 26, 1965, issued Oct. 28, 1969.
22. R.R. Ernst, in *The Applications of Computer Techniques in Chemical Research*, The Institute of Petroleum, London, 1972, p.61.
23. O.W. Sørensen, G.W. Eich, M.H. Levitt, G. Bodenhausen, and R.R. Ernst, Prog. NMR Spectrosc. **16**, 163 (1983).
24. J.B.J. Fourier, 'Theorie analytique de la chaleur', Firmin Didot, Père et fils, Paris, 1822.
25. I.J. Lowe and R.E. Norberg, Phys. Rev. **107**, 46 (1957).
26. N. Wiener, M.I.T. Radiation Lab., Rep.V-16S, Apr.6, 1942; N. Wiener, *Non-Linear Problems in Random Theory*, Wiley, New York, 1958.
27. R.H. Varian, US Patent No. 3.287.629 'Gyromagnetic resonance methods and apparatus', filed Aug. 29, 1956, issued Nov. 22, 1966.
28. H. Primas, Helv. Phys. Acta **34**, 36 (1961).
29. R.R. Ernst and H. Primas, Helv. Phys. Acta **36**, 583 (1963).
30. R.R. Ernst, J. Chem. Phys. **45**, 3845 (1966).
31. R.R. Ernst, Mol. Phys. **16**, 241 (1969).
32. R. Kaiser, J. Magn. Reson. **3**, 28 (1970).

33. R.R. Ernst, J. Magn. Reson. **3**, 10 (1970).
34. D. Ziessow and B. Blümich, Ber. Bunsenges. Phys. Chem. **78**, 1169 (1974); B. Blümich and D. Ziessow, J. Chem. Phys. **78**, 1059 (1983).
35. B. Blümich, Bull. Magn. Reson. **7**, 5 (1985).
36. J. Dadok and R.F. Sprecher, J. Magn. Reson. **13**, 243 (1974).
37. R.K. Gupta, J.A. Ferretti, and E.D. Becker, J. Magn. Reson. **13**, 275 (1974).
38. J.A. Ferretti and R.R. Ernst, J. Chem. Phys. **65**, 4283 (1976).
39. B.L. Tomlinson and H.D.W. Hill, J. Chem. Phys. **59**, 1775 (1973).
40. M.H. Levitt and R. Freeman, J. Magn. Reson. **33**, 473 (1979).
41. M.H. Levitt, Prog. NMR Spectrosc. **18**, 61 (1986).
42. R.L. Vold, J.S. Waugh, M.P. Klein, and D.E. Phelps, J. Chem. Phys. **48**, 3831 (1968).
43. R. Freeman and H.D.W. Hill, in *Dynamic NMR Spectroscopy* (eds. L.M. Jackman and F.A. Cotton), p.131, Academic Press, New York, 1975.
44. S. Forsén and R.A. Hoffman, J. Chem. Phys. **39**, 2892 (1963).
45. H.C. Torrey, Phys. Rev. **75**, 1326 (1949); **76**, 1059 (1949).
46. E.L. Hahn, Phys. Rev. **76**, 145 (1949).
47. E.L. Hahn, Phys. Rev. **80**, 297 (1950).
48. E.L. Hahn, Phys. Rev. **80**, 580 (1950).
49. M. Emshwiller, E.L. Hahn, and D. Kaplan, Phys. Rev. **118**, 414 (1960).
50. S.R. Hartmann and E.L. Hahn, Phys. Rev. **128**, 2042 (1962).
51. M.B. Comisarow and A.G. Marshall, Chem. Phys. Lett. **25**, 282 (1974) ibid **26**, 489 (1974).
52. J.C. McGurk, H. Mäder, R.T. Hofmann, T.G. Schmalz, and W.H. Flygare, J. Chem. Phys. **61**, 3759 (1974).
53. E.g. M.K. Bowman, in *Modern Pulsed and Continuous-Wawe Electron Spin Resonance*, ed. L. Kevan and M.K. Bowman, p. 1, J. Wiley, New York, 1990.
54. M. Karplus, J. Chem. Phys. **30**, 11 (1959).
55. J.H. Noggle and R.E. Schirmer, *The Nuclear Overhauser Effect*, Academic Press, New York, 1971.
56. K. Wüthrich, *NMR of Proteins and Nucleic Acids*, Wiley Interscience, New York, 1986.
57. S. Yatsiv, Phys. Rev. **113**, 1522 (1952).
58. W.A. Anderson and R. Freeman, J. Chem. Phys. **37**, 85 (1962).
59. R. Freeman and W.A. Anderson, J. Chem. Phys. **37**, 2053 (1962).
60. R.A. Hoffman and S. Forsén, Prog. NMR Spectrosc. **1**, 15 (1966).
61. J. Jeener, Ampère International Summer School, Basko Polje, Jugoslavia, 1971, unpublished.
62. R.R. Ernst, *VIth International Conference on Magnetic Resonance in Biological Systems*, Kandersteg, Switzerland, 1974, unpublished.
63. W.P. Aue, E. Bartholdi, and R.R. Ernst, J. Chem. Phys. **64**, 2229 (1976).
64. J. Jeener, B.H. Meier, and R.R. Ernst, J. Chem. Phys. **71**, 4546 (1979).
65. B.H. Meier and R.R. Ernst, J. Am. Chem. Soc. **101**, 6641 (1979).
66. S. Macura and R.R. Ernst, Mol. Phys. **41**, 95 (1980).
67. Anil Kumar, R.R. Ernst, and K. Wüthrich, Biochem. Biophys. Res. Commun. **95**, 1 (1980).
68. M.P. Williamson, T.F. Havel, and K. Wüthrich, J. Mol. Biol. **182**, 295 (1985).
69. A.D. Kline, W. Braun, and K. Wüthrich, J. Mol. Biol. **189**, 377 (1986).
70. B.A. Messerle, A. Schäffer, M. Vasák, J.H.R. Kägi, and K. Wüthrich, J. Mol. Biol. **214**, 765 (1990).
71. G. Otting, Y.Q. Qian, M. Billeter, M. Müller, M. Affolter, W.J. Gehring, and K. Wüthrich, EMBO J. **9**, 3085 (1990).
72. T.F. Haveland and K. Wüthrich, Bull. Math. Biol. **46**, 673 (1984).
73. W. Braun and N. Gō, J. Mol. Biol. **186**, 611 (1985).

74. R. Kaptein, E.R.P. Zuiderweg, R.M. Scheek, R. Boelens, and W.F. van Gunsteren, J. Mol. Biol. **182**, 179 (1985).
75. G.M. Clore, A.M. Gronenborn, A.T. Brünger, and M. Karplus, J. Mol. Biol. **186**, 435 (1985).
76. Y. Huang, S. Macura, and R.R. Ernst, J. Am. Chem. Soc. **103**, 5327 (1981).
77. G.W. Eich, G. Bodenhausen, and R.R. Ernst, J. Am. Chem. Soc. **104**, 3731 (1982).
78. P.H. Bolton and G. Bodenhausen, Chem. Phys. Lett. **89**, 139 (1982).
79. Spectra recorded by C. Griesinger, see R.R. Ernst, Chimia **41**, 323 (1987).
80. L. Braunschweiler and R.R. Ernst, J. Magn. Reson. **53**, 521 (1983).
81. D.G. Davis and A. Bax, J. Am. Chem. Soc. **107**, 2821 (1985).
82. A.A. Bothner-By, R.L. Stephens, J. Lee, C.O. Warren, and R.W. Jeanloz, J. Am. Chem. Soc. **106**, 811 (1984).
83. R. Brüschweiler, B. Roux, M. Blackledge, C. Griesinger, M. Karplus, and R.R. Ernst, J. Am. Chem. Soc. **114**, 2289 (1992).
84. C. Griesinger, G. Otting, K. Wüthrich, and R.R. Ernst, J. Am. Chem. Soc. **110**, 7870 (1988).
85. J. Briand and R.R. Ernst, Chem. Phys. Lett. **185**, 276 (1991).
86. S. Vega, T.W. Shattuck, and A. Pines, Phys. Rev. Lett. **37**, 43 (1976).
87. S. Vega and A. Pines, J. Chem. Phys. **66**, 5624 (1977).
88. A. Wokaun and R.R. Ernst, Mol. Phys. **36**, 317 (1978).
89. L. Braunschweiler, G. Bodenhausen, and R.R. Ernst, Mol. Phys. **48**, 535 (1983).
90. A. Bax, R. Freeman, and S.P. Kempsell, J. Am. Chem. Soc. **102**, 4849 (1980).
91. A. Bax, R. Freeman, and S.P. Kempsell, J. Magn. Reson. **41**, 349 (1980).
92. U. Piantini, O.W. Sørensen, and R.R. Ernst, J. Am. Chem. Soc. **104**, 6800 (1982).
93. N. Müller, G. Bodenhausen, K. Wüthrich, and R.R. Ernst, J. Magn. Reson. **65**, 531 (1985).
94. C. Radloff and R.R. Ernst, Mol. Phys. **66**, 161 (1989).
95. A. Wokaun and R.R. Ernst, Chem. Phys. Lett. **52**, 407 (1977).
96. G. Bodenhausen, H. Kogler, and R.R. Ernst, J. Magn. Reson. **58**, 370 (1984).
97. M.H. Levitt and R.R. Ernst, Chem. Phys. Lett. **100**, 119 (1983).
98. M.H. Levitt and R.R. Ernst, J. Chem. Phys. **83**, 3297 (1985).
99. C. Griesinger, O.W. Sørensen, and R.R. Ernst, J. Am. Chem. Soc. **107**, 6394 (1985).
100. C. Griesinger, O.W. Sørensen, and R.R. Ernst, J. Chem. Phys. **85**, 6837 (1986).
101. C. Griesinger, O.W. Sørensen, and R.R. Ernst, J. Magn. Reson. **75**, 474 (1987).
102. A. Bax and R. Freeman, J. Magn. Reson. **44**, 542 (1981).
103. B.U. Meier and R.R. Ernst, J. Magn. Reson. **79**, 540 (1988).
104. A.A. Maudsley and R.R. Ernst, Chem. Phys. Lett. **50**, 368 (1977).
105. G. Bodenhausen and R. Freeman, J. Magn. Reson. **28**, 471 (1977).
106. L. Müller, J. Am. Chem. Soc. **101**, 4481 (1979).
107. M. Ernst, C. Griesinger, R.R. Ernst, and W. Bermel, Mol. Phys. **74**, 219 (1991).
108. M.H. Levitt, O.W. Sørensen, and R.R. Ernst, Chem. Phys. Lett. **94**, 540 (1983).
109. H. Kogler, O.W. Sørensen, G. Bodenhausen, and R.R. Ernst, J. Magn. Reson. **55**, 157 (1983).
110. H.D. Plant, T.H. Mareci, M.D. Cockman, and W.S. Brey, 27th Exp. NMR Conference, Baltimore, MA 1986.
111. G.W. Vuister and R. Boelens, J. Magn. Reson. **73**, 328 (1987).
112. C. Griesinger, O.W. Sørensen, and R.R. Ernst, J. Magn. Reson. **73**, 574 (1987).

113. C. Griesinger, O.W. Sørensen, and R.R. Ernst, J. Am. Chem. Soc. **109**, 7227 (1987).
114. H. Oschkinat, C. Griesinger, P. Kraulis, O.W. Sørensen, R.R. Ernst, A.M. Gronenborn, and G.M. Clore, Nature (London) **332**, 374 (1988).
115. G.W. Vuister, R. Boelens, and R. Kaptein, J. Magn. Reson. **80**, 176 (1988).
116. C. Griesinger, O.W. Sørensen, and R.R. Ernst, J. Magn. Reson. **84**, 14 (1989).
117. E.R.P. Zuiderweg and S.W. Fesik, Biochemistry **28**, 2387 (1989).
118. D. Marion, P.C. Driscoll, L.E. Kay, P.T. Wingfield, A. Bax, A.M. Gronenborn, and G.M. Clore, Biochemistry **28**, 6150 (1989).
119. S. Boentges, B.U. Meier, C. Griesinger, and R.R. Ernst, J. Magn. Reson. **85**, 337 (1989).
120. O.W. Sørensen, J. Magn. Reson. **89**, 210 (1990).
121. L.E. Kay, G.M. Clore, A. Bax, and A.M. Gronenborn, Science **249**, 411 (1990).
122. Z.L. Mádi, C. Griesinger, and R.R. Ernst, J. Am. Chem. Soc. **112**, 2908 (1990).
123. R. Brüschweiler, M. Blackledge, and R.R. Ernst, J. Biomol. NMR **1**, 3 (1991).
124. W. Burgermeister, T. Wieland, and R. Winkler, Eur. J. Biochem. **44**, 311 (1974).
125. H. Kessler, M. Klein, A. Müller, K. Wagner, J.W. Bats, K. Ziegler, and M. Frimmer, Angew. Chem. **98**, 1030 (1986); H. Kessler, A. Müller, and K.H. Pook, Liebig Ann. Chem. 903 (1989); H. Kessler, J.W. Bats, J. Lantz, and A. Müller, Liebig Ann. Chem. 913 (1989); J. Lantz, H. Kessler, W.F. van Gunsteren, H.J. Berendsen, R.M. Scheek, R. Kaptein, and J. Blaney, Proc. 20th Eur. Pept. Symp. 1989, p. 438 (Ed. G. Jung, E. Bayer).
126. R.R. Ernst, M. Blackledge, S. Boentges, J. Briand, R. Brüschweiler, M. Ernst, C. Griesinger, Z.L. Mádi, T. Schulte-Herbrüggen, and O.W. Sørensen, in *Proteins, Structure, Dynamics, Design*, ed. V. Renugopalakrishnan, P.R. Carey, I.C.P. Smith, S.G. Huang, and A.C. Storer, ESCOM, Leiden, 1991.
127. P.C. Lauterbur, Nature **242**, 190 (1973).
128. Anil Kumar, D. Welti, and R.R. Ernst, J. Magn. Reson. **18**, 69 (1975).
129. W.A. Edelstein, J.M.S. Hutchison, G. Johnson, and T.W. Redpath, Phys. Med. Biol. **25**, 751 (1980).
130. P. Mansfield, A.A. Maudsley, and T. Baines, J. Phys. **E9**, 271 (1976).
131. P.C. Lauterbur, D.M. Kramer, W.V. House, and C.-N. Chen, J. Am. Chem. Soc. **97**, 6866 (1975).

Chemistry 1992

RUDOLPH A. MARCUS

*for his contributions to the theory of electron transfer reactions
in chemical systems*

THE NOBEL PRIZE IN CHEMISTRY

Speech by Professor Lennart Eberson of the Royal Swedish Academy of Sciences.
Translation from the Swedish text.

Your Majesties, Your Royal Highnesses, Ladies and Gentlemen,

The 1992 Nobel Prize in Chemistry is being awarded to Professor Rudolph Marcus for *his contributions to the theory of electron transfer reactions in chemical systems*. To understand the background of his achievements, we must transport ourselves back to the period around 1950, when chemistry looked completely different than it does today. In those days, it was still difficult to determine the structure of chemical compounds, and even more difficult to make theoretical calculations of the rate of chemical reactions.

Reaction rate is a fundamental concept in chemistry. A mixture of chemical compounds undergoes changes, or chemical reactions, at different rates. Today we can measure reaction rates using virtually any time scale from quadrillionths of a second to thousands of years. By the late 19th century, Sweden's Svante Arrhenius, later a Nobel Laureate, had shown that the rate of a chemical reaction can be described in terms of the requirement for a reacting system to cross an energy barrier. The size of this barrier was easy to determine experimentally. Calculating it was a formidable problem.

In the years after 1945, a new technique for determining reaction rates had been developed: the radioactive tracer technique. By substituting a radioactive isotope for a given atom in a molecule, new types of reactions could be studied. One such reaction was the transfer of an electron between metal ions in different states of oxidation, for example between a bivalent and a trivalent iron ion in an aqueous solution. This turned out to be a slow reaction, that is, it took place over a period of hours, something highly unexpected by the chemists of that day. Compared with an atomic nucleus, an electron is a very light particle. How could the slowness of its movement between iron ions be explained?

This problem led to lively discussion around 1950. Marcus became interested when he happened to read through some papers from a symposium on electron transfer reactions, where the American chemist Willard Libby had suggested that a well-known spectroscopic principle known as the Franck-Condon principle might apply to the movement of an electron between two molecules. Marcus realized that this ought to create an energy barrier, which might explain the slow electron transfer between bivalent and trivalent iron in an aqueous solution. To enable the two iron ions to exchange an electron, a number of water molecules in their surroundings must be rearranged. This increases the energy of the system temporarily, and at some point the electron can jump without violating the restrictions of the Franck-Condon principle.

In 1956, Marcus published a mathematical model for this type of reaction, based on classic theories of physical chemistry. He was able to calculate the size of the energy barrier, using simple quantities such as ionic radii and ionic charges. He later extended the theory to cover electron transfer between different kinds of molecules and derived simple mathematical expressions known as "the quadratic equation" and "the cross-equation." These could be tested empirically and led to new experimental programs in all branches of chemistry. The Marcus theory greatly contributed to our understanding of such widely varying phenomena as the capture of light energy in green plants, electron transfer in biological systems, inorganic and organic oxidation and reduction processes and photochemical electron transfer.

The quadratic equation predicts that electron transfer reactions will occur more slowly the larger the driving force of the reaction is. This phenomenon received its own name, "the inverted region." To a chemist, the phenomenon is just as unexpected as when a skier finds himself gliding more slowly down a slope the steeper it is. In 1965, Marcus himself suggested that certain chemiluminescent reactions ("cold light") might serve as an example of the inverted region. Only after 1985, however, could further examples of such reactions be demonstrated. The most improbable prediction in his theory had thereby been verified.

Professor Marcus,
In the space of a few minutes, I have tried to trace and explain the origins of the theory of electron transfer that carries your name. Your theory is a unifying factor in chemistry, promoting understanding of electron transfer reactions of biochemical, photochemical, inorganic and organic nature and thereby contributing to science as a whole. It has led to the development of many new research programs, demonstrating the lasting impact of your work. In recognition of your contribution to chemistry, the Royal Swedish Academy of Sciences has decided to confer upon you this year's Nobel Prize in Chemistry.

Professor Marcus, I have the honor and pleasure to extend to you the congratulations of the Royal Swedish Academy of Sciences and to ask you to receive your Prize from the hands of His Majesty the King.

Rudolph A. Marcus

RUDOLPH A. MARCUS

My first encounters with McGill University came when I was still in a baby carriage. My mother used to wheel me about the campus when we lived in that neighborhood and, as she recounted years later, she would tell me that I would go to McGill. There was some precedent for my going there, since two of my father's brothers received their M.D.'s at McGill.

I have always loved going to school. Since neither of my parents had a higher education, my academic "idols" were these two paternal uncles and one of their uncles, my great-uncle, Henrik Steen (né Markus). My admiration for him, living in faraway Sweden, was not because of a teol.dr. (which he received from the University of Uppsala in 1915) nor because of the many books he wrote — I knew nothing of that — but rather because he was reputed to speak 13 languages. I learned decades later that the number was only 9! Growing up, mostly in Montreal, I was an only child of loving parents. I admired my father's athletic prowess — he excelled in several sports — and my mother's expressive singing and piano playing.

My interest in the sciences started with mathematics in the very beginning, and later with chemistry in early high school and the proverbial home chemistry set. My education at Baron Byng High School was excellent, with dedicated masters (boys and girls were separate). I spent the next years at McGill University, for both undergraduate and, as was the custom of the time, graduate study. Our graduate supervisor, Carl A. Winkler, specialized in rates of chemical reactions. He himself had received his Ph.D. as a student of Cyril Hinshelwood at Oxford. Hinshelwood was later the recipient of the Nobel Prize for his work on chemical kinetics. Winkler brought to his laboratory an enthusiastic joyousness in research and was much loved by his students.

During my McGill years, I took a number of math courses, more than other students in chemistry. Upon receiving a Ph.D. from McGill University in 1946, I joined the new post-doctoral program at the National Research Council of Canada in Ottawa. This program at NRC later became famous, but at the time it was still in its infancy and our titles were Junior Research Officers. The photochemistry group was headed by E.W.R. Steacie, an international figure in the study of free-radical reactions and a major force in the development of the basic research program at NRC. I benefitted from the quality of his research on gas phase reaction rates. Like my research on chemical reaction rates in solution at McGill (kinetics of nitration), it was experimental in nature. There were no theoretical chemists in Canada at the time, and as students I don't think we ever considered how or where theories were conceived.

About 1948 a fellow post-doctoral at NRC, Walter Trost, and I formed a two-man seminar to study theoretical papers related to our experimental work. This adventure led me to explore the possibility of going on a second post-doctoral, but in theoretical work, which seemed like a radical step at the time. I had a tendency to break the glass vacuum apparatus, due to a still present impetuous haste, with time-consuming consequences. Nevertheless, the realization that breaking a pencil point would have far less disastrous consequences played little or no role, I believe, in this decision to explore theory!

I applied in 1948 to six well-known theoreticians in the U.S. for a postdoctoral research fellowship. The possibility that one of them might take on an untested applicant, an applicant hardly qualified for theoretical research, was probably too much to hope for. Oscar K. Rice at the University of North Carolina alone responded favorably, subject to the success of an application he would make to the Office of Naval Research for this purpose. It was, and in February 1949 I took the train south, heading for the University of North Carolina in Chapel Hill. I was impressed on arrival there by the red clay, the sandy walks, and the graciousness of the people.

After that, I never looked back. Being exposed to theory, stimulated by a basic love of concepts and mathematics, was a marvelous experience. During the first three months I read everything I could lay my hands on regarding reaction rate theory, including Marcelin's classic 1915 theory which came within one small step of the Transition State Theory of 1935. I read numerous theoretical papers in German, a primary language for the "chemical dynamics" field in the 1920s and 1930s, attended my first formal course in quantum mechanics, given by Nathan Rosen in the Physics Department, and was guided by Oscar in a two-man weekly seminar in which I described a paper I had read and he pointed out assumptions in it that I had overlooked. My life as a working theorist began three months after this preliminary study and background reading, when Oscar gently nudged me toward working on a particular problem.

Fortunately for me, Oscar's gamble paid off. Some three months later, I had formulated a particular case of what was later entitled by B. Seymour Rabinovitch, RRKM theory ("Rice-Ramsperger-Kassel-Marcus"). In it, I blended statistical ideas from the RRK theory of the 1920s with those of the transition state theory of the mid-1930s. The work was published in 1951. In 1952 I wrote the generalization of it for other reactions. In addition, six months after arrival in Chapel Hill, I was also blessed by marriage to Laura Hearne, an attractive graduate student in sociology at UNC. She is here with me at this ceremony. Our three sons, Alan, Kenneth and Raymond, and two daughters-in-law are also present today.

In 1951, I attempted to secure a faculty position. This effort met with little success (35 letters did not yield 35 no's, since not everyone replied!). Very fortunately, that spring I met Dean Raymond Kirk of the Polytechnic Institute of Brooklyn at an American Chemical Society meeting in Cleveland, which I was attending primarily to seek a faculty position. This meeting with Dean Kirk, so vital for my subsequent career, was arranged by Seymour Yolles, a graduate student at UNC in a course I taught during Rice's illness. Seymour had been a student at Brooklyn Poly and learned, upon accidentally encountering Dr. Kirk, that Kirk was seeking new faculty. After a subsequent interview at Brooklyn Poly, I was hired, and life as a fully independent researcher began.

I undertook an experimental research program on both gas phase and solution reaction rates, wrote the 1952 RRKM papers, and wondered what to do next in theoretical research. I felt at the time that it was pointless to continue with RRKM since few experimental data were available. Some of our experiments were intended to produce more.

After some minor pieces of theoretical study that I worked on, a student in my statistical mechanics class brought to my attention a problem in polyelectrolytes. Reading everything I could about electrostatics, I wrote two papers on that topic in 1954/55. This electrostatics background made me fully ready in 1955 to treat a problem I had just read about on electron transfers. I comment on this next period on electron transfer research in my Nobel Lecture. About 1960, it became clear that it was best for me to bring the experimental part of my research program to a close—there was too much to do on the theoretical aspects—and I began the process of winding down the experiments. I spent a year and a half during 1960—61 at the Courant Mathematical Institute at New York University, auditing many courses which were, in part, beyond me, but which were, nevertheless, highly instructive.

In 1964, I joined the faculty of the University of Illinois in Urbana-Champaign and I never undertook any further experiments there. At Illinois, my interests in electron transfer continued, together with interests in other aspects of reaction dynamics, including designing "natural collision coordinates", learning about action-angle variables, introducing the latter into molecular collisions, reaction dynamics, and later into semiclassical theories of collisions and of bound states, and spending much of my free time in the astronomy library learning more about classical mechanics, celestial mechanics, quasiperiodic motion, and chaos. I spent the academic year of 1975—76 in Europe, first as Visiting Professor at the University of Oxford and later as a Humboldt Awardee at the Technical University of Munich, where I was first exposed to the problem of electron transfer in photosynthesis.

In 1978, I accepted an offer from the California Institute of Technology to come there as the Arthur Amos Noyes Professor of Chemistry. My semiclassical interlude of 1970—80 was intellectually a very stimulating one, but it involved for me less interaction with experiments than had my earlier work on unimolecular reaction rates or on electron transfers. Accordingly, prompted by the extensive experimental work of my colleagues at Caltech in these fields of unimolecular reactions, intramolecular dynamics and of electron transfer processes, as well as by the rapidly growing experimental work in both broad areas world-wide, I turned once again to those particular topics and to the many new types of studies that were being made. Their scope and challenge continues to grow to this day in both fields. Life would be indeed easier if the experimentalists would only pause for a little while!

There was a time when I had wondered about how much time and energy had been lost doing experiments during most of my stay at Brooklyn Poly — experiments on gas phase reactions, flash photolysis, isotopic exchange electron transfer, bipolar electrolytes, nitration, and photoelectrochemistry, among others — and during all of my stay at NRC and at McGill. In retrospect, I realized that this experimental background heavily flavored my attitude and interests in theoretical research. In the latter I drew, in most but not all cases, upon experimental findings or puzzles for theoretical problems to study. The growth of experiments in these fields has served as a continually rejuvenating influence. This interaction of experiment and theory, each stimulating the other, has been and continues to be one of the joys of my experience.

Honors received for the theoretical work include the Irving Langmuir, the Peter Debye, and the Theoretical Chemistry Awards of the American Chemical Society (1978, 1988, 1997), the Willard Gibbs, Theodore William Richards, and Pauling Medals, and the Remsen, Edgar Fahs Smith and Auburn-Kosolapoff Awards, from various sections of the ACS (1988, 1990, 1991, 1991, 1991, 1996), the Robinson and the Centenary Medals of the Faraday Division of the Royal Society of Chemistry (1982, 1988), Columbia University's Chandler Medal (1983) and Ohio State's William Lloyd Evans Award (1990), Lavoisier Medal of the Société Française de Chimie (1994), a Professorial Fellowship at University College, Oxford (1975 to 1976) and a Visiting Professorship in Theoretical Chemistry at Oxford during that period, the Visiting Linnett Professorship at the University of Cambridge in 1996, the Wolf Prize in Chemistry (1985), the National Medal of Science (1989), the Hirschfelder Prize in Chemistry (1993), election to the National Academy of Sciences (1970), the American Academy of Arts and Sciences (1973), the American Philosophical Society (1990), honorary membership in the Royal Society of Chemistry (1991), the International Society of Electrochemistry (1994), and the Korean Chemical Society (1996), foreign membership in the Royal Society (London) (1987) and in the Royal Society of Canada (1993), and elected to Honorary Fellowship at University College, Oxford (1995) and Honorary Professorships at Fudan University, Shanghai and at the Institute of Chemistry, Chinese Academy of Sciences, Beijing, China (1995). Honorary degrees were conferred by the University of Chicago and by Göteburg, Polytechnic, McGill, and Queen's Universities (1983, 1986, 1987, 1988, 1993, 1993), by the Universities of New Brunswick, Oxford and North Carolina (1993, 1995, 1996), and by Yokohama National University (1996). A commemorative issue of the Journal of Physical Chemistry was published in 1986.

ELECTRON TRANSFER REACTIONS IN CHEMISTRY: THEORY AND EXPERIMENT

Nobel Lecture, December 8, 1992

by

RUDOLPH A. MARCUS

Noyes Laboratory of Chemical Physics, California Institute of Technology, MS 127-72, Pasadena, CA 91125, USA

ELECTRON TRANSFER EXPERIMENTS SINCE THE LATE 1940s

Since the late 1940s, the field of electron transfer processes has grown enormously, both in chemistry and biology. The development of the field, experimentally and theoretically, as well as its relation to the study of other kinds of chemical reactions, represents to us an intriguing history, one in which many threads have been brought together. In this lecture, some history, recent trends, and my own involvement in this research are described.

The early experiments in the electron transfer field were on "isotopic exchange reactions" (self-exchange reactions) and, later, "cross reactions." These experiments reflected two principal influences. One of these was the availability after the Second World War of many radioactive isotopes, which permitted the study of a large number of isotopic exchange electron transfer reactions, such as

$$Fe^{2+} + Fe^{*3+} \rightarrow Fe^{3+} + Fe^{*2+}, \tag{1}$$

and

$$Ce^{3+} + Ce^{*4+} \rightarrow Ce^{4+} + Ce^{*3+}, \tag{2}$$

in aqueous solution, where the asterisk denotes a radioactive isotope.

There is a two-fold simplicity in typical self-exchange electron transfer reactions (so-called since other methods beside isotopic exchange were later used to study some of them): (1) the reaction products are identical with the reactants, thus eliminating one factor which usually influences the rate of a chemical reaction in a major way, namely the relative thermodynamic stability of the reactants and products; and (2) no chemical bonds are broken or formed in *simple* electron transfer reactions. Indeed, these self-exchange reactions represent, for these combined reasons, the simplest class of reactions in chemistry. Observations stemming directly from this simplicity were to have major consequences, not only for the electron transfer field but also, to a lesser extent, for the study of other kinds of chemical reactions as well (cf Shaik *et al*, ref. 2).

A second factor in the growth of the electron transfer field was the introduction of new instrumentation, which permitted the study of the rates of rapid chemical reactions. Electron transfers are frequently rather fast, compared with many reactions which undergo, instead, a breaking of chemical bonds and a forming of new ones. Accordingly, the study of a large body of fast electron transfer reactions became accessible with the introduction of this instrumentation. One example of the latter was the stopped-flow apparatus, pioneered for inorganic electron transfer reactions by N. Sutin. It permitted the study of bimolecular reactions in solution in the millisecond time scale (a fast time scale at the time). Such studies led to the investigation of what has been termed electron transfer "cross reactions," i.e., electron transfer reactions between two different redox systems, as in

$$Fe^{2+} + Ce^{4+} \rightarrow Fe^{3+} + Ce^{3+}, \tag{3}$$

which supplemented the earlier studies of the self-exchange electron transfer reactions. A comparative study of these two types of reaction, self-exchange and cross-reactions, stimulated by theory, was also later to have major consequences for the field and, indeed, for other areas.

Again, in the field of electrochemistry, the new post-war instrumentation in chemical laboratories led to methods which permitted the study of fast electron transfer reactions at metal electrodes. Prior to the late 1940s only relatively slow electrochemical reactions, such as the discharge of an H_3O^+ ion at an electrode to form H_2, had been investigated extensively. They involved the breaking of chemical bonds and the forming of new ones.

Numerous electron transfer studies have now also been made in other areas, some depicted in Figure 1. Some of these investigations were made possible by a newer technology, lasers particularly, and now include studies in the picosecond and subpicosecond time regimes. Just recently, (non-laser) nanometer-sized electrodes have been introduced to study electrochemical processes that are still faster than those hitherto investigated. Still other recent investigations, important for testing aspects of the electron transfer theory at electrodes, involve the new use of an intervening ordered adsorbed monolayer of long chain organic compounds on the electrode to facilitate the study of various effects, such as varying the metal-solution potential difference on the electrochemical electron transfer rate.

In some studies of electron transfer reactions in solution there has also been a skillful blending of these measurements of chemical reaction rates with various organic or inorganic synthetic methods, as well as with site-directed mutagenesis, to obtain still further hitherto unavailable information. The use of chemically modified proteins to study the distance dependence of electron transfer, notably by Gray and coworkers, has opened a whole new field of activity.

The interaction of theory and experiment in these many electron transfer fields has been particularly extensive and exciting, and each has stimulated

Developments in Electron Transfer Reactions

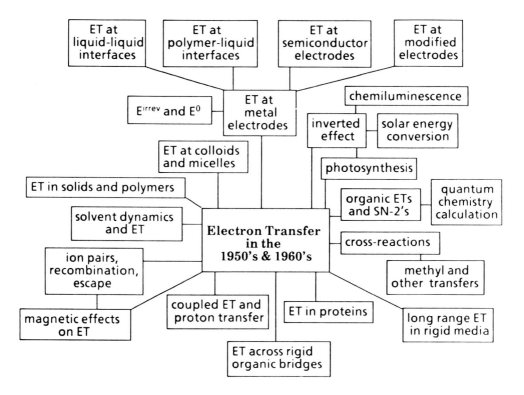

Figure 1. Examples of topics in the electron transfer field (Marcus and Siddarth, ref. 2).

the other. The present lecture addresses the underlying theory and this interaction.

THE EARLY EXPERIENCE

My own involvement in the electron transfer field began in a rather circuitous way. In an accompanying biographical note I have commented on my earlier background, which was in experimental measurements of reaction rates as a chemistry graduate student at McGill University (1943–46) and as a post-doctoral associate at the National Research Council of Canada (1946–49). A subsequent post-doctoral study at the University of North Carolina (1949–51) on the theory of reaction rates resulted in what is now known in the literature as RRKM theory (Rice, Ramsperger, Kassel, Marcus).

This unimolecular reaction field reflects another long and extensive interaction between theory and experiment. RRKM theory enjoys widespread use and is now usually referred to in the literature only by its acronym (or by the texts written about it, ref. 4), instead of by citation of the original articles.

After the theoretical post-doctoral, I joined the faculty of the Polytechnic Institute of Brooklyn in 1951 and wondered what theoretical research to do next after writing the RRKM papers (1951–52). I remember vividly how a friend of mine, a colleague at Brooklyn Poly, Frank Collins, came down to my office every day with a new idea on the liquid state transport theory which he was developing, while I, for theoretical research, had none. Perhaps this gap in not doing anything immediately in the field of theory was, in retrospect, fortunate: In not continuing with the study of the theory of unimolecular reactions, for which there were too few legitimate experimental data at the time to make the subject one of continued interest, I was open for investigating quite different problems in other areas. I did, however, begin a program of experimental studies in gas phase reactions, prompted by my earlier studies at NRC and by the RRKM work.

In the biographical note I have also recalled how a student in my statistical mechanics class in this period (Abe Kotliar) asked me about a particular problem in polyelectrolytes. It led to my writing two papers on the subject (1954–55), one of which required a considerable expansion in my background in electrostatics, so as to analyze different methods for calculating the free energy of these systems: In polyelectrolyte molecules, it may be recalled, the ionic charges along the organic or inorganic molecular backbone interact with each other and with the solvent. In the process, I read the relevant parts of the texts that were readily available to me on electrostatics (Caltech's Mason and Weaver's was later to be particularly helpful!). When shortly thereafter I encountered some papers on electron transfer, a field entirely new to me, I was reasonably well prepared for treating the problems which lay ahead.

DEVELOPING AN ELECTRON TRANSFER THEORY

Introduction

My first contact with electron transfers came in 1955 as a result of chancing upon a 1952 symposium issue on the subject in the Journal of Physical Chemistry. An article by Bill Libby caught my eye—a use of the Franck-Condon principle to explain some experimental results, namely, why some isotopic exchange reactions which involve electron transfer between pairs of small cations in aqueous solution, such as reaction (1), are relatively slow, whereas electron transfers involving larger ions, such as $Fe(CN)_6^{3-}$ – $Fe(CN)_6^{4-}$ and MnO_4^- – MnO_4^{2-}, are relatively fast.

Libby explained this observation in terms of the Franck-Condon principle, as discussed below. The principle was used extensively in the field of spectroscopy for interpreting spectra for the excitation of the molecular electronic-vibrational quantum states. An application of that principle to chemical reaction rates was novel and caught my attention. In that paper Libby gave a "back-of-the-envelope" calculation of the resulting solvation

energy barrier which slowed the reaction. However, I felt instinctively that
even though the idea—that somehow the Franck—Condon principle was
involved—seemed strikingly right, the calculation itself was incorrect. The
next month of study of the problem was, for me, an especially busy one. To
place the topic in some perspective I first digress and describe the type of
theory that was used for other types of chemical reaction rates at the time
and continues to be useful today.

Reaction rate theory
Chemical reactions are often described in terms of the motion of the atoms
of the reactants on a potential energy surface. This potential energy surface
is really the electronic energy of the entire system, plotted versus the
positions of all the atoms. A very common example is the transfer of an
atom or a group B from AB to form BC

$$AB + C \rightarrow A + BC. \tag{4}$$

An example of reaction (4) is the transfer of an H, such as in $IH + Br \rightarrow I + HBr$, or the transfer of a CH_3 group from one aromatic sulfonate to
another. To aid in visualizing the motion of the atoms in this reaction, this
potential energy function is frequently plotted as constant energy contours
in a space whose axes are chosen to be two important relative coordinates
such as, in reaction (4), a scaled AB bond length and a scaled distance from
the center of mass of AB to C, as in Figure 2.

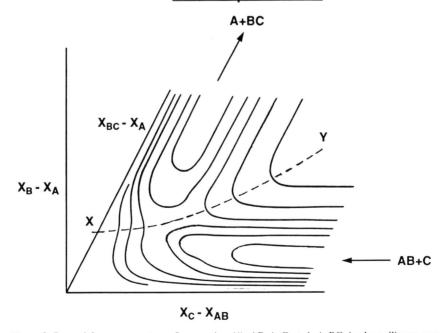

Potential Energy Contours for an Atom or Group Transfer

Figure 2. Potential energy contours for reaction (4), $AB + C \rightarrow A + BC$, in the collinear case.

A point representing this reacting system begins its trajectory in the lower right region of the figure in a valley in this plot of contours, the "valley of the reactants." When the system has enough energy, appropriately distributed between the various motions, it can cross the "mountain pass" (saddle-point region) separating the initial valley from the products' valley in the upper left, and so form the reaction products. There is a line in the figure, XY, analogous to the "continental divide" in the Rocky Mountains in the U.S., which separates systems which could spontaneously flow into the reactants' valley from those which could flow into the products' one. In chemists' terminology this line represents the "transition state" of the reaction.

In transition state theory a quasi-equilibrium between the transition state and the reactants is frequently postulated, and the reaction rate is then calculated using equilibrium statistical mechanics. A fundamental dynamical basis, which replaces this apparently *ad hoc* but common assumption of transition state theory and which is perhaps not as well known in the chemical literature as it deserves to be, was given many years ago by the physicist and one-time chemical engineer, Eugene Wigner (1938). He used a classical mechanical description of the reacting system in the many-dimensional space (of coordinates and momenta). Wigner pointed out that the quasi-equilibrium would follow as a dynamical consequence if each trajectory of a moving point representing the reacting system in this many-dimensional space did not recross the transition state (and if the distribution of the reactants in the reactants' region were a Boltzmann one). In recent times, the examination of this recrossing has been a common one in classical mechanical trajectory studies of chemical reactions. Usually, recrossings are relatively minor, except in nonadiabatic reactions, where they are readily treated (cf discussion, later).

In practice, transition state theory is generalized so as to include as many coordinates as are needed to describe the reacting system. Further, when the system can "tunnel" quantum mechanically through the potential energy barrier (the "pass") separating the two valleys, as for example frequently happens at low energies in H-transfer reactions, the method of treating the passage across the transition state region needs, and has received, refinement. (The principal problem encountered here has been the lack of "dynamical separability" of the various motions in the transition state region.)

Electron transfer theory. Formulation
In contrast to the above picture, we have already noted that in simple electron transfer reactions no chemical bonds are broken or formed and so a somewhat different picture of the reaction is needed for the electron transfer reaction.

In his 1952 symposium paper, Libby noted that when an electron is transferred from one reacting ion or molecule to another, the two new molecules or ions formed are in the wrong environment of the solvent molecules, since the nuclei do not have time to move during the rapid electron jump: in reaction (1) a Fe^{2+} ion would be formed in some configuration of the many nearby dipolar solvent molecules that was appropriate to the original Fe^{3+} ion. Analogous remarks apply to the newly formed Fe^{3+} ion in the reaction. On the other hand, in reactions of "complex ions," such as those in the $Fe(CN)_6^{-3} - Fe(CN)_6^{-4}$ and $MnO_4^- - MnO_4^{2-}$ self-exchange reactions, the two reactants are larger, and so the change of electric field in the vicinity of each ion, upon electron transfer, would be smaller. The original solvent environment would therefore be less foreign to the newly formed charges, and so the energy barrier to reaction would be less. In this way Libby explained the faster self-exchange electron transfer rate for these complex ions. Further confirmation was noted in the ensuing discussion in the symposium: the self-exchange $Co(NH_3)_6^{3+} - Co(NH_3)_6^{2+}$ reaction is very slow, and it was pointed out that there was a large difference in the equilibrium Co-N bond lengths in the $3+$ and the $2+$ ions, and so each ion would be formed in a very "foreign" configuration of the vibrational coordinates, even though the ions are "complex ions."

After studying Libby's paper and the symposium discussion, I realized that what troubled me in this picture for reactions occurring in the dark was that energy was not conserved: the ions would be formed in the wrong high-energy environment, but the only way such a non-energy conserving event could happen would be by the absorption of light (a "vertical transition"), and not in the dark. Libby had perceptively introduced the Franck-Condon principle to chemical reactions, but something was missing.

In the present discussion, as well as in Libby's treatment, it was supposed that the electronic interaction of the reactants which causes the electron transfer is relatively weak. That view is still the one that seems appropriate today for most of these reactions. In this case of weak-electronic interaction, the question becomes: how does the reacting system behave in the dark so as to satisfy both the Franck-Condon principle and energy conservation? I realized that fluctuations had to occur in the various nuclear coordinates, such as in the orientation coordinates of the individual solvent molecules and indeed in any other coordinates whose most probable distribution for the products differs from that of the reactants. With such fluctuations, values of the coordinates could be reached which satisfy both the Franck-Condon and energy conservation conditions and so permit the electron transfer to occur in the dark.

For a reaction such as reaction (1), an example of an initial and final configuration of the solvent molecules is depicted in Figure 3. Fluctuations from the original equilibrium ensemble of configurations were ultimately needed, prior to the electron transfer, and were followed by a relaxation to the equilibrium ensemble for the products, after electron transfer.

The theory then proceeded as follows. The potential energy U_r of the entire system, reactants plus solvent, is a function of the many hundreds of relevant coordinates of the system, coordinates which include, among others, the position and orientation of the individual solvent molecules (and

Electron Transfer in Solution

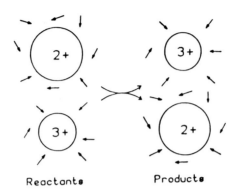

Figure 3. Typical nuclear configurations for reactants, products, and surrounding solvent molecules in reaction (1). The longer $M-OH_2$ bond length in the $+2$ state is indicated schematically by the larger ionic radius. (Sutin, ref. 2)

hence of their dipole moments, for example), and the vibrational coordinates of the reactants, particularly those in any inner coordination shell of the reacting ions. (E.g., the inner coordination shell of an ion such as Fe^{2+} or Fe^{3+} in water is known from EXAFS experiments to contain six water molecules.) No longer were there just the two or so important coordinates that were dominant in reaction (4).

Similarly, after the electron transfer, the reacting molecules have the ionic charges appropriate to the reaction products, and so the relevant potential energy function U_p is that for the products plus solvent. These two potential energy surfaces will intersect if the electronic coupling which leads to electron transfer is neglected. For a system with N coordinates this intersection occurs on an $(N\text{-}1)$ dimensional surface, which then constitutes in our approximation the transition state of the reaction. The neglected electronic coupling causes a well-known splitting of the two surfaces in the vicinity of their intersection. A schematic profile of the two potential energy surfaces in the N-dimensional space is given in Figure 4. (The splitting is not shown.)

Due to the effect of the previously neglected electronic coupling and the coupling between the electronic motion and the nuclear motion near the intersection surface S, an electron transfer can occur at S. In classical terms, the transfer at S occurs at fixed positions and momenta of the atoms, and so the Franck-Condon principle is satisfied. Since U_r equals U_p at S, energy is also conserved. The details of the electron transfer depend on the extent of electronic coupling and how rapidly the point representing the system in this N-dimensional space crosses S. (It has been treated, for example, using as an approximation the well-known one-dimensional Landau-Zener expression for the transition probability at the near-intersection of two potential energy curves.)

Potential Energy Surfaces, Profile

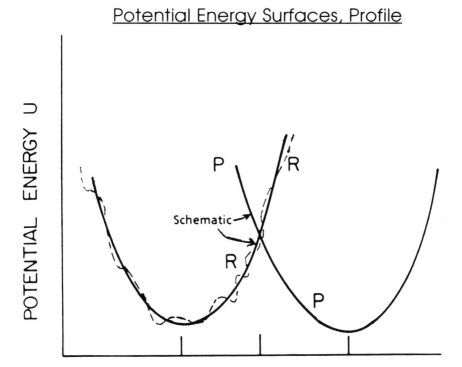

Figure 4. Profile of potential energy surfaces for reactants plus environment, *R*, and for products plus environment, *P*. Solid curves: schematic. Dashed curves: schematic but slightly more realistic. The typical splitting at the intersection of U_r and U_p is not shown in the Figure (Marcus and Siddarth, ref. 2).

When the splitting caused by the electronic coupling between the electron donor and acceptor is large enough at the intersection, a system crossing *S* from the lower surface on the reactants' side of *S* continues onto the lower surface on the products' side, and so an electron transfer in the dark has then occurred. When the coupling is, instead, very weak, ("nonadiabatic reactions") the probability of successfully reaching the lower surface on the products' side is small and can be calculated using quantum mechanical perturbation theory, for example, using Fermi's "Golden Rule," an improvement over the 1-dimensional Landau-Zener treatment.

Thus, there is some difference and some similarity with a more conventional type of reaction such as reaction (4), whose potential energy contour plots were depicted in Figure 2. In both cases, fluctuations of coordinates are needed to reach the transition state, but since so many coordinates can now play a significant role in the electron transfer reaction, because of the major and relatively abrupt change in charge distribution on passing through the transition state region, a rather different approach from the conventional one was needed to formulate the details of the theory.

Electron transfer theory. Treatment

In the initial paper (1956) I formulated the above picture of the mechanism of electron transfer and, to make the calculation of the reaction rate tractable, treated the solvent as a dielectric continuum. In the transition state the position-dependent dielectric polarization $P_u(r)$ of the solvent, due to the orientation and vibrations of the solvent molecules, was not the one in equilibrium with the reactants' or the products' ionic charges. It represented instead, some macroscopic fluctuation from them. The electronic polarization for the solvent molecules, on the other hand, can rapidly respond to any such fluctuations and so is that which is dictated by the reactants' charges and by the instantaneous $P_u(r)$.

With these ideas as a basis, what was then needed was a method of calculating the electrostatic free energy G of this system with its still un-known polarization function $P_u(r)$. I obtained this free energy G by finding a reversible path for reaching this state of the system. Upon then minimizing G, subject to the constraint imposed by the Franck-Condon principle (reflected in the electron transfer occurring at the intersection of the two potential energy surfaces), I was able to find the unknown $P_u(r)$ and, hence, to find the G for the transition state. That G was then introduced into transition state theory and the reaction rate calculated.

In this research I also read and was influenced by a lovely paper by Platzmann and Franck (1952) on the optical absorption spectra of halide ions in water and later by work of physicists such as Pekar and Frohlich (1954) on the closely related topic of polaron theory. As best as I can recall now, my first expressions for G during this month of intense activity seemed rather clumsy, but then with some rearrangement a simple expression emerged that had the right "feel" to it and that I was also able to obtain by a somewhat independent argument. The expression also reduced reassuringly to the usual one, when the constraint of arbitrary $P_u(r)$ was removed. Obtaining the result for the mechanism and rate of electron transfer was indeed one of the most thrilling moments of my scientific life.

The expression for the rate constant k of the reaction is given by

$$k = A\exp\left(\frac{-\Delta G^*}{k_B T}\right),$$
(5a)

where ΔG^*, in turn, is given by

$$\Delta G^* = \frac{\lambda}{4}\left(1 + \frac{\Delta G^0}{\lambda}\right)^2.$$
(5b)

The A in Eq. (5a) is a term depending on the nature of the electron transfer reaction (e.g., bimolecular or intramolecular), ΔG^0 is the standard free energy of reaction (and equals zero for a self-exchange reaction), λ is a "reorganization term," composed of solvational (λ_o) and vibrational (λ_i) components.

$$\lambda = \lambda_o + \lambda_i$$
(6)

In a two-sphere model of the reactants, λ_o was expressed in terms of the two ionic radii a_1 and a_2 (including in the radius any inner coordination shell), the center-to-center separation distance R of the reactants, the optical (D_{op}) and static (D_s) dielectric constants of the solvent, and the charge transferred Δe from one reactant to the other:

$$\lambda_o = (\Delta e)^2 \left(\frac{1}{2a_1} + \frac{1}{2a_2} - \frac{1}{R} \right) \left(\frac{1}{D_{op}} - \frac{1}{D_s} \right) \tag{7}$$

For a bimolecular reaction, work terms, principally electrostatic, are involved in bringing the reactants together and in separating the reaction products, but are omitted from Eq. (5) for notational brevity. The expression for the vibrational term λ_i is given by

$$\lambda_i = \frac{1}{2} \sum_j k_j (Q_j^r - Q_j^p)^2 \tag{8}$$

where Q_j^r and Q_j^p are equilibrium values for the jth normal mode coordinate Q, and k_j is a reduced force constant $2k_j^r k_j^p / (k_j^r + k_j^p)$ k_j^r being the force constant for the reactants and k_j^p being that for the products. (I introduced a "symmetrization" approximation for the vibrational part of the potential energy surface, to obtain this simple form of Eqs. (5) to (8), and tested it numerically.)

In 1957 I published the results of a calculation of the λ_i arising from a stretching vibration in the innermost coordination shell of each reactant, (the equation used for λ_i was given in the 1960 paper). An early paper on the purely vibrational contribution using chemical bond length coordinates and neglecting bond-bond correlation had already been published for self-exchange reactions by George and Griffiths in 1956.

I also extended the theory to treat electron transfers at electrodes, and distributed it as an Office of Naval Research Report in 1957, the equations being published later in a journal paper in 1959. I had little prior knowledge of the subject, and my work on electrochemical electron transfers was facilitated considerably by reading a beautiful and logically written survey article of Roger Parsons on the equilibrium electrostatic properties of electrified metal-solution interfaces.

In the 1957 and 1965 work I showed that the electrochemical rate constant was again given by Eqs. (5)–(8), but with A now having a value appropriate to the different "geometry" of the encounter of the participants in the reaction. The $1/2a_2$ in Eq. (7) was now absent (there is only one reacting ion) and R now denotes twice the distance from the center of the reactant's charge to the electrode (it equals the ion-image distance). A term $e\eta$ replaced the ΔG^o in Eq.(5b), where e is the charge transferred between the ion and the electrode, and η is the activation overpotential, namely the metal-solution potential difference, relative to the value it would have if the rate constants for the forward and reverse reactions were equal. These rate constants are equal when the minima of the two G curves in Figure 5 have the same height.

Free Energy Curves

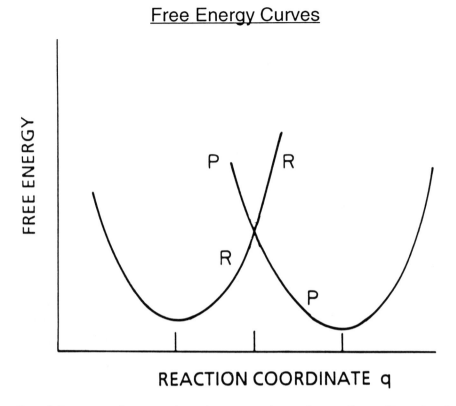

Figure 5. Free energy of reactants plus environment vs. the reaction coordinate q (R curve), and free energy of products plus environment vs. reaction coordinate q (P curve). The three vertical lines on the abscissa denote, from left to right, the value for the reactants, for the transition state, and for the products. (Marcus and Siddarth, ref. 2).

When $|e\eta| < \lambda$, most electrons go into or out of quantum states in the metal that are near the Fermi level. However, because of the continuum of states in the metal, the inverted effect (described below) was now predicted to be absent for this process, i.e., the counterpart of Eq. (5b) is applicable only in the region $|e\eta| < \lambda$: In the case of an intrinsically highly exothermic electron transfer reaction at an electrode, the electron can remove the immediate "exothermicity" by (if entering) going into a high unoccupied quantum state of the metal, or (if leaving) departing from a low occupied quantum state, each far removed from the Fermi level. (The inverted region effect should, however, occur for the electron transfer when the electrode is a narrow band semiconductor.)

After these initial electron transfer studies, which were based on a dielectric continuum approximation for the solvent outside the first coordination shell of each reactant, I introduced a purely molecular treatment of the reacting system. Using statistical mechanics, the solvent was treated as a collection of dipoles in the 1960 paper, and later in 1965 a general charge distribution was used for the solvent molecules and for the reactants. At the same time I found a way in this 1960 paper of introducing rigorously a global reaction coordinate in this many-dimensional (*N*) coordinate space of the reacting system. The globally defined coordinate so introduced was equivalent to using $U_p - U_r$, the potential energy difference between the products plus solvent (U_p) and the reactants plus solvent (U_r) (cf A. Warshel, 1987). It was, thereby, a coordinate defined everywhere in this *N*-dimensional space.

The free energy G_r of a system containing the solvent and the reactants, and that of the corresponding system for the products, G_p, could now be defined along this globally defined reaction coordinate. (In contrast, in reactions such as that depicted by Figure 2, it is customary, instead, to define a reaction coordinate locally, namely, in the vicinity of a path leading from the valley of the reactants through the saddle point region and into the valley of the products.)

The potential energies U_r and U_p in the many-dimensional coordinate space are simple functions of the vibrational coordinates but are complicated functions of the hundreds of relevant solvent coordinates: there are many local minima corresponding to locally stable arrangements of the solvent molecules. However, I introduced a "linear response approximation," in which any hypothetical change in charge of the reactants produces a proportional change in the dielectric polarization of the solvent. (Recently, I utilized a central limit theorem to understand this approximation better — beyond simple perturbation theory, and plan to submit the results for publication shortly.) With this linear approximation the free energies G_r and G_p became simple quadratic functions of the reaction coordinate.

Such an approach had major consequences. This picture permitted a depiction of the reaction in terms of parabolic free energy plots in simple and readily visualized terms, as in Figure 5. With them the trends predicted from the equations were readily understood. It was also important to use the free energy curves, instead of oversimplified potential energy profiles, because of the large entropy changes which occur in many electron transfer cross-reactions, due to changes in strong ion-polar solvent interactions. (The free energy plot is legitimately a one-coordinate plot while the potential energy plot is at most a profile of the complicated U_r and U_p in *N*-dimensional space.)

With the new statistical mechanical treatment of 1960 and 1965 one could also see how certain relations between rate constants initially derivable from the dielectric continuum-based equations in the 1956 paper could also be valid more generally. The relations were based, in part, on Equations (5) and (initially via (7) and (8)) on the approximate relation

$$\lambda_{12} \cong \frac{1}{2} (\lambda_{11} + \lambda_{22}) \tag{9}$$

where λ_{12} is the λ for the cross-reaction and the λ_{11} and λ_{22} are those of the self-exchange reactions.

Predictions

In the 1960 paper I had listed a number of theoretical predictions resulting from these equations, in part to stimulate discussion with experimentalists in the field at a Faraday Society meeting on oxidation-reduction reactions, where this paper was to be presented. At the time I certainly did not anticipate the subsequent involvement of the many experimentalists in testing these predictions. Among the latter was one which became one of the most widely tested aspects of the theory, namely, the "cross-relation." This expression, which follows from Eqs. (5) and (9), relates the rate constant k_{12} of a cross-reaction to the two self-exchange rate constants, k_{11} and k_{22}, and to the equilibrium constant K_{12} of the reaction.

$$k_{12} \cong (k_{11}k_{22}K_{12}f_{12})^{1/2}, \tag{10}$$

where f_{12} is a known function of k_{11}, k_{22} and K_{12} and is usually close to unity.

Another prediction in the 1960 paper concerned what I termed there the inverted region: In a series of related reactions, similar in λ but differing in ΔG^o, a plot of the activation free energy ΔG^* vs. ΔG^o is seen from Eq. (5b) to first decrease as ΔG^o is varied from 0 to some negative value, vanish at $\Delta G^o = -\lambda$, and then increase when ΔG^o is made still more negative. This initial decrease of ΔG^* with increasingly negative ΔG^o is the expected trend in chemical reactions and is similar to the usual trend in "Bronsted plots" of acid or base catalyzed reactions and in "Tafel plots" of electrochemical reactions. I termed that region of ΔG^o the "normal" region. However, the prediction for the region where $-\Delta G^o > \lambda$, the "inverted region," was the unexpected behavior, or at least unexpected until the present theory was introduced.

This inverted region is also easily visualized using Figures 6 and 7: Successively making ΔG^o more negative, by lowering the products' G curve vertically relative to the reactant curve, decreases the free energy barrier ΔG^* (given by the intersection of the reactants' and products' curves): that barrier is seen in Figure 6 to vanish at some ΔG^o and then to increase again.

Other predictions dealt with the relation between the electrochemical and the corresponding self-exchange electron transfer rates, the numerical estimate of the reaction rate constant k and, in the case of non-specific solvent effects, the dependence of the reaction rate on solvent dielectric properties. The testing of some of the predictions was delayed by an extended sabbatical in 1960−61, which I spent auditing courses and attending seminars at the nearby Courant Mathematical Institute.

Comparisons of Experiment and Theory

Around 1962, during one of my visits to Brookhaven National Laboratory, I showed Norman Sutin the 1960 predictions. Norman had either measured via his stopped-flow apparatus or otherwise knew rate constants and equilibrium constants which permitted the cross-relation Eq. (10) to be tested. There were about six such sets of data which he had available. I remember vividly the growing sense of excitement we both felt as, one by one, the

The Inverted Region Effect

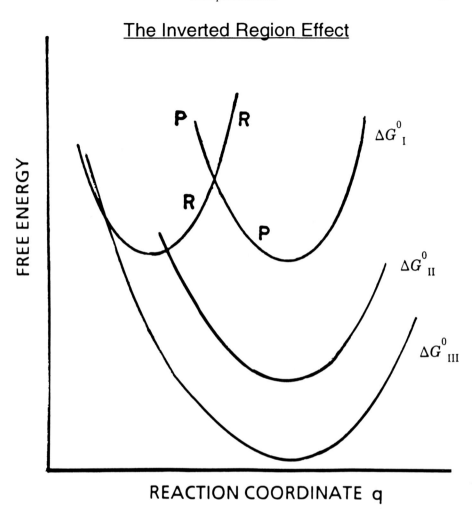

Figure 6. Plot of the free energy *G* versus the reaction coordinate *q*, for reactants' (R) and products' (P), for three different values of ΔG^o, the cases I to III indicated in Figure 7 (Marcus and Siddarth, ref. 2).

The Inverted Region Effect

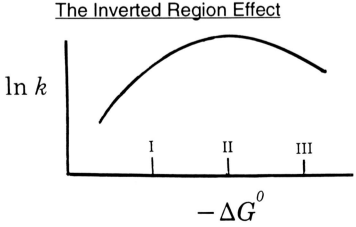

Figure 7. Plot of ln k_r vs $-\Delta G^o$. Points I and III are in the normal and inverted regions, respectively, while point II, where ln k_r is a maximum, occurs at $-\Delta G^o = \lambda$ (Marcus and Siddarth, ref. 2).

observed k_{12}'s more or less agreed with the predictions of the relation. I later collected the results of this and of various other tests of the 1960 predictions and published them in 1963. Perhaps by showing that the previously published expressions were not mere abstract formulae, but rather had concrete applications, this 1963 paper, and many tests by Sutin and others, appear to have stimulated numerous subsequent tests of the cross-relation and of the other predictions. A few examples of the cross-relation test are given in Table 1.

The encouraging success of the experimental tests given in the 1963 paper suggested that the theory itself was more general than the approximations (e.g., solvent dipoles, unchanged force constants) used in 1960 and stimulated me to give a more general formulation (1965). The latter paper also contains a unified treatment of electron transfers in solution and at metal electrodes, and served, thereby, to generalize my earlier (1957) treatment of the electrochemical electron transfers.

The best experimental evidence for the inverted region was provided in 1984 by Miller, Calcaterra and Closs, almost 25 years after it was predicted.

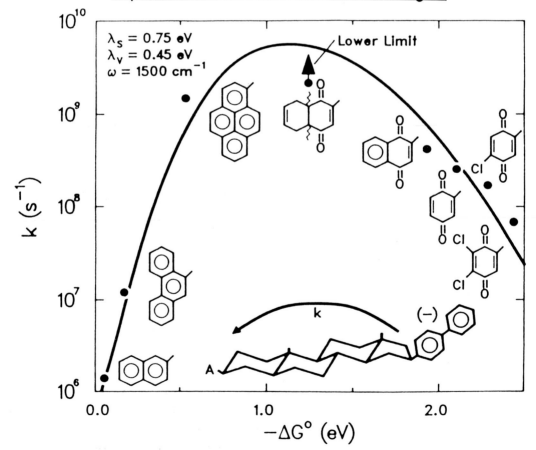

Figure 8. Inverted region effect in chemical electron transfer reactions. (Miller, *et al*, ref. 3).

This successful experimental test, which was later obtained for other electron transfer reactions in other laboratories, is reproduced in Figure 8. Possible reasons for not observing it in the earlier tests are several-fold and have been discussed elsewhere.

Previously, indirect evidence for the inverted region had been obtained by observing that electron transfer reactions with a very negative ΔG^* may result in chemiluminescence: when the G_r and G_p curves intersect at a high ΔG^* because of the inverted region effect, there may be an electron transfer to a more easily accessible G_p curve, one in which one of the products is electronically excited and which intersects the G_r curve in the normal region at a low ΔG^*, as in Figure 9. Indeed, experimentally in some reactions 100 % formation of an electronically excited state of a reaction product has been observed by Bard and coworkers, and results in chemiluminescence.

Another consequence of Eq. (5) is the linear dependence of $k_B T \ln k$ on $- \Delta G^o$ with a slope of $1/2$, when $|\Delta G^o / \lambda|$ is small, and a similar behavior at electrodes, with ΔG^o replaced by $e\eta$ the product of the charge transferred

Formation of Electronically Excited Products

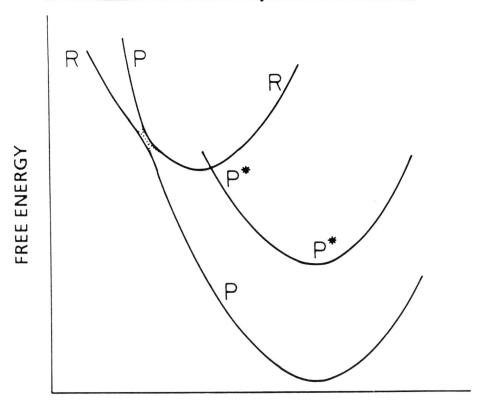

REACTION COORDINATE q

Figure 9. A favored formation of an electronically excited state of the products (Marcus and Siddarth, ref. 2).

and the activation overpotential. Extensive verification of both these results has been obtained. More recently, the curvature of plots of ln k vs. $e\eta$, expected from these equations, has been demonstrated in several experiments. The very recent use of ordered organic molecular monolayers on electrodes, either to slow down the electron transfer rate or to bind a redox-active agent to the electrode, but in either case to avoid or minimize diffusion control of the fast electron transfer processes, has considerably facilitated this study of the curvature in the ln k vs. $e\eta$ plot.

Comparison of experiment and theory has also included that of the absolute reaction rates of the self-exchange reactions, the effect on the rate of varying the solvent, an effect sometimes complicated by ion pairing in the low dielectric constant media involved, and studies of the related problem of charge transfer spectra, such as

$$DA + h\upsilon \rightarrow D^+A^- \tag{11}$$

Here, the frequency of the spectral absorption maximum υ_{max} is given

$$h\upsilon_{max} = \lambda + \Delta G^o. \tag{12}$$

Comparisons with Eq: (12), using Eq.(7) for λ, have included those of the effects of separation distance and of the solvent dielectric constant.

Comparisons have also been made of the self-exchange reaction rates in solution with the rates of the corresponding electron transfer reactions at

Electrochemical vs Self-Exchange Rate Constants

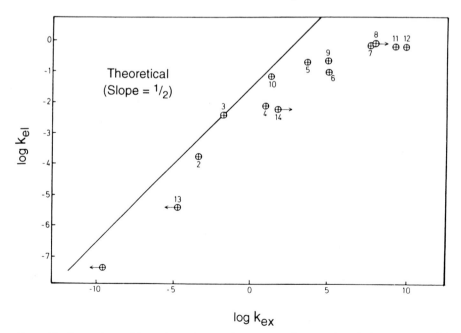

Figure 10. Comparison of isotopic exchange electron transfer rates in solution, covering 20 orders of magnitude, with rates of corresponding electron transfers at metal electrodes. (Cannon, ref. 2).

electrodes. An example of the latter is the plot given in Figure 10, where the self-exchange rates are seen to vary by some twenty orders of magnitude. The discrepancy at high *k*'s is currently the subject of some reinvestigation of the fast electrode reaction rates, using the new nanotechnology. Most recently, a new type of interfacial electron transfer rate has also been measured, electron transfer at liquid-liquid interfaces. In treating the latter, I extended the "cross relation" to this two-phase system. It is clear that much is to be learned from this new area of investigation. (The study of the transfer of ions across such an interface, on the other hand, goes back to the time of Nernst and of Planck, around the turn of the century.)

Other Applications and Extensions
As noted in Figure 1, one aspect of the electron transfer field has been its continued and, indeed, ever-expanding growth in so many directions. One of these is in the biological field, where there are now detailed experimental and theoretical studies in photosynthetic and other protein systems. The three-dimensional structure of a photosynthetic reaction center, the first membrane protein to be so characterized, was obtained by Deisenhofer, Michel and Huber, who received the Nobel Prize in Chemistry in 1988 for this work. A bacterial photosynthetic system is depicted in Figure 11, where the protein framework holding fast the constituents in this reaction center is not shown.

The Reaction Center

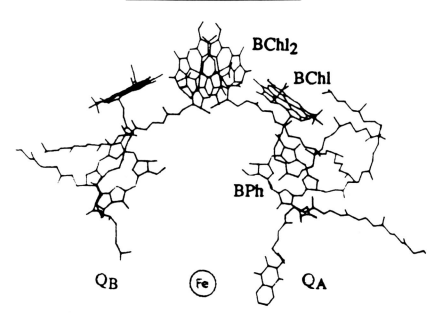

Figure 11. Redox-active species involved in the initial charge separation for a photosynthetic bacterium (cf Deisenhofer *et al*, ref. 3, and Yeates *et al*, ref. 3), with labels added, to conform to the present text; they include a missing Q_B.

In the photosynthetic system there is a transfer of electronic excitation from "antenna" chlorophylls (not shown in Figure 11) to a special pair $BChl_2$. The latter then transfers an electron to a pheophytin BPh within a very short time (~ 3 picoseconds) and from it to a quinone Q_A in 200 psec and thence to the other quinone Q_B. (Other chemical reactions then occur with these separated charges at each side of the membrane, bridged by this photosynthetic reaction center.)

To avoid wasting the excitation energy of the $BChl_2^*$ unduly it is necessary that the $-\Delta G^o$ of this first electron transfer to BPh be small. (It is only about 0.25 eV out of an overall excitation energy of $BChl_2^*$ of 1.38 eV.) In order that this electron transfer also be successful in competing with two wasteful processes, the fluorescence and the radiationless transition of $BChl_2^*$, it is also necessary that ΔG^* for that first electron transfer step be small and hence, by Eq. (5b), that the λ be small. The size of the reactants is large, and the immediate protein environment is largely nonpolar, so leading to a small λ (cf Eq.(7)). Nature appears, indeed, to have constructed a system with this desirable property.

Furthermore, to avoid another form of wasting the energy, it is also important that an unwanted back electron transfer reaction from the BPh^- to the $BChl_2^+$ not compete successfully with a second forward electron transfer step from BPh^- to Q_A. That is, it is necessary that the back transfer, a "hole-electron recombination" step, be slow, even though it is a very highly exothermic process ($\sim 1.1eV$). It has been suggested that the small λ ($\sim 0.25eV$) and the resulting inverted region effect play a significant role in providing this essential condition for the effectiveness of the photosynthetic reaction center.

There is now a widespread interest in synthesizing systems which can mimic the behavior of nature's photosynthetic systems, and so offer other routes for the harnessing of solar energy. The current understanding of how nature works has served to provide some guidelines. In this context, as well as that of electron transfer in other proteins, there are also relevant experiments in long range electron transfer. Originally the studies were of electron transfer in rigid glasses and were due to Miller and coworkers. More recently the studies have involved a donor and receptor held together by synthetically made rigid molecular bridges. The effect of varying the bridge length has been studied in the various systems. A theoretical estimate of the distance dependence of electron transfers in a photosynthetic system was first made by Hopfield, who used a square barrier model and an approximate molecular estimate of the barrier height.

Recently, in their studies of long range electron transfer in chemically modified proteins, Gray and coworkers have studied systematically the distance or site dependence of the electronic factor, by attaching an appropriate electron donor or acceptor to a desired site. For each such site the reactant chosen should be such that $-\Delta G^o \approx \lambda$, i.e., which has a k at the maximum of the $\ln k$ vs. $-\Delta G^o$ curve (cf Eq. (5)). The value of k then no longer depends on a ΔG^*. Since ΔG^* is distance-dependent (cf Eq.(7)), it is particularly desirable to make $\Delta G^* \approx 0$, so that the relative k's at the various sites now reflect only the electronic factor. Dutton and coworkers have treated data similarly for a number of reactions by using, where possible, the k at the maximum of each ln k vs. ΔG^0 curve. Of particular interest in such studies is whether there is a simple exponential decrease of the electronic factor on the separation distance between donor and acceptor, or whether there are deviations from this monotonic behavior, due to local structural factors.

In a different development, the mechanism of various organic reactions has been explored by several investigators, notably by Eberson (ref. 2), in the light of current electron transfer theory. Other organic reactions have been explored by Shaik and Pross, in their analysis of a possible electron transfer mechanism vs. a conventional mechanism, and by Shaik *et al* (ref.2).

Theoretical calculations of the donor-acceptor electronic interactions, initially by McConnell and by Larsson, and later by others, our group among them, have been used to treat long-range electron transfer. The methods have recently been adapted to large protein systems. In our studies with Siddarth we used an "artificial intelligence" searching technique to limit the number of amino acids used in the latter type of study.

Another area of much current activity in electron transfers is that of solvent dynamics, following the pioneering treatment for general reactions by Kramers (1940). Important later developments for electron transfer were made by many contributors. Solvent dynamics affects the electron transfer reaction rate when the solvent is sufficiently sluggish. As we showed recently with Sumi and Nadler, the solvent dynamics effect can also be modified significantly, when there are vibrational (λ_i) contribution to λ.

Computational studies, such as the insightful one of David Chandler and coworkers on the $Fe^{2+} + Fe^{3+}$ self-exchange reaction, have also been employed recently. Using computer simulations they obtained a verification of the parabolic G curves, even for surprisingly high values of the fluctuation in G. They also extended their studies to dynamical and quantum mechanical effects of the nuclear motion. Studies of the quantum mechanical effects on the nuclear motion on electron transfer reactions were initiated in 1959 by Levich and Dogonadze, who assumed a harmonic oscillator model for the polar solvent medium and employed perturbation theory. Their method was related to that used for other problems by Huang and Rhys (1951) and Kubo and Toyozawa (1954).

There were important subsequent developments by various authors on these quantum effects, including the first discussion of quantum effects for the vibrations of the reactants by Sutin in 1962 and the important work of Jortner and coworkers in 1974−75, who combined a Levich and Dogonadze type approach to treat the high frequency vibrations of the reactants with the classical expression which I described earlier for the polar medium. These quantum effects have implications for the temperature dependence of k, among other effects. Proceeding in a different (classical) direction Saveant recently showed how to extend Eq. (5b) to reactions which involved the rupture of a chemical bond by electron transfer and which he had previously studied experimentally: $M(e) + RX \rightarrow M + R + X^-$, where R is an alkyl group, X a halide and M a metal electrode.

A particularly important early development was that by Taube in the 1950s; he received the Nobel Prize for his work in 1983. Taube introduced the idea of different mechanisms for electron transfer — outer sphere and inner sphere electron transfers, which he had investigated experimentally. His experimental work on charge transfer spectra of strongly interacting systems (''Creutz-Taube'' ion, 1959, 1973) and of weakly interacting ones has been similarly influential. Also notable has been Hush's theoretical work on charge transfer spectra, both of intensities and absorption maxima (1967), which supplemented his earlier theoretical study of electron transfer rates (1961).

There has been a "spin-off" of the original electron transfer theory to other types of chemical reactions as well. In particular, the ΔG^* vs ΔG^o relation and the cross-relation have been extended to these other reactions, such as the transfer of atoms, protons, or methyl groups. (Even an analog of Eqs. (5b) and (9), but for binding energies instead of energy barriers has been introduced to relate the stability of isolated protonbound dimers AHB^+ to those of AHA^+ and BHB^+!)

Since the transfer of these nuclei involves strong electronic interactions, it is not well represented by intersecting parabolic free energy curves, and so a different theoretical approach was needed. For this purpose I adapted (1968) a "bond-energy-bond-order" model of H. Johnston, in order to treat the problem for a reaction of the type given by Eq.(4). The resulting simple expression for ΔG^* is similar to Eq.(5), when $|\Delta G^o/\lambda|$ is not large ($<1/2$), but differs from it in not having any inverted region. It has the same λ property as that given by Eq.(9), and has resulted in a cross-relation analogous to Eq. (10). The cross-relation has been tested experimentally for the transfer of methyl groups by E. Lewis, and the ΔG^* vs ΔG^o relation has been used or tested for other transfers by Albery and by Kreevoy and their coworkers, among others.

It is naturally gratifying to see one's theories used. A recent article, which showed the considerable growth in the use of papers such as the 1956 and 1964 articles (ref. 5), points up the impressive and continued vitality of the field itself. The remarks above on many areas of electron transfer and on the spin-off of such work on the study of other types of reactions represent a necessarily brief picture of these broad-based investigations.

ACKNOWLEDGMENTS

My acknowledgments are to my many fellow researchers in the electron transfer field, notably Norman Sutin, with whom I have discussed so many of these matters for the past thirty or more years. I also thank my students and post-doctorals, whose presence was a constant source of stimulation to me, both in the electron transfer field and in the other fields of research

Table I. Comparison of Calculated and Experimental k_{12} Values[a]

Reaction	k_{12}, M^{-1} sec^{-1}	
	Observed	Calculated
$IrCl_6^{2-}$ + $W(CN)_8^{4-}$	6.1×10^7	6.1×10^7
$IrCl_6^{2-}$ + $Fe(CN)_6^{4-}$	3.8×10^5	7×10^5
$IrCl_6^{2-}$ + $Mo(CN)_8^{4-}$	1.9×10^6	9×10^5
$Mo(CN)_8^{3-}$ + $W(CN)_8^{4-}$	5.0×10^6	4.8×10^6
$Mo(CN)_8^{3-}$ + $Fe(CN)_6^{4-}$	3.0×10^4	2.9×10^4
$Fe(CN)_6^{3-}$ + $W(CN)_8^{4-}$	4.3×10^4	6.3×10^4
Ce^{IV} + $W(CN)_8^{4-}$	$>10^5$	4×10^8
Ce^{IV} + $Fe(CN)_6^{4-}$	1.9×10^6	8×10^6
Ce^{IV} + $Mo(CN)_6^{4-}$	1.4×10^7	1.3×10^7
L-Co[(−)PDTA]$^{2-}$ + $Fe(bipy)_3^{3+}$	8.1×10^4	$\geq 10^5$
L-Fe[(−)PDTA]$^{2-}$ + $Co(EDTA)^-$	1.3×10^1	1.3×10^1
L-Fe[(−)PDTA]$^{2-}$ + $Co(ox)_3^{3-}$	2.2×10^2	1.0×10^2
$Cr(EDTA)^{2-}$ + $Fe(EDTA)^-$	$\geq 10^6$	10^9
$Cr(EDTA)^{2-}$ + $Co(EDTA)^-$	$\simeq 3 \times 10^5$	4×10^7
$Fe(EDTA)^{2-}$ + $Mn(CyDTA)^-$	$\simeq 4 \times 10^5$	6×10^6
$Co(EDTA)^{2-}$ + $Mn(CyDTA)$	9×10^{-1}	2.1
$Fe(PDTA)^{2-}$ + $Co(CyDTA)^-$	1.2×10^1	1.8×10^1
$Co(terpy)_2^{2+}$ + $Co(bipy)_3^{3+}$	6.4×10	3.2×10
$Co(terpy)_2^{2+}$ + $Co(phen)_3^{3+}$	2.8×10^2	1.1×10^2
$Co(terpy)_2^{2+}$ + $Co(bipy)(H_2O)_4^{3+}$	6.8×10^2	6.4×10^4
$Co(terpy)_2^{2+}$ + $Co(phen)(H_2O)_4^{3+}$	1.4×10^3	6.4×10^4
$Co(terpy)_2^{2+}$ + $Co(H_2O)_6^{3+}$	7.4×10^4	2×10^{10}
$Fe(phen)_3^{3+}$ + MnO_4^-	6×10^3	4×10^3
$Fe(CN)_6^{4-}$ + MnO_4^-	1.3×10^4	5×10^3
$V(H_2O)_6^{2+}$ + $Ru(NH_3)_6^{3+}$	1.5×10^{3} [a]	4.2×10^3
$Ru(en)_3^{2+}$ + $Fe(H_2O)_6^{3+}$	8.4×10^4	4.2×10^5
$Ru(NH_3)_6^{2+}$ + $Fe(H_2O)_6^{3+}$	3.4×10^5	7.5×10^6
$Fe(H_2O)_6^{2+}$ + $Mn(H_2O)_6^{3+}$	1.5×10^4	3×10^4

[a] Bennett, ref. 3.

which we have explored. In its earliest stage and for much of this period this research was supported by the Office of Naval Research and also later by the National Science Foundation. The support of both agencies continues to this day and I am very pleased to acknowledge its value and timeliness here.

In my Nobel lecture, I concluded on a personal note with a slide of my great uncle, Henrik Steen (né Markus), who came to Sweden in 1892. He received his doctorate in theology from the University of Uppsala in 1915, and was an educator and a prolific writer of pedagogic books. As I noted in the biographical sketch in Les Prix Nobel, he was one of my childhood idols. Coming here, visiting with my Swedish relatives — some thirty or so descendants — has been an especially heartwarming experience for me and for my family. In a sense I feel that I owed him a debt, and that it is most fitting to acknowledge that debt here.

REFERENCES

Some of my relevant articles, largely from the 1956−65 period, are listed in ref.1 below, and some general references which review the overall literature are listed in ref. 2. Several additional references for the Table and for the Figures are given in ref. 3. Classic texts on unimolecular reactions are given in ref. 4.

1. R. A. Marcus, J. Chem. Phys. **24,** 966 (1956); *ibid.,* **24,** 979 (1956); *ibid.,* **26,** 867 (1957); *ibid,* **26,** 872 (1957); Trans. N. Y. Acad. Sci. **19,** 423 (1957); ONR Technical Report No. 12, Project NR 051-331 (1957), reproduced in *Special Topics in Electrochemistry, P.* A. Rock, ed., Elsevier, New York, 1977, p. 181; Can. J. Chem. **37,** 155 (1959); Discussions Faraday Soc. **29,** 21 (1960); J. Phys. Chem. **67,** 853, 2889 (1963); J. Chem. Phys. **38,** 1858 (1963); ibid., **39,** 1734 (1963); Ann. Rev. Phys. Chem. **15,** 155 (1964); J.Chem.Phys. **43,** 679 (1965); ibid., **43,** 1261 (1965); ibid., **43,** 2654 (1965), (corr.) **52,** 2803 (1970); J. Phys. Chem. **72,** 891 (1968)

2. R. A. Marcus and N. Sutin, Biochim. Biophys. Acta, **811,** 265 (1985); J.R. Bolton, N. Mataga and G. McLendon (eds.) Adv. Chem. Ser., **228,** (1991), assorted articles; M.D. Newton and N. Sutin, Ann. Rev. Phys. Chem., **35,** 437 (1984); N. Sutin, Prog. Inorg. Chem., **30,** 441 (1983); M.D. Newton, Chem. Rev., **91,** 767 (1991); J. Ulstrup, *Charge Transfer Processes in Condensed Media,* Springer, New York, 1979; R.D. Cannon, *Electron Transfer Reactions,* Butterworths, London, 1980; L. Eberson, *Electron Transfer Reactions in Organic Chemistry,* Springer, New York, 1987; M.A. Fox and M. Chanon (eds.), *Photoinduced Electron Transfer,* Elsevier, New York, 1988, 4 vols.; M.V. Twigg (ed.), *Mechanisms of Inorganic and Organometallic Reactions,* vol. 7, 1991, Chaps. 1 and 2, and earlier volumes; R.A. Marcus and P. Siddarth, in *Photoprocesses in Transition Metal Complexes, Biosystems and Other Molecules: Experiment and Theory,* E. Kochanski, ed., Kluwer, Norwall, Massachusetts, 1992, p. 49; S.S. Shaik, H.B. Schlegel and S. Wolfe, *Theoretical Aspects of Physical Organic Chemistry,* J. Wiley, New York, 1992; N. Sutin, Pure & Applied Chem. **60,** 1817 (1988); Assorted articles in Chem. Revs. **92,** No. 3 (1992); R.A. Marcus Commemorative Issue, J. Phys. Chem. **90,** (1986).

3. L. E. Bennett, Prog.Inorg.Chem. **18,** 1 (1973); J.R. Miller, L.T. Calcaterra and G.L. Closs, J.Am.Chem.Soc. **106,** 3047 (1984); J. Deisenhofer, O. Epp, K. Miki, R. Huber and H. Michel, J.Mol.Biol. **180,** 385 (1984); J. Deisenhofer and H. Michel, Angew.Chem.Int.Ed.Engl. **28,** 829 (1989); T.G. Yeates, H. Komiya, D.C. Rees, J.P. Allen and G. Feher, Proc.Nat.Acad.Sci. **84,** 6438 (1987).

4. P. J. Robinson and H.A. Holbrook, *Unimolecular Reactions,* J.Wiley, New York 1972; W. Forst, *Theory of Unimolecular Reactions,* Academic Press, New York, 1973; cf also the very recent text, R.G. Gilbert and S.C. Smith, *Theory of Unimolecular and Recombination Reactions,* Blackwells, Oxford, 1990.

5. Science Watch, **3,** No.9, November,1992, p.8.

Chemistry 1993

KARY B. MULLIS

for his invention of the polymerase chain reaction (PCR) method

and

MICHAEL SMITH

for his fundamental contributions to the establishment of oligonucleotide-based, site-directed mutagenesis and its development for protein studies

THE NOBEL PRIZE IN CHEMISTRY

Speech by Professor Carl-Ivar Brändén of the Royal Swedish Academy of Sciences.
Translation from the Swedish text.

Your Majesties, Your Royal Highnesses, Ladies and Gentlemen,

Our genetic material, which gives every living organism its unique characteristics, is built up from large and complex DNA molecules, each comprising hundreds of millions of atoms. For a long time it was believed that these molecules were outside the realm of the chemical laboratory and that their manipulation could only be achieved through the complex machinery of a living cell. This year's Nobel Laureates in Chemistry, Kary B. Mullis and Michael Smith, have drastically changed this notion by providing research tools that chemists can use outside a living cell to amplify and specifically modify any given gene.

The foundations of the conceptual framework for this development were laid down by the discovery of the double-helical nature of the DNA structure for which the Nobel Prize in Physiology or Medicine was given to Francis Crick, James Watson and Maurice Wilkins in 1962. The four building blocks of the DNA molecule, the nucleotides, are arranged in a specific order along the molecule to form a genetic code for the sequence of amino acids in the corresponding protein molecules. This genetic information is present in a complementary fashion in the two strands of the DNA molecule. Hence, one strand can serve as a template for the synthesis of the second strand. Mutations occur when the sequence of nucleotides is changed. In living cells such mutations occur randomly and evolution is therefore a trial and error process. It can easily be calculated that for a given gene only a tiny fraction of all possible combinations of these nucleotides have been made and tested during the 3.5 billion years that life has been present on Earth. Consequently, there is scope for the design of novel and interesting protein molecules provided we have the knowledge and methodology to obtain and amplify existing genes and to make appropriate changes to them.

In the 1970s, Michael Smith developed a general method for producing mutations in a gene, not in a random fashion but at specific positions determined in advance from the sequence of the nucleotides in the gene. This method of site-directed mutagenesis has created completely new opportunities to study the properties of protein molecules: how they function as catalysts or as signal transmitters through membranes, which factors determine how they fold into specific three-dimensional structures and how they interact with other molecules in the cell. Such protein engineering is also of importance in modern biotechnology and drug design. Novel antibodies have been created that can kill certain cancer cells. Plants that produce proteins enriched in essential amino acids are being field tested and in the future this method might produce engineered wheat and corn flour that has the same nutritional value as meat.

Isolation and amplification of a specific gene was one of the outstanding problems in DNA technology, including site-directed mutagenesis, until 1985 when Kary Mullis presented the Polymerase Chain Reaction, now commonly known as PCR. Using this method it is possible to amplify and isolate in a test tube a specific DNA segment within a background of a complex gene pool. In this repetitive process the number of copies of the specific DNA segment doubles during each cycle. In a few hours it is possible to achieve more than 20 cycles, which produces over a million copies.

The PCR method has already had a profound influence on basic research in biology. Cloning and sequencing of genes as well as site-directed mutagenesis have been facilitated and made more efficient. Genetic and evolutionary relationships are easily studied by the PCR method even from ancient fossils containing only fragments of DNA. Biotechnology applications of PCR are numerous. In addition to being an indispensable research tool in drug design, the PCR method is now used in diagnosis of viral and bacterial infections including HIV. The method is so sensitive that it is used in forensic medicine to analyze the DNA content of a drop of blood or a strand of hair.

Dr. Kary Mullis and Professor Michael Smith,
On behalf of the Royal Swedish Academy of Sciences I wish to convey to you our warmest congratulations for your outstanding accomplishments and ask you to receive the Nobel Prize from the hands of his Majesty the King.

Kary B. Mullis

KARY B. MULLIS

My father Cecil Banks Mullis and mother, formerly Bernice Alberta Barker grew up in rural North Carolina in the foothills of the Blue Ridge Mountains. My dad's family had a general store, which I never saw. My grandparents on his side had already died before I started noticing things. My mother's parents were close to me all during my childhood, and her father Albert stopped by to see me in a non-substantial form on his way out of this world in 1986. I was living in California. "Pop" died at 92 and wondering what was happening to me out in Californa, stopped by Kensington for a couple days. My house afforded a view of San Francisco and the Golden Gate Bridge. His visit was an odd experience. Not at all frightening. I have cultivated the curious things in life and found this one pleasant. "Pop" and I sat in the evenings in my kitchen and I told him about the contemporary California world while we drank beer. I drank his for him as it appeared that although he was very much there for me, he was not there at all for the beer. Many of my friends when I told them of this thought it fanciful. (I think it more likely than much of our math today and at least half of our physics, both of which I like).

Until I was five my immediate family lived near my grandfather's farm where my mother had grown up, and with the exception of a few modern conveniences, had not changed a lot over the years.

My grandfather milked several cows twice a day and supplied the neighbours with dairy products. He liked to go visiting around the county on saturdays and he also enjoyed the neighbours when they came by once a week with their empty milk jars. He walked them out to their cars and hung over the driver's side window until they drove off. The road was two tire tracks on well mown grass between barbed wire fences, cows off to the right, alfalfa or sometimes corn to the left.

I remember mostly the summers. My mother and aunts presided out on the big screened back porch shelling peas, stringing beans, peeling apples, pears, and peaches. The peaches were peeled with a special machine that had a hand crank and left a spiraling groove on what was left of the peach. The peels went to the pigs. Everything else went into steaming Mason jars which would go down into the earthen floored cellar. Down there in the dark, and it was always a little moist, were spiders in abundance and magnificent biodiversity. My brothers, and my cousins, and I ventured into the cellar once in a while to inspect the sweet potatoes and the hibernating jars. No one wanted to stay there alone ever, and mostly we played in the woods, the swamp, the orchards, the barn, the granary, which had wasps, and the woodshed, which also had wasps and, like the barn, allegedly, snakes.

We tortured the cows. We sliced apples and slipped them onto the electric fence that contained them in the newer parts of the pasture. Cows like apples and they kept trying. We watched the chickens pecking at the black mud around their chicken house. We heard the squeal of young pigs being castrated by my grandfather and the veternarian, but we weren't allowed to watch. We heard stories from our moms about balls of fire during thunder storms streaming up the drain pipe that led down to the chicken yard and dancing out of the sink onto the grey floor of the back porch. All the scorched marks had been sanded and painted over by the time we heard about it, and sadly it never happened while we were there. But there were thunderstorms. Rain would come down from a cloudburst in the summer afternoons and the woods would explode with thunder. Our moms would keep us inside and out of the draft from any windows. We . . . wanted to see those fireballs.

We could play in the attic. Even in the day there was not enough light to keep us calm in the attic, and there were animal-skin coats and unfamiliar garments that lurked in the closets. There was a horrible picture of Teddy Roosevelt killing a bear. Very bloody. And there were black widow spiders waiting for us always, down in their funnel shaped webs in all the dark corners. It was a thrilling place during a thunderstorm and, like the hay loft of the barn, a place where my pre-adolescent sexuality concerning my cousin Judy, who was one month my senior, would come a little more sharply into focus. We were only nine or ten, but it was there already with it's pressing curiosity. We sometimes kissed. My techniques have improved, but not the thrill.

When my great-grandmother died she was almost a hundred and we were glad to see her go because every time she would come over to my grandmother's house, she would try to kiss all of us. She looked almost a hundred and, heartless, cruel, mindless little children that we were, she repulsed us. She grabbed us anyway and kissed us until she was through. They put her body in a metal casket with gauzy curtains and left it in the living room near the grandfather's clock, which announced the hours with a number of resonant bongs and marked the half-hours with a single chilling tone. Her body was there for three days until the service on Sunday at Mt Zion Baptist Church. We dared each other to go in and look at her. The adults were uneffected and took their regular meals right in the next room. We found it difficult to sleep. The clock seemed more alive than usual.

My great-grandmother, as I learned from Judy much later, when we were adults, had been an unusual woman in Saw Mills, North Carolina. She lived just a bit on the wild side. She gave birth to my grandmother out of wedlock following an affair with a railroad man named Stowe. We never heard much about him. "Nanny", as we called our great-grandmother, was tolerated by the community because she was the only person for miles around who knew the rudiments of medicine. She provided medical care to livestock, for which she had been trained, but also to people for whom she was the only alternative on her side of the Catawba River. She also ran the post office in

Granite Falls. She was the first postmistress anyone had heard about, and rural North Carolinians at the time were not in the mood for new customs, but they accepted what they couldn't avoid. And granite does fall.

When my grandfather, "Pop", James Albert Barker, son of Cary Barker from Cary, N.C. decided to marry Nanny's illegitimate daughter, Princess Escoe Miller, his father gave him a piece of land to farm and tolerated his choice of bride. My given name derives from Cary with a slight change of spelling that my mother thought practical so as to keep my initialed name from being the same as my Dad's, C. B. Mullis. She probably never imagined that I would be living far away before it ever mattered.

The rest of my life has passed quite suddenly. Around ten or twelve I fell into the inevitable logarithms of time. It seems to go faster and faster. I wonder now why we have to have Christmas so often.

I went to high school in Columbia. I met my first wife, Richards, whom I married while I was working on a B. S. in chemistry at Georgia Tech. She bore Louise and I studied. I learned most of the useful technical things, math, physics, chemistry, that I now use, during those four years. I did little else, except to play with Louise and change her diapers at night. We moved to Berkeley, Californa in 1966. I did my Ph. D. in biochemistry under J. B. Neilands and there I learned the rest, the non-technical things. After that, it happened so quickly that it's hard to really talk about in the wake of my grandparents' farm.

Except for Cynthia and our boys.

I met Cynthia while I was in Kansas for three years. She's the very special daughter of an old grain trading family and a pathologist, David Gibson. Cynthia encouraged me to write and brought Christopher and Jeremy into the world. I left her, some say foolishly, when we were living in California in about 1981.

I was working for Cetus, making oligonucleotides. They were heady times. Biotechnology was in flower and one spring night while the California buckeyes were also in flower I came across the polymerase chain reaction. I was driving with Jennifer Barnett to a cabin I had been building in northern California. She and I had worked and lived together for two years. She was an inspiration to me during that time as only a woman with brains, in the bloom of her womanhood, can be. That morning she had no idea what had just happened. I had an inkling. It was the first day of the rest of my life.

From there it's a single sentence. I worked as a consultant, got the Nobel Prize, and have now turned to writing. It is 1994.

Biography: Dr. Kary B. Mullis

Born

December 28, 1944 Lenoir, North Carolina

Children

R. Louise Mullis, 1965

Christopher W. Mullis 1977

Jeremy G. Mullis 1980

Education

Georgia Institute of Technology, Atlanta, Georgia
B.S. Chemistry, 1966

University of California, Berkeley
Ph. D. Biochemistry, Advisor John B. Neilands, 1972

University of Kansas Medical School
Postdoctoral fellow, Pediatric Cardiology 1973−5

University of California, San Francisco
Postdoctoral fellow, Pharmaceutical Chemistry 1977−9

Employment

Cetus Corporation, Emeryville, CA 1979−86
Xytronyx Corporation, San Diego 1986−8
Private Consultant Biotechnology 1987−92
Writer 1993−

Some significant awards

Preis Biochemische Analytik 1990
Gairdner Foundation International Award 1991
California Scientist of the Year 1992
Robert Koch Prize 1992
Japan Prize 1993
Nobel Prize, Chemistry 1993

THE POLYMERASE CHAIN REACTION

Nobel Lecture, December 8, 1993

by

KARY B. MULLIS

6767 Neptune Place, Apt. 4, La Jolla, CA 92037, USA

In 1944 Erwin Schroedinger, stimulated intellectually by Max Delbruck, published a little book called *What is Life?* It was an inspiration to the first of the molecular biologists, and has been, along with Delbruck himself, credited for directing the research during the next decade that solved the mystery of how "like begat like."

Max was awarded this Prize in 1969, and rejoicing in it, he also lamented that the work for which he was honored before all the peoples of the world was not something which he felt he could share with more than a handful. Samuel Beckett's contributions in literature, being honored at the same time, seemed to Max somehow universally accessible to anyone. But not his. In his lecture here Max imagined his imprisonment in an ivory tower of science.

"The books of the great scientists," he said, "are gathering dust on the shelves of learned libraries. And rightly so. The scientist addresses an infinitesimal audience of fellow composers. His message is not devoid of universality but it's universality is disembodied and anonymous. While the artist's communication is linked forever with it's original form, that of the scientist is modified, amplified, fused with the ideas and results of others, and melts into the stream of knowledge and ideas which forms our culture. The scientist has in common with the artist only this: that he can find no better retreat from the world than his work and also no stronger link with his world than his work."

Well, I like to listen to the wisdom of Max Delbruck. Like my other historical hero, Richard Feynman, who also passed through here, Max had a way of seeing directly into the core of things and clarifying it for the rest of us.

But I am not convinced with Max that the joy of scientific creation must remain completely mysterious and unexplainable, locked away from all but a few esoterically informed colleagues. I lean toward Feynman in this matter. I think Feynman would have said, if you can understand it, you can explain it.

So I'm going to try to explain how it was that I invented the polymerase chain reaction. There's a bit of it that will not easily translate into normal language. If that part weren't of some interest to more than a handful of people here, I would just leave it out. What I will do instead is let you know when we get to that and also when we are done with it. Don't trouble yourself over it. It's esoteric and not crucial. I think you can understand what it felt like to invent PCR without following the details.

In 1953, when Jim Watson and Francis Crick published the structure of DNA, Schroedinger's little book and I were eight years old. I was too young to notice that mankind had finally understood how it might be that "like begat like." The book had been reprinted three times. I was living in Columbia S.C., where no one noticed that we didn't have a copy. But my home was a few blocks away from an undeveloped wooded area with a creek, possums, racoons, poisonous snakes, dragons, and a railroad track. We didn't need a copy. It was a wilderness for me and my brothers, an unknown and unregimented place to grow up. And if we got bored of the earth, we could descend into the network of storm drains under the city. We learned our way around that dark, subterranean labyrinth. It always frightened us. And we always loved it.

By the time Watson and Crick were being honored here in Stockholm in 1962, I hade been designing rockets with my adolescent companions for three years. For fuel, we discovered that a mixture of potassium nitrate and sugar could be very carefully melted over a charcoal stove and poured into a metal tube in a particular way with remarkable results. The tube grew larger with our successive experiments until it was about four feet long. My mother grew more cautious and often her head would appear out of an upstairs window and she would say things that were not encouraging. The sugar was reluctantly furnished from her own kitchen, and the potassium nitrate we purchased from the local druggist.

Back then in South Carolina young boys seeking chemicals were not immediately suspect. We could even buy dynamite fuse from the hardware with no questions asked. This was good, because we were spared from early extinction on one occasion when our rocket exploded on the launch pad, by the very reliable, slowly burning dynamite fuses we could employ, coupled with our ability to run like the wind once the fuse had been lit. Our fuses were in fact much improved over those which Alfred Nobel must have used when he was frightening his own mother. In one of our last experiments before we became so interested in the maturing young women around us that we would not think deeply about rocket fuels for another ten years, we blasted a frog a mile into the air and got him back alive. In another, we inadvertently frightened an airline pilot, who was preparing to land a DC-3 at Columbia airport. Our mistake.

At Dreher High School, we were allowed free, unsupervised access to the chemistry lab. We spent many an afternoon there tinkering. No one got hurt and no lawsuits resulted. They wouldn't let us in there now. Today, we would be thought of as a menace to society. If I'm not mistaken, Alfred Nobel for a time was not allowed to practice his black art on Swedish soil.

Sweden, of course, was then and still is a bit ahead of the United States in these matters.

I never tired of tinkering in labs. During the summer breaks from Georgia Tech, Al Montgomery and I built an organic synthesis lab in an old chicken house on the edge of town where we made research chemicals to sell. Most of them were noxious or either explosive. No one else wanted to make them, somebody wanted them, and so their production became our domain. We suffered no boredom and no boss. We made enough money to buy new equipment. Max Gergel, who ran Columbia Organic Chemicals Company, and who was an unusually nice man, encouraged us and bought most of our products, which he resold. There were no government regulators to stifle our fledgling efforts, and it was a golden age, but we didn't notice it. We learned a lot of organic chemistry.

By the time I left Georgia Tech for graduate school in biochemistry at the University of California at Berkeley, the genetic code had been solved. DNA did not yet interest me. I was excited by molecules. DNA before PCR was long and stringy, not really molecular at all. Six years in the biochemistry department didn't change my mind about DNA, but six years of Berkeley changed my mind about almost everything else.

I was in the laboratory of Joe Neilands who provided his graduate students with a place to work and very few rules. I'm not even sure that Joe knew any rules except the high moral ground of social responsibility and tolerance. Not knowing that the department did have rules, I took astrophysics courses instead of molecular biology, which I figured I could learn from my molecular biologist friends. I published my first scientific paper in *Nature,* in 1968. It was a sophomoric astrophysical hypothesis called "The Cosmological Significance of Time Reversal." I think *Nature* is still embarrassed about publishing it, but it was immensely useful to me when it came time for my qualifying examination. The committee would decide whether or not I would be allowed to take a Ph. D, without having taken molecular biology. And my paper in *Nature,* helped them to justify a "yes." In retrospect, the membership of that committee is intriguing.

Don Glaser, who received this Prize in physics in 1960 at age 34, would later be one of the founders of Cetus Corporation, where I was working when I invented PCR. Henry Rapaport, who discovered psoralens would be the scientific advisor to my department at Cetus, and would co-author two patents with me. Alan Wilson, now sadly passed away, would be the first researcher outside of Cetus to employ PCR. And Dan Koshland would be the editor of *Science* when my first PCR paper was rejected from that journal and also the editor when PCR was three years later proclaimed Molecule of the Year. I passed. None of us, I think, as we walked out of that room, had any conscious inkling of the way things would turn out among us.

In Berkeley it was a time of social upheaval and Joe Neilands was the perfect mentor to see his people through it with grace. We laughed a lot over tea at four every afternoon around a teakwood table that Joe had brought from home and oiled once a month. Our lab had an ambience that was special. I decided to become a neurochemist. Joe was the master of microbial iron transport molecules. It wasn't done like that in most labs, where the head of the lab would prefer that you help advance his career by elaborating on some of his work. Not so with Neilands. As long as I wrote a thesis and got a degree, he didn't care what else I did, and I stayed in his lab happily, following my own curiosity even if it carried me into music courses, for as long as Joe thought we could get away with it. The department was paying me a monthly stipend from the NIH, and eventually, Joe knew, I would have to leave.

After six years I headed east with a Ph. D. and confidence in my education. My wife of a few months went to Kansas to go to medical school and I followed her there. That was 1972.

I had made no professional plans that would work in Kansas, so I decided to become a writer. I discovered pretty quickly that I was far too young. I didn't know anything yet about tragedy, and my characters were flat. I didn't know how to describe a mean spririt in terms someone else could believe.

So I had to get a job as a scientist. I found one at the medical school working with two pediatric cardiologists and a pathologist. It was a very fortunate accident. For one thing pediatricians are always the nicest doctors, and for another thing these doctors were very special: Leone Mattioli, whose wife could cook, Agostino Molteni and Richard Zakheim. For two years I did medical research, learned how to appreciate Old World values from two Italians and a New York Jew, and learned human biology for the first time.

Marriage over, I returned to Berkeley, working for a time in a restaurant and then at the University of California at San Francisco killing rats for their brains. I saw Max Delbruck talk, but I don't think I understood the significance of who he was, nor was I influenced to go into molecular biology by him. I was working on the enkephalins.

But then there was a seminar describing the synthesis and cloning of a gene for somatostatin. That impressed me. For the first time I realized that significant pieces of DNA could be synthesized chemically and that they were likely to be very exciting. I started studying DNA synthesis in the library. And I started looking for a job making DNA molecules.

Cetus hired me in the fall of 1979. I worked long hours and enjoyed it immensely. DNA synthesis was much more fun than killing rats, and the San Francisco Bay Area was a good place to be doing it. There were a number of biotechnology companies and several academic groups working on improving the synthesis methods for DNA. Within two years, there was a machine in my lab from Biosearch of San Rafael, California, turning out oligonucleotides much faster than the molecular biologists at Cetus could use them. I started playing with the oligonucleotides to find out what they could do.

The lab next door to me was run by Henry Erlich and was working on methods for detecting point mutations. We had made a number of oligonucleotides for them. I started thinking about their problem and proposed an idea of my own which they ended up calling oligomer restriction. It worked as long as the target sequence was fairly concentrated, like a site on a purified plasmid, but it didn't work if the site was relatively rare, like a single copy gene in human DNA.

I apologize to those of you who just got lost, but I do have to say a few things now that are going to be difficult. I will get back to the story in a few minutes.

The oligomer restriction method also relied on the fact that the target of interest contained a restriction site polymorphism, which kept it from being universally applicable to just any point mutation. I started thinking about doing some experiments wherein an oligonucleotide hybridized to a specific site could be extended by DNA polymerase in the presence of only dideoxynucleoside triphosphates. I reasoned that if one of the dideoxynucleoside triphosphates in each of four aliquots of a reaction was radioactive then a analysis of the aliquots on a gel could indicate which of the dideozynucleoside triphosphates had added to the hybridized oligonucleotide and therefore which base was adjacent to the three prime end of the oligonucleotide. It would be like doing Sanger sequencing at a single base pair.

On human DNA, it would not have worked because the oligonucleotide would not have specifically bound to a single site. On a DNA as complex as human DNA it would have bound to hundreds or thousands of sites depending on the sequence involved and the conditions used. What I needed to make this work was some method of raising the relative concentration of the specific site of interest. What I needed was PCR, but I had not considered that possibility. I knew the difference numerically between five thousand base pairs as in a plasmid and three billion base pairs as in the human genome, but somehow it didn't strike me as sharply as it should have. My ignorance served me well. I kept on thinking about my experiment without realizing that it would never work. And it turned into PCR.

One Friday night I was driving, as was my custom, from Berkeley up to Mendocino where I had a cabin far away from everything off in the woods. My girlfriend, Jennifer Barnett, was asleep. I was thinking. Since oligonucleotides were not that hard to make anymore, wouldn't it be simple enough to put two of them into the reaction instead of only one such that one of them would bind to the upper strand and the other to the lower strand with their three prime ends adjacent to the opposing bases of the base pair in question. If one were made longer than the other then their single base extension products could be separated on a gel from each other and one could act as a control for the other. I was going to have to separate them on a gel anyway from the large excess of radioactive nucleosidetriphosphate. What I would hope to see is that one of them would pick up one radioactive nucleotide and the other would pick up its complement. Other combinations would indicate that something had gone wrong. It was not a perfect control, but it would not require a lot of effort. It was about to lead me to PCR.

I liked the idea of a control that was nearly free in terms of cost and effort. And also, it would help use up the oligonucleotides that my lab could now make faster than they could be used.

As I drove through the mountains that night, the stalks of the California buckeyes heavily in blossom leaned over into the road. The air was moist and cool and filled with their heady aroma.

Encouraged by my progress on the thought experiment I continued to think about it and about things that could possibly go wrong. What if there were deoxynucleoside triphosphates in the DNA sample, for instance? What would happen? What would happen, I reasoned, is that one or more of them would be added to the oligonucleotide by the polymerase prior to the termination of chain elongation by addition of the dideoxynucleoside triphosphate, and it could easily be the wrong dideoxynucleoside triphosphate and it surely would result in an extension product that would be the wrong size, and the results would be spurious. It would not do. I needed a way to insure that the sample was free from contamination from deoxynucleoside triphosphates. I could treat the sample before the extension reaction with bacterial alkaline phosphatase. The enzyme would degrade any triphosphates present down to nucleosides which would not interfere with the main reaction, but then I would need to deactivate the phosphatase before adding the dideoxynucleoside triphosphates and everyone knew at that time that BAP, as we called it, was not irreversibly denaturable by heat. The reason we knew this was that the renaturation of heat denatured BAP had been demonstrated in classic experiments that had shown that a protein's shape was dictated by it's sequence. In the classical experiments the renaturation had been performed in a buffer containing lots of zinc. What had not occurred to me or apparently many others was that BAP could be irreversibly denatured if zinc was omitted from the buffer, and that zinc was not necessary in the buffer if the enzyme was only going to be used for a short time and had its own tightly bound zinc to begin with. There was a product on the market at the time called matBAP wherein the enzyme was attached to an insoluble matrix which could be filtered out of a solution after it had been used. The product sold because people were of the impression that you could not irreversibly denature BAP. We'd all heard about, but not read, the classic papers.

This says something about the arbitrary way that many scientific facts get established, but for this story, it's only importance is that, had I known then that BAP could be heat denatured irreversibly, I may have missed PCR. As it was, I decided against using BAP, and tried to think of another way to get rid of deoxynucleoside triphosphates. How about this, I thought? What if I leave out the radioactive dideoxynucleoside triphosphates, mix the DNA sample with the oligonucleotides, drop in the polymerase and wait? The polymerase should use up all the deoxynucleoside triphosphates by adding them to the hybridized oligonucleotides. After this was complete I could heat the mixture, causing the extended oligonucleotides to be removed from the target, then cool the mixture allowing new, unextended oligonucleotides to hybridize. The extended oligonucleotides would be far outnumbered by the vast excess of unextended oligonucleotides and therefore would not rehybridize to the target to any great extent. Then I would add the dideoxynucleoside triphosphate mixtures, and another aliquot of polymerase. And now things would work.

But what if the oligonucleotides in the original extension reaction had been extended so far they could now hybridize to unextended oligonucleotides of the opposite polarity in this second round. The sequence which they hade been extended into would permit that. What would happen?

EUREKA!!!! The result would be exactly the same only the signal strength would be doubled.

EUREKA again!!!! I could do it intentionally, adding my own deoxynucleoside triphosphates, which were quite soluble in water and legal in California.

And again, EUREKA!!!! I could do it over and over again. Every time I did it I would double the signal. For those of you who got lost, we're back!

I stopped the car at mile marker 46,7 on Highway 128. In the glove compartment I found some paper and a pen. I confirmed that two to the tenth power was about a thousand and that two to the twentieth power was about a million, and that two to the thirtieth power was around a billion, close to the number of base pairs in the human genome. Once I had cycled this reaction thirty times I would be able to the sequence of a sample with an immense signal and almost no background.

Jennifer wanted to get moving. I drove on down the road. In about a mile it occurred to me that the oligonucleotides could be placed at some arbitrary distance from each other, not just flanking a base pair and that I could make an arbitrarily large number of copies of any sequence I chose and what's more, most of the copies after a few cycles would be the same size. That size would be up to me. They would look like restriction fragments on a gel. I stopped the car again.

"Dear Thor!," I exclaimed. I had solved the most annoying problems in DNA chemistry in a single lightening bolt. Abundance and distinction. With two oligonucleotides, DNA polymerase, and the four nucleosidetriphosphates I could make as much of a DNA sequence as I wanted and I could make it on a fragment of a specific size that I could distinguish easily.

Somehow, I thought, it had to be an illusion. Otherwise it would change DNA chemistry forever. Otherwise it would make me famous. It was too easy. Someone else would have done it and I would surely have heard of it. We would be doing it all the time. What was I failing to see? "Jennifer, wake up. I've thought of something incredible."

She wouldn't wake up. I had thought of incredible things before that somehow lost some of their sheen in the light of day. This one could wait till morning. But I didn't sleep that night. We got to my cabin and I starting drawing little diagrams on every horizontal surface that would take pen, pencil or crayon until dawn, when with the aid of a last bottle of good Mendocino county cabernet, I settled into a perplexed semiconsciousness.

Afternoon came, including new bottles of celebratory red fluids from

Jack's Valley Store, but I was still puzzled, alternating between being absolutely pleased with my good luck and clever brain, and being mildly annoyed at myself and Jennifer Barnett, for not seeing the flaw that must have been there. I had no phone at the cabin and there were no other biochemists besides Jennifer and me in Anderson Valley. The conundrum which lingered throughout the week-end and created an unprecedented desire in me to return to work early was compelling. If the cyclic reactions which by now were symbolized in various ways all over the cabin really worked, why had I never heard of them being used? If they had been used, I surely would have heard about it and so would everybody else including Jennifer, who was presently sunning herself by the pond taking no interest in the explosions that were rocking my brain.

Why wouldn't these reactions work?

Monday morning I was in the library. The moment of truth. By afternoon it was clear. For whatever reasons, there was nothing in the abstracted literature about succeeding or failing to amplify DNA by the repeated reciprocal extension of two primers hybridized to the separate strands of a particular DNA sequence. By the end of the week I had talked to enough molecular biologists to know that I wasn't missing anything really obvious. No one could recall such a process ever being tried.

However, shocking to me, not one of my friends or colleagues would get excited over the potential for such a process. True. I was always having wild ideas, and this one maybe looked no different than last week's. But it WAS different. There was not a single unknown in the scheme. Every step involved had been done already. Everyone agreed that you could extend a primer on a DNA template, everyone knew you could melt double stranded DNA. Everyone agreed that what you could do once, you could do again. Most people didn't like to do things over and over, me in particular. If I had to do a calculation twice, I preferred to write a program instead. But no one thought it was impossible. It could be done, and there was always automation. The result on paper was so obviously fantastic, that even I had little irrational lapses of faith that it would really work in a tube, and most everyone who could take a moment to talk about it with me, felt compelled to come up with some reason why it wouldn't work. It was not easy in that post-cloning, pre-PCR year to accept the fact that you could have all the DNA you wanted. And that it would be easy.

I had a directory full of untested ideas in the computer. I opened a new file and named this one polymerase chain reaction. I didn't immediately try an experiment, but all summer I kept talking to people in and out of the company. I described the concept around August at an in-house seminar. Every Cetus scientist had to give a talk twice a year. But no one had to listen. Most of the talks were dry descriptions of labor performed and most of the scientists left early without comment.

One or two technicians were interested, and on the days when she still loved me, Jennifer, thought it might work. On the increasingly numerous days when she hated me, my ideas and I suffered her scorn together.

I continued to talk about it, and by late summer had a plan to amplify a 400-bp fragment from Human Nerve Growth Factor, which Genentech had cloned and published in *Nature*. I would start from whole human placental DNA from Sigma. taking a chance that the cDNA sequence had derived from a single exon. No need for a cDNA library. No colonies, no nothing. It would be dramatic. I would shoot for the moon. Primers were easy to come by in my lab, which made oligonucleotides for the whole company. I entered the sequences I wanted into the computer and moved them to the front of the waiting list.

My friend Ron Cook, who had founded Biosearch, and produced the first successful commercial DNA synthesis machine, was the only person I remember during that summer who shared my enthusiasm for the reaction. He knew it would be good for the oligonucleotide business. Maybe that's why he believed it. Or maybe he's a rational chemist with an intact brain. He's one of my best friends now, so I have to disqualify myself from claiming any really objective judgement regarding him. Perhaps I should have followed his advice, but then things would have worked out differently and I probably wouldn't be here on the beach in La Jolla writing this, which I enjoy. Maybe I would be rich in Tahiti. He suggested one night at his house that since no one at Cetus had taken it seriously, I should resign my job, wait a little while, make it work, write a patent, and get rich. By rich he wasn't imagining $300 000 000. Maybe one or two. The famous chemist Albert Hofmann was at Ron's that night. He had invented LSD in 1943. At the time he didn't realize what he had done. It only dawned on him slowly, and then things worked their way out over the years like no one would have ever predicted, or could have controlled by forethought and reason.

I responded weakly to Ron's suggestion. I hade already described the idea at Cetus, and if it turned out to be commercially successful they would have lawyers after me forever. Ron was not sure that Cetus had rights on my ideas unless they were directly related to my duties. I wasn't sure about the law, but I was pretty happy working at Cetus and assumed innocently that if the reaction worked big time I would be amply rewarded by my employer.

The subject of PCR was not yet party conversation, even among biochemists, and it quickly dropped. Albert being there was much more interesting, even to me. He had given a fine talk that afternoon at Biosearch.

Anyhow, my problems with Jennifer were not getting any better. That night was no exception to the trend. I drove home alone feeling sad and unsettled, not in the mood for leaving my job, or any big change in what was left of stability in my life. PCR seemed distant and very small compared to our very empty house.

In September I did my first experiment. I like to try the easiest possibilities first. So one night I put human DNA and the nerve growth factor primers in a little screw-cap tube with an O-ring and a purple top. I boiled for a few minutes, cooled, added about 10 units of DNA polymerase, closed the tube and left it at 37°. It was exactly midnight on the ninth of September. I poured a cold Becks into a 400-ml beaker and contemplated my notebook for a few minutes before leaving the lab.

Driving home I figured that the primers would be extended right away, and I hoped that at some finite rate the extension products would come unwound from their templates, be primed and re-copied, and so forth. I did not relish the idea of heating, cooling, adding polymerase over and over again, and held this for a last resort method of accomplishing the chain reaction. I was thinking of DNA:DNA interactions as being reversible with all the ramifications thereof. I wasn't concerned about the absolute rate of dissociation, because I didn't care how long the reaction took as long as nobody had to do anything. I assumed there would always be some finite concentration of single strands, which would be available for priming by a relatively high concentration of primer with pseudo-first order kinetics.

For a reaction with the potential which I dreamed of for this one, especially in light of the absence of anything else that could do the same thing, time was only a very secondary consideration. Would it work at all was important. The next most important thing was, would it be easy to do? Then came time.

At noon the next day I went to the lab to take a 12-hour sample. There was no sign by ethidium bromide of any 400-bp bands. I could have waited another hundred years as I had no idea what the absolute rates might be. But I succumbed slowly to the notion that I couldn't escape much longer the unpleasant prospect of cycling the reaction between single stranded temperatures and double stranded temperatures. This also meant adding the thermally unstable polymerase after every cycle.

For three months I did sporadic experiments while my life at home and in the lab with Jennifer was crumbling. It was slow going. Finally, I retreated from the idea of starting with human DNA, I wasn't even absolutely sure that the Genentech sequence from *Nature* that I was using was from a single exon. I settled on a target of more modest proportions, a short fragment from pBR322, a purified plasmid. The first successful experiment happened on December 16th. I remember the date. It was the birthday of Cynthia, my former wife from Kansas City, who had encouraged me to write fiction and bore us two fine sons. I had strayed from Cynthia eventually to spend two tumultuous years with Jennifer. When I was sad for any other reason, I would also grieve for Cynthia. There is a general place in your brain, I think, reserved for "melancholy of relationships past." It grows and prospers as life progresses, forcing you finally, against your grain, to listen to country music.

And now as December threatened Christmas, Jennifer, that crazy, wonderful woman chemist, had dramatically left our house, the lab, headed to New York and her mother, for reasons that seemed to have everything to do with me but which I couldn't fathom. I was beginning to learn tragedy. It differs a great deal from pathos, which you can learn from books. Tragedy is personal. It would add strength to my character and depth someday to my writing. Just right then, I would have preferred a warm friend to cook with. Hold the tragedy lessons. December is a rotten month to be studying your love life from a distance.

I celebrated my victory with Fred Faloona, a young mathematician and a wizard of many talents whom I had hired as a technician. Fred had helped me that afternoon set up this first successful PCR reaction, and I stopped by his house on the way home. As he had learned all the biochemistry he knew directly from me he wasn't certain whether or not to believe me when I informed him that we had just changed the rules in molecular biology. "Okay, Doc, if you say so." He knew I was more concerned with my life than with those cute little purple-topped tubes.

In Berkeley it drizzles in the winter. Avocados ripen at odd times and the tree in Fred's front yard was wet and sagging from a load of fruit. I was sagging as I walked out to my little silver Honda Civic, which never failed to start. Neither Fred, empty Becks bottles, nor the sweet smell of the dawn of the age of PCR could replace Jenny. I was lonesome.

MICHAEL SMITH

I was born on April 26th, 1932 at 65 St. Heliers Road, South Shore, Blackpool, England in the house of my maternal grandmother, Mary Martha Armstead, having been delivered by the District Nurse, Ms. Parkinson, a lady who I can remember from my infant and juvenile days in her uniform and navy blue raincoat on her bicycle doing her rounds and visiting schools for health inspections. My parents, Mary Agnes Smith and Rowland Smith, both had to work since their early teens, she in the holiday boarding house of her mother and he in his father's market garden in Marton Moss, a village on the south side of Blackpool, just north of Saint Anne's-on-Sea. I went to the local school, Marton Moss Church of England School for 6 years from the age of 5. My mother attended the local church, Saint Nicolas, and consequently I attended that church and its Sunday School. My only prizes from the Sunday School were "for attendance", so I presume my atheism, which developed when I left home to attend university, although latent, was discernible.

During my last year at elementary school, 1943, I sat for the "Eleven-plus" examination which was used in the English schools in those days. In principle, of course, it was an invidious system designed to identify the approximately 20% of the school population that would be offered an academic education and the 80% who would be obligated to take a secondary education that terminated with no further academic options at age 15 (of course, there was the alternative of private schooling, but that was not an option if you were the child of poor parents, as was I). I was lucky enough to obtain a scholarship to the local private school, Arnold School. I did not, at the time, consider this to be luck. I did not want to go to Arnold School because the pupils were considered to be snobs and I thought that I would be ostracized by my friends in Marton Moss. Luckily, my mother insisted, and I went to Arnold School. I cannot say it was the happiest time in my life (I was no good at sports, and proficiency in sport is important in private school life. And I hated the war-time meals that were provided at lunch, as well as the prefect who insisted that I eat the awful food). But the schooling was first-rate, and in this I flourished, although not equally well in all subjects. Clearly, science was my metier, and I was lucky to have a chemistry teacher, Sidney Law, who stimulated my interest in chemistry and who took a personal interest in me (he told me I should read a better newspaper than the one to which my parents subscribed and, as a consequence, I became a life-long reader of the Manchester Guardian. That, in turn, stimulated me to become a reader of the New Yorker as soon as I came to North America, another life-long addiction).

The seven years from 1943 to 1950 were also a time when I became a boy scout. That was a piece of luck. The headmaster at Arnold School, Mr. Holdgate, at the end of my first term, sent me to a dentist, Mr. Paterson, in the hope that he could correct my protruding front teeth, about which I had been teased by my schoolmates. Mr. Paterson did not correct the problem with my teeth but he did introduce me to a wonderful scoutmaster, Mr. Barnes, who inducted me into the happy world of camping and outdoorsmanship which provided me with enjoyment and vacations throughout my secondary school years and right up to the present. An enjoyment that explains why I have a particular delight in living amid the rugged outdoors and beauty of British Columbia.

The second World War impinged on the lives of many of us who were alive at the time. Blackpool, as it turned out, was a very safe place, being in the northwest of England and distant from the targets for German bombing. The large number of hotels and boarding houses in this seaside resort were used to house military trainees, mainly for the airforce. And my father, working his father's market garden, grew primarily food crops rather than his preference, chrysanthemums. Occasionally, bombers, presumably diverted from their primary targets of Manchester and Liverpool, would try to bomb the new factory behind our house that produced Wellington bombers. Usually, they hit the market gardeners' greenhouses which showed up better at night. And I remember one night, alone in the house with my baby brother Robin, when a stick of bombs fell on either side of the house.

I was not proficient in Latin and so was not able to go to Oxford or Cambridge. However, I did enter the first-rate chemistry honours program at the University of Manchester in 1950, where the professors were E.R.H. Jones and M.G. Evans, and graduated in 1953, with the financial support of a Blackpool Education Committee Scholarship. I had hoped to get a first-class degree, but only got a 2(i)! I was very disappointed. However, I still was able to obtain a State Scholarship which supported me throughout my graduate studies until I finished my Ph.D. degree in 1956. My supervisor was H.B. Henbest. He was an outstanding young organic chemist, and I was glad to have him as a supervisor of my work on cyclohexane diols. However, we did not have a particularly warm relationship. I was socially shy and moody and was probably quite hard to understand.

The last year of our graduate studies saw me and my classmates writing to various American professors seeking post-doctoral fellowships. I had no luck in obtaining my desire of a fellowship on the west coast of the United States, but I heard, in the summer of 1956, that a young scientist in Vancouver, Canada, Gobind Khorana, might have a fellowship to work on the synthesis of biologically important organo-phosphates. While I knew this kind of chemistry was much more difficult than the cyclohexane stereochemistry in which I was trained, I wrote to him and was awarded a fellowship after an interview in London with the Director of the British

Columbia Research Council, Dr. G.M. Shrum.

I arrived in Vancouver in September 1956. My first project was to develop a general, efficient procedure for the chemical synthesis of nucleoside-5' triphosphates based on the synthesis of ATP by Khorana in 1954. This study led to more extensive investigations of the reactions of carbodiimides with acids, including phosphoric acid esters and to a general procedure for the preparation of nucleoside-3',5' cyclic phosphates, a class of compounds whose existence and great biological significance had only recently been discovered. One particular pleasure of that period was the development of the methoxyl-trityl family of protecting groups for nucleoside-5'-hydroxyl groups (one synthesis of trimethoxytritanol erupted and left a large orange stain on the laboratory ceiling); this class of protecting group is still in use in modern automated syntheses of DNA and RNA fragments.

In 1960, the Khorana group, including myself, newly married (I have three children, Tom, Ian and Wendy. My wife Helen and I separated in early 1983), moved to the Institute for Enzyme Research at the University of Wisconsin. There I worked on the synthesis of ribo-oligonucleotides, that most challenging of chemical problems for a nucleic acid chemist. Early in 1961, I began to realize that it was time to move on. Helen and I wanted to return to the West Coast of North America, and I accepted a position with the Fisheries Research Board of Canada Laboratory in Vancouver. I enjoyed my time there because of the opportunity it presented to learn about marine biology and I was able to sustain my interest in nucleic acid chemistry because of the award of a U.S. National Institutes of Health Grant, which led to a new synthetic method for nucleoside-3',5' cyclic phosphates. However, the atmosphere of the laboratory, although based on the campus of the University of British Columbia, was not really conducive to, or supportive of, academic research. Hence, in 1966, I was very glad that Dr. Marvin Darrach, then Head of the Department of Biochemistry, offered to nominate me for the position of Medical Research Associate of the Medical Research Council of Canada. This award, which provided salary support, allowed me to become a faculty member of the Department, my academic home ever since, except for sabbaticals at Rockefeller University, the Laboratory of Molecular Biology of the Medical Research Council in Cambridge and Yale University. The Council also has provided research grant support throughout my academic career.

In 1981, Ben Hall and Earl Davie, of the University of Washington, invited me to be a scientific cofounder of a new biotechnology company, Zymos, which was funded by the Seattle venture capital group, Cable and Howse. One of the first contractors was the Danish pharmaceutical company, Novo, who asked Zymos to develop a process for producing human insulin in yeast. After a considerable cooperative effort by Zymos and Novo researchers a successful process was developed. In 1988, the pharmaceutical company, now Noro-Nodisk, purchased outright the biotechnology company, now named ZymoGenetics. I am pleased that, although I no longer have any involvement, ZymoGenetics has subsequently expanded and has continued research on a wide variety of potential protein pharmaceuticals.

In 1986, I was asked by the then Dean of Science at the University of British Columbia, Dr. R.C. Miller, Jr., to establish a new interdisciplinary institute, the Biotechnology Laboratory. I decided that it was time for me to start paying back for the thirty years of fun that I had been able to have in research. I have very much enjoyed recruiting and helping to get established the group of young faculty members that constitute the core of the Biotechnology Laboratory. I also have enjoyed being Scientific Leader of the National Network of Centres of Excellence in Protein Engineering that was funded in 1990. It has been very satisfying, in this case, to see established scientists, working in the various subdisciplines of biochemistry, come together in nation-wide collaborations to solve important problems in protein structure-function analysis and to work with Canadian industry in improving technology transfer which has been less than optimal in the past.

One difficult chore was presented to me in 1991 when I became Acting Director of the Biomedical Research Centre, a privately funded research institute on the Campus of the University of British Columbia. Its source of funding disappeared; therefore I had the responsibilities of managing the Centre on a tight budget, negotiating future funding from the Provincial Government, and helping to ensure the transfer of ownership to the University. This task was made difficult because many of the staff had been led to believe that I was trying to take over and subvert the activities of the Centre. This misguided belief helped the problems of the Centre to become a public political football in an election year. However, funding was negotiated, the University took over the ownership and I was able to step down after 12 months with the Centre and its mission intact.

I look forward to shedding all my administrative responsibilities in another couple of years and returning to my first scientific love, working at the bench and having more time for sailing and for skiing.

SYNTHETIC DNA AND BIOLOGY

Nobel Lecture, December 8, 1993

by

MICHAEL SMITH

Biotechnology Laboratory and Department of Biochemistry, University of British Columbia, Vancouver, B.C. V6T 1W5, Canada

> *"Most of the significant work has been summarized in a number of reviews and articles. In these there was, of necessity, a good deal of simplification and omission of detail... With the passage of time, even I find myself accepting such simplified accounts."*
>
> F. Sanger (1988)

INTRODUCTION

I had the good fortune to arrive in the laboratory of Gobind Khorana, in September 1956, just one month after he had made the accidental discovery of the phosphodiester method for the chemical synthesis of deoxyribo-oligonucleotides (Khorana *et al.*, 1956), a synthetic approach whose full exploitation led to elucidation of the genetic code and the first total synthesis of a gene (Khorana, 1961, 1969, 1979). An identifying characteristic of the Khorana approach to the production of synthetic polynucleotides for the solution of biological problems has been a willingness to use both chemical and enzymatic tools as appropriate (Chambers *et al.*, 1957; Khorana, 1979) and this catholic approach is the foundation of all the molecular technology employed in modern molecular genetics (Ausubel *et al.*, 1987; Sambrook *et al.*, 1989; Watson *et al.*, 1993). It had a major impact on my thinking.

During my post-doctoral time in the Khorana group, I was mainly involved in small molecule synthesis; nucleoside-5' triphosphates (Smith and Khorana, 1958) and nucleoside-3',5' cyclic phosphates (Smith *et al.*, 1961). However, I was able to contribute to the technology of polynucleotide synthesis by the development of the methoxytrityl series of 5'-hydoxyl protecting groups (Smith *et al.*, 1961) and with a first approach to the chemical synthesis of ribo-oligonucleotides (Smith and Khorana, 1959).

In 1961, I left the Khorana group to join the Fisheries Research Board of Canada Vancouver Laboratory. Whilst much of my research effort was directed at studies on salmonid physiology and endocrinology, it was possible to continue studies on the chemistry of phosphodiester synthesis and this resulted in a new method for nucleoside-3'5' cyclic phosphate synthesis by alkaline transesterification under anhydrous conditions, a reaction which obviated the need for chemical protecting groups on the heterocyclic bases (Smith, 1964; Borden and Smith, 1966).

In 1966, I became a faculty member in the Department of Biochemistry of the University of British Columbia, which has been my academic home ever since. Studies on oligonucleotide synthesis continued, being directed at the reaction of deoxyribonucleoside phosphorofluoridates in anhydrous alkaline conditions (von Tigerstrom and Smith, 1970). Although this method, in obviating the need for base protecting groups, represented a significant advance, it did not have the generality required of a universal synthetic method (von Tigerstrom *et al.*, 1975a, 1975b).

An approach to deoxyribo-oligonucleotide synthesis which did prove to be versatile, simple and useful, although inefficient, involved the extension of short primers using *E. coli* polynucleotide phosphorylase, in the presence of Mn^{+2} and NaCl, with doxyribonucleosise-5' diphosphates as substrate (Gillam and Smith, 1972, 1974, 1980; Gillam *et al.*, 1978). This procedure, which made possible the synthesis of deoxyribo-oligonucleotides up to 12 to 13 nucleotides in length, proved to be a significant breakthrough for our small group because it allowed us to undertake a number of fairly ambitious molecular biological projects at a time, between 1970 and 1980, when oligonucleotides were not generally accessible. Of course, it has been displaced as a routine synthetic procedure with the advent of automated chemical synthesis (which still uses the methoxytrityl protecting group) with nucleoside-3' phosphoroamidites as the key intermediates (Adams *et al.*, 1983; Atkinson and Smith, 1984; McBride and Caruthers, 1983).

MODEL STUDIES ON OLIGONUCLEOTIDE DOUBLE HELIX STABILITY AND SPECIFICITY

In 1968, planned studies on the biosynthesis of salmonid protamine (Ingles *et al.*, 1966) and on the sequence of dA,dT-rich crab DNA (Astell *et al.*, 1969) became impossible for personal or for technological reasons. Casting around for a new project, I suggested to Caroline Astell, who was studying for her Ph.D., a series of model studies using chemically synthesized oligonucleotides directed at defining the stability and specificity of oligonucleotide duplexes of defined sequence. The objective was to see if a synthetically accessible oligonucleotide could be a tool for identifying and isolating a specific messenger RNA based on Watson-Crick hydrogen-bonding. From the statistical point of view it should be possible, since the length required of an oligonucleotide in order for it to be unique in a given genome size ranges from 9 nucleotides for phage lambda with a genome of 4.6×10^4 base pairs to 18 nucleotides for the higher plants, e.g. *A. cepa* with a genome of 1.5×10^{10} base pairs (Table 1). Also relevant was the work of Michelson and Monny (1967) and of Niyogi and Thomas (1968) on interactions of ribo-oligonucleotides with ribo-polynucleotides which indicated that stable dou-

Table 1. DNA content of haploid genomes and oligonucleotide length required for unique recognition.

Organism	Base pairs	Oligonucleotide length (N)*
Viruses		
Phage lambda	4.6×10^4	9
Bacteria		
D. pneumoniae	1.7×10^6	11
E. coli	4.1×10^6	12
Fungi		
S. cerevisiae	1.7×10^7	13
N. crassa	2.1×10^7	13
Plants		
C. reinhardtii	10^8	14
A. thaliana	10^8	14
Z. mays	6.6×10^9	17
A. cepa	1.5×10^{10}	18
Animals		
C. elegans	10^8	14
D. melanogaster	1.3×10^8	14
B. rerio	10^9	16
M. musculus	2.2×10^9	17
H. sapiens	3.3×10^9	17

* (N) is unique when $4^N \geq 2 \times$ Base Pairs.

ble helices can be formed with as few as seven base pairs and, equally importantly, which demonstrated that there is a significant increase in duplex stability for each additional base pair up to about 16 base pairs (Figure 1). The work of Gilham with random length oligothymidylates covalently attached to cellulose demonstrated their potential power for affinity chromatography (Gilham, 1962; Gilham and Robinson, 1964) which led to an important procedure for isolation of eukaryote mRNAs (Aviv and Leder, 1972). However, although data was available on duplex stability involving *mixtures* of oligonucleotides of defined length with DNA (McConaughy and McCarthy, 1967; Niyogi, 1969), there was no systematic data obtained with pure oligonucleotides of defined sequence and length.

Our approach was to prepare deoxyribo-oligonucleotide-celluloses using simple oligonucleotides of defined length and sequence synthesized by the Khorana method and to use these oligonucleotide-celluloses, in the form of thermally-eluted columns, to establish the stability of duplexes with a variety of complementary, or partially complementary deoxyribo- and ribo-oligonucleotides (Astell *et al.*, 1973; Astell and Smith, 1971, 1972; Gillam *et al.*, 1975). The results and conclusions of this time-consuming but critical series

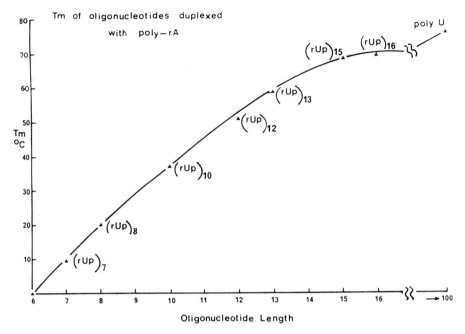

Fig. 1: The relationship between oligouridylate length and the *Tm* of duplexes with polyadeny-late in 0.02 M NaCl, 0.01 M sodium cacodylate, pH 7.0. Drawn from data of Michelson and Monny (1967).

of experiments has been summarized in more detail elsewhere (Smith, 1983) and only some highlights will be discussed here.

Firstly, it was clear that, under appropriate conditions of ionic strength, stable duplexes containing as few as six dA-dT base-pairs can be formed (Figure 2). Further, additional base-pairs increase duplex stability significantly, but in a non-linear manner with the incremental increase in stability diminishing with increased length. Also, a single base-pair mismatch of dT with dT or of dT with dG decreases duplex stability by the same amount (equivalent to removing about two base-pairs), but duplex formation is still possible at low temperature (Figure 2). In addition, it was obvious that oligonucleotide duplexes form rapidly at low temperatures, conditions under which denatured double strand DNA does not reanneal.

These experiments convinced me that it would be possible to use a synthetic oligonucleotide to identify an RNA or a DNA containing the exact Watson-Crick complement of its sequence and to differentiate it from similar but not identical sequences. An opportunity to demonstrate this principle was provided by mutants in the lysozyme-encoding locus of phage T4 (Doel and Smith, 1973). Also, at a later date, the data made me realize that it should be possible to use short synthetic deoxyribo-oligonucleotides as specific mutagens and to differentiate between a point mutant and wild-type DNA.

Experiments on the interaction of deoxyribo-oligonucleotide-celluloses and ribo-oligonucleotides confirmed earlier studies on the relative stabili-

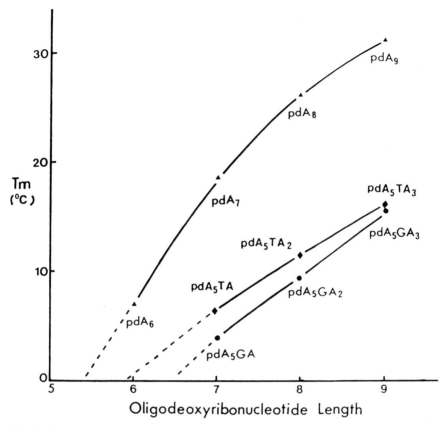

Fig. 2: The *Tm* values of the duplexes with cellulose-pdT$_9$ of complementary oligodeoxyribonu-cleotides containing a single base-pair mismatch in 1 M NaCl, 0.01 M sodium phosphate, pH 7.0. Drawn from data of Gillam, Waterman, and Smith (1975).

ties of DNA-DNA and DNA-RNA double strands (Chamberlin, 1965; Smith, 1983). They also demonstrated a further important principle: a dT, rG base pair is relatively stable (Figure 3). Differences in the stabilities of dT, dG and dT, rG base pairs (compare Figures 2 and 3), presumably, reflect differences in the base pairing allowed in the B-forms and the A-forms of double helix (Watson *et al.*, 1987). The observation has two practical implications. First, the most specific and stringent interaction of an oligonucleotide probe will be the one between a *deoxyribo-oligonucleotide* and *DNA*. Second, one cannot assume that duplex structures that are formed by RNA will be formed by DNA or by deoxyribo-oligonucleotides. In connection with this, I have always been puzzled by the widespread acceptance of the idea that deoxyinosine is precocious in base pairing because inosine behaves in a precocious manner in t-RNA duplex interactions. Studies on deoxyinosine, in fact, indicate that it functions as a specific analog of deoxyguanosine, although it does not self-aggregate as does deoxyguanosine (Ausubel *et al.*, 1987; Kornberg, 1980; Sambrook *et al.*, 1989).

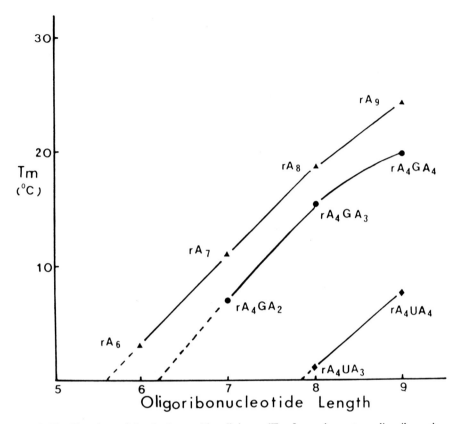

Fig. 3: The *Tm* values of the duplexes with cellulose-pdT$_9$ of complementary oligoribonucleo-tides containing a single base-pair mismatch in 1 M NaCl, 0.01 M sodium phosphate, pH 7.0. Drawn from data of Giliiam, Waterman, and Smith (1975).

The above empirical studies, and other related ones not discussed here (Smith, 1983), provided the direct inspiration for and commitment to our use of synthetic deoxyribo-oligonucleotides as probes for isolation of specific DNA fragments, as primers for the direct sequencing of double-stranded DNA, as primers for precise definition of the ends of mRNAs, as specific mutagens and as tools for the identification and isolation of point mutations.

DEOXYRIBO-OLIGONUCLEOTIDES AS DNA PROBES

When the previously described model studies were commenced, in 1968, the intent was to develop a method for isolation of specific m-RNAs by affinity chromatography. However, the development of the full panoply of DNA cloning technologies in the early 1970's (Watson *et al.*, 1993) made it clear that the prime target for work with synthetic doxyribo-oligonucleotides should be a method for monitoring gene isolation. Discussions with Dr. Benjamin D. Hall made me aware of the important studies of F. Sherman on double-frameshift mutants of yeast cytochrome *c* (Stewart and

Sherman, 1974). Knowledge of the amino-acid sequence of these mutants allowed the unambiguous prediction of the N-terminal coding sequence and, hence, provided a specific target for an oligonucleotide of defined sequence. A second attractive feature about a yeast gene as a first target was the fact that the length of oligonucleotide required for a specific probe, 13 nucleotides (Table 1), was close to the limit for our enzymatic method of synthesis (Gillam and Smith, 1972, 1980; Gillam *et al.*, 1978). The required probe was synthesized using the enzymatic method (Gillam *et al.*, 1977) and was used to isolate the gene, albeit with some difficulty (Montgomery *et al.*, 1978), thus clearly validating the principle that a synthetic oligonucleotide could function as a probe for gene isolation. The subsequent generalization of the method using mixtures of oligonucleotides (Wallace *et al.*, 1981; Suggs *et al.*, 1981) has made this approach to gene isolation into a powerful tool in molecular genetics (Ausubel *et al.*, 1987; Sambrook *et al.*, 1989).

SEQUENCING DOUBLE-STRANDED DNA USING OLIGONUCLEOTIDE PRIMERS

At the time when the iso-1-cytochrome *c* gene was isolated, the chemical method for DNA sequence determination (Maxam and Gilbert, 1977) and the chain terminating enzymatic method (Sanger *et al.*, 1977) had newly been developed, the former being applicable to double-stranded DNA and the latter to single-stranded DNA. The sequences of the ends of the Eco RI-Hind III DNA fragment containing the cytochrome *c* coding sequence were obtained using the chemical method (Montgomery *et al.*, 1978). I decided to investigate the possibility of directly sequencing double-stranded DNA by the enzymatic method using synthetic oligonucleotide primers, based on my conviction that an oligonucleotide should hybridize *at low temperature* with a complementary DNA strand of a denatured DNA double helix in solution, even though the second DNA strand was present and capable of producing a thermodynamically much more stable structure. The critical question was whether the second DNA strand would displace a short oligonucleotide from a complex with the complementary DNA strand under the conditions of the Sanger enzymatic sequencing method (Sanger, 1981). The first experiment was completely successful, an enormously exciting event. As a consequence, the sequence of the gene was completed by "walking" along the sequence using a series of short oligonucleotide primers 9 or 10 nucleotides in length (Smith *et al.*, 1979). I believe that the full potential of this approach to gene sequencing has yet to be realized.

Another important application of oligonucleotide-primed sequencing of double-stranded DNA is the precise identification of point mutations produced by classical genetic techniques at a given locus. A particularly good example was in the analysis of a large number of independently isolated point mutations in a yeast suppressor t-RNA gene (Koski *et al.*, 1980; Kurjan *et al.*, 1980).

DEFINING THE ENDS OF mRNAs PRECISELY

It is important to exactly define the sequences at the 5'- and 3'-ends of m-RNAs. For the 3'-end of an m-RNA, this can be done by oligonucleotide-primed sequencing using reverse transcriptase for the Sanger sequencing method with RNA enriched for the desired m-RNA as template. The family of 12 deoxyribo-oligonucleotides consisting of a tract of dT-residues terminated at the 3'-end by two other deoxyribonucleotides can be annealed to the 3'-end of polyadenylated eukaryote m-RNA. If the transcribed sequence has a unique 3'-end, only one of the twelve oligonucleotides should act as a specific primer. This strategy was used to define the sequence of the unique 3'-end of bovine growth hormone m-RNA (Sasavage *et al.*, 1980) and to demonstrate and precisely define the microheterogeneity at the 3'-end of the bovine prolactin m-RNA transcript (Sasavage *et al.*, 1982). The iso-1-cytochrome *c* proved to have a unique 3'-end (Boss *et al.*, 1981) while the 5'-end of the same m-RNA is very disperse indeed as was shown by oligonucleotide primed extension using the m-RNA as template (McNeil and Smith, 1985). There are no other methods for defining the 5'- and 3'-ends of m-RNAs with this precision, a degree of precision which is essential to the accurate definition of transcription initiation and termination signals (McNeil and Smith, 1985; Sasavage *et al.*, 1982).

OLIGONUCLEOTIDE-DIRECTED MUTAGENESIS

> *"I strongly believe that, through the products of organic synthesis, it will be possible to gain influence over the development of organisms and to produce changes that surpass all that can be achieved by conventional breeding."*
>
> Emil Fischer (1917)

> *"The ignis fatuus of Genetics has been the specific mutagen, the reagent that would penetrate to a given gene, recognize it, and modify it in a specific way."*
>
> J. Lederberg (1959)

Initiation of the project to isolate the yeast cytochrome *c* gene made me realize that I needed to learn about DNA sequence determination. I was lucky to be able to spend a year, starting in the fall of 1975, in Fred Sanger's laboratory working with the "plus-minus" sequencing method as part of the team which was attacking the sequence of the *E. coli* phage ∅X174 (Sanger and Coulson, 1975; Sanger *et al.*, 1978). There are three pairs of overlapping genes in the 5386 nucleotide genome (Barrell *et al.*, 1976; Brown and Smith, 1977; Gillam *et al.*, 1985) all genes being encoded on the same DNA strand, with each overlapping pair using different codon reading frames. An integral essential component of defining the position and reading frame of these genes was the availability of nonsense mutants suppressible by amber or ochre suppressors. This use of precisely located mutants of a particular phenotype highlighted the need for a very specific mutagenic method that would define the target to a specific base-pair in the genome and introduce a predetermined change with sufficiently high efficiency to allow genomic screening in order to identify phenotypically silent mutants. Given that the DNA of phage ⌀X174 is single stranded and with the knowledge of the complete sequence of its genome, our earlier studies, which demonstrated that small oligonucleotides as short as 7 nucleotides in length could form stable duplexes at low temperature even with a mismatch (Figure 2), suggested that oligonucleotide-directed mutagenesis should be possible. There was additional useful information. It was known that point mutants could be reverted, albeit with low efficiency, by annealing mutant phage ⌀X174 DNA with fragments from the complementary strand of wild-type DNA prior to transfection (Hutchison and Edgell, 1971; Weisbeek and van de Pol, 1970). However, fragments very much longer than those that we could readily synthesize were needed and the low efficiency precluded genotypic screening (Hutchison and Edgell, 1971; Weisbeek and van de Pol, 1970). In discussing these issues, Clyde Hutchison (who, also, was spending one year in Fred Sanger's group and whose biological knowledge of ⌀X174 was invaluable to the sequencing project) and I realized that the studies of Kornberg and Goulian (Goulian, 1968a; Goulian *et al.*, 1967, 1973; Kornberg, 1980) provided an obvious route to a mutagenic method since they had demonstrated that an oligonucleotide as short as nine nucleotides in length could act as a primer for *E. coli* DNA polymerase I on a circular single strand template and that the product could be converted to a closed circular duplex by enzymatic ligation. It was also known that most of the primer molecule is excised from the product (Goulian, 1968b), presumably due to the 5'-exonuclease activity of *E. coli* DNA polymerase I (Klett *et al.*, 1968). With this information and our knowledge of the stability of duplexes involving a mismatched oligonucleotide, we decided to use a 12 nucleotide oligomer, with a centrally positioned single nucleotide mismatch, as primer and with ⌀X174 DNA as template and *E. coli* DNA polymerase I in which the 5'-exonuclease had been inactivated by subtilisin (Brutlag *et al.*, 1969; Klenow and Henningsen, 1970; Klenow *et al.*, 1971) to construct a closed circular double-stranded DNA with the oligonucleotide in one strand. Transfection of *E. coli* with this DNA should produce wild-type and mutant phage. The specific mutations chosen for the first experiment were the production and reversion of a known nonsense mutation, *am3*, in the lytic function, gene E, of ⌀X174, since convenient phenotypic screens were available. The mutations involved the interconversion of a Trp codon, TGG, and an amber codon, TAG, by G-T and A-C mismatches (Hutchison *et al.*, 1978). In the first experiments, mutation was achieved but at low efficiency. This was increased to a very encouraging level of about 15% after removal of incompletely closed duplexes by adsorption to nitro cellulose or treatment with a single-strand specific nuclease under conditions where a single base-pair mismatch was not degraded (Hutchison *et al.*,

1978). Further studies directed at optimizing the mutagenic conditions resulted in efficiencies of up to 39% with a 12 nucleotide mutagenic primer and significant mutation with a heptamer mutagenic primer (Gillam and Smith, 1979a). The wisdom of using subtilisin-treated DNA polymerase is evident from comparison of these results with those of another study which used intact *E. coli* DNA polymerase I (Razin *et al.*, 1978).

Further studies on the mutagenesis of phage $\emptyset$X174 demonstrated that, in addition to the two transition mutations of the first experiments, it was possible to produce transversion mutations and single nucleotide deletions, also using very short oligonucleotides (Smith and Gillam, 1981). Clearly, the method was highly specific, efficient and general within the context of the genome of $\emptyset$X174. It was time to further generalize the method so that it could be applied to any cloned fragment of DNA. The icosahedral bacteriophages, such as $\emptyset$X174, are not potential vectors for recombinant DNA because the amount of DNA that can be accommodated in the viral particle is limited. The filamentous bacteriophages, on the other hand, can accommodate additional DNA incorporated into the genome. At the time when Mark Zoller and I were planning the generalization of the method, there were those that argued in favour of a methodology applied to double-stranded recombinant plasmids. It seemed to us that the biology of a filamentous bacteriophage, which produces single-stranded circular DNA and then purifies it as phage particles, argued overpoweringly in favour of phage DNA as vector. Subsequent developments in recombinant phage biology and in the construction of phagemids (Ausubel *et al.*, 1987; Sambrook *et al.*, 1989), have vindicated this strategic decision, although viable procedures for plasmid mutagenesis are available (Smith, 1985; Zoller, 1991, 1992). Oligonucleotide mutagenesis with filamentous phage DNA as template proved to be more difficult than with $\emptyset$X174 DNA, presumably because of a greater degree of secondary structure which inhibited the production of full length double-stranded circles. Consequently, new procedures for enhancing the efficiency of the mutagenic procedure had to be developed (Zoller and Smith, 1982, 1983, 1984, 1987). The power of the methodology was graphically illustrated by our first involvement in protein engineering in collaboration with Greg Winter and Alan Fersht (Winter *et al.*, 1982). This, of course, presaged the enormous use of the methodology in protein structure-function analysis that has become a feature of modern biochemistry.

GENOTYPIC SELECTION OR SCREENING FOR MUTANTS

Because the immediate product of the enzymatic incorporation of a mutagenic oligonucleotide into a double-stranded recombinant DNA is a heteroduplex, biological replication of the DNA will produce mutated and unmutated progeny DNAs, providing there is no asymmetry in the mechanism of DNA replication or providing that there is not an asymmetric mismatch repair mechanism. In fact, it is possible to achieve mutagenic efficiencies of close to 50 % using standard procedures (Gillam and Smith, 1979a; Zoller and Smith, 1984). With this level of efficiency, screening by DNA sequence determination is the most effective procedure for detecting the desired mutant DNA. However, often the yield of mutant is considerably less than 50 %. It is my opinion that the predominant reason for this is the use of template DNA that is contaminated with random oligonucleotides resulting from bacterial DNA degradation; others ascribe it to biological mismatch repair favouring the sequence of the template DNA (Smith, 1985). Whatever the cause of the reduced yield of mutant, procedures other than sequence determination are required for mutant identification. The ones that are available fall into two categories, those using the mutagenic oligonucleotide as a tool for identifying mutant DNA and those which select for progeny derived from the newly synthesized, oligonucleotide-containing DNA strand.

The significantly enhanced stability of a perfectly matched oligonucleotide duplex relative to one with a mismatch (Figure 2) suggested that a mutagenic oligonucleotide should be able to identify mutant complementary DNA. This principle was used to develop a method for *selection* of mutant DNA. In this procedure the mixture of wild-type and mutant single-stranded phage DNA produced by oligonucleotide mutagenesis is used as a template for double-stranded DNA synthesis using the mutagenic oligonucleotide as primer under conditions when it should not form a mismatched duplex (Gillam and Smith, 1979b). Following transfection into *E. coli*, this procedure can select mutant DNA with close to 100 % efficiency.

An alternative procedure is to use the mutagenic oligonucleotide as a probe to screen for mutant DNA in plaques produced by *E. coli* with mutagenized phage (Zoller and Smith, 1982, 1983). Again, this procedure is very reliable in the identification of mutant DNA. The same principle is used in the identification of point mutations in human DNA (Conner *et al.*, 1983) where, of course, the complexity of the human genome requires a probe containing a minimum of 17 nucleotides (Table 1).

A number of procedures for selection of progeny derived from the DNA strand containing the mutagenic oligonucleotide have been developed (Smith, 1985). Two of these, one based on *in vivo* selection and the other on *in vitro* selection have proved most popular and are widely used. *In vivo* selection involves the use of a template DNA where about 1% of the deoxythymidine residues have been replaced by deoxyuridine residues. This template is incorporated into double-stranded DNA where the mutagenic oligonucleotide is part of the *in vitro* synthesized second strand. On transfection of an appropriate *E. coli* host, the deoxyuridine strand is selectively degraded by the bacterium's DNA damage repair system, resulting in progeny derived primarily from the mutant strand (Kunkel, 1985). *In vitro* selection involves the use of a normal recombinant phage template DNA with the mutagenic oligonucleotide incorporated in the second strand which contains enzymatically synthesized thiophosphate inter-nucleotide linkages. These are resistant to cleavage by restriction endonuclease which, as a consequence, nicks the template. The nicked template then is partially degraded with an exonuclease and repaired using a DNA polymerase and DNA ligase. The result is a double-stranded DNA which, predominantly, is mutant in both strands and which produces, predominantly, mutant recombinant phage on transfection of *E. coli* (Taylor *et al.*, 1985).

CONCLUSION

This concludes my account of our studies on the use of synthetic dioxyribo-oligonucleotides as tools for characterizing naturally occurring nucleic acids, which is as accurate as I can make it, bearing in mind the caveat of Fred Sanger cited at the beginning of this article. Clearly, many other applications of oligonucleotide in biology exist, notably the polymerase chain reaction (Mullis, 1994) and its multitudinous variations, but also in double strand DNA synthesis and in the use of oligonucleotide probes *in vitro* and *in vivo*. In the latter area there is the potential for development of a whole new chemistry directed at pharmaceuticals which block the expression of specific genes and at diagnostic agents (Buchardt *et al.*, 1993; Crooke, 1993).

With regard to our own work, it is very satisfying that the model studies started in 1968 provided the inspiration for so many applications and scientific collaborations (only a few of which are cited in this article). The potential of oligonucleotide-directed site-specific mutagenesis was apparent from the outset; to quote:

> *"This new method of mutagenesis has considerable potential . . . to define the role of . . . origins of DNA replication, promoters and the sequences for ribosome-building sites. . . . specific mutation within protein structural genes . . . will allow precise studies of protein structure-function relationships."*
>
> Hutchison *et al.* (1978)

However, we could not have anticipated the explosion of gene isolations, the improvements in DNA sequence determination methodology and the advances in the chemistry of nucleic acids synthesis that have occurred since 1978. This has resulted in an amazing increase in the use of site-directed mutagenesis as an analytical tool in biochemistry and biology. And it has been accompanied by continual improvements in the basic methodologies and versatility of site-directed mutagenesis (Smith, 1985; Zoller, 1991, 1992) and the initiation of new scientific publications such as Protein Engineering and Protein Science. It would not be too much of an exaggeration to say that the prediction of Emil Fischer in 1917 has been fulfilled and the dilemma posed by Joshua Lederberg in 1959 has been resolved.

ACKNOWLEDGMENTS

My career in nucleic acid research would not exist without the example and inspiration of Gobind Khorana. Also, I am indebted to the many young scientists who have been members of my group over the past 30 years, but especially to Caroline Astell, Shirley Gillam, Patricia Jahnke and Mark Zoller, who were the individuals primarily responsible for the studies discussed in this article, as well as to Tom Atkinson who undertook the arduous task of manual chemical synthesis of oligonucleotides in the time immediately prior to the introduction of automated synthesizers.

Two crucial, indispensable collaborators, whose biological expertise and commitment were essential to these studies, were Ben Hall in the studies on the isolation of the yeast cytochrome *c* gene and Clyde Hutchison in the development of oligonucleotide-directed mutagenesis. And my year in Fred Sanger's laboratory was wonderful in its excitment and in its timeliness.

Research funding support by the U.S. National Institutes of Health was essential in allowing me to continue synthetic nucleotide chemistry in the years immediately after I left the Khorana group. Since 1966, the Medical Research Council of Canada has continuously provided me with the salary support and research grants that have been the stable and very much appreciated foundation for my more speculative research activities. I have also received, for more limited periods, research support from the British Columbia Health Research Foundation and from the National Cancer Institute of Canada.

For the past four years I have been Scientific Leader of the Canadian Protein Engineering Network of Centres of Excellence, a country-wide collaboration on protein structure-function analysis, generously funded by the Network of Centres of Excellence program.

Finally, firstly at the British Columbia Research Council, then at the Fisheries Board of Canada Vancouver Laboratory, mostly in the Department of Biochemistry and more recently in the Biotechnology Laboratory, the campus of the University of British Columbia has been my scientific home since 1956. I am very grateful for the environment that it has provided, especially in the Department of Biochemistry where the studies described in this article were carried out.

REFERENCES

Adams, S.P., Kavka, K.S., Wykes, E.J., Holder, S.B. and Galluppi, G.R. (1983). Hindered dialkylamino nucleoside phosphite reagents in the synthesis of two DNA 51-mers. J. Am. Chem. Soc. *105*: 661–663.

Astell, C.R., Suzuki, D.T., Klett, R.P., Smith, M. and Goldberg, I.H. (1969). The intracellular location of the adenine- and thymine-rich component of deoxyribonucleate in testicular cells of the crab, *Cancer productus*. Exptl. Cell Res. *54*: 3–10.

Astell, C.R. and Smith, M. (1971). Thermal elution of complementary sequences of nucleic acids from cellulose columns with covalently attached oligonucleotides of known length and sequence. J. Biol. Chem. *246*: 1944–1946.

Astell, C.R. and Smith, M. (1972). Synthesis and properties of oligonucleotide-cellulose columns. Biochemistry *11*: 4114–4120.

Astell, C.R., Doel, M.T., Jahnke, P.A. and Smith, M. (1973). Further studies on the properties of oligonucleotide cellulose columns. Biochemistry *12*: 5068–5074.

Atkinson, T. and Smith, M. (1984). Solid-phase synthesis of oligodeoxyribo- nucleotides by the phosphite-triester method. In: Oligonucleotide Synthesis. A Practical Approach (M.J. Gait, ed.). IRL Press, Oxford. pp. 35–81.

Ausubel, F.M., Brent, R, Kingston, R.E., Moore, D.D., Seidman, J.G., Smith, J.A. and Struhl, K. (1987). Current Protocols in Molecular Biology, (2 volumes, supplemented quarterly). Wiley Interscience, New York, N.Y.

Aviv, H. and Leder, P. (1972). Purification of biologically active globin messenger RNA by chromatography on oligothymidylic acid-cellulose. Prov. Natl. Acad. Sci. USA *69*: 1408–1412.

Barrell, B.G., Air, G.M. and Hutchison, C.A. III (1976). Overlapping genes in bacteriophage ØX174. Nature *264*: 34–41.

Borden, R.K. and Smith, M. (1966). Preparation of nucleoside-3',5' cyclic phosphates in strong base. J. Org. Chem. *31*: 3247–3253.

Boss, J.M., Gillam, S., Zitomer, R.S. and Smith, M. (1981). Sequence of the yeast iso-1-cytochrome *c* mRNA. J. Biol. Chem. *256*: 12958–12961.

Brown, N.L. and Smith, M. (1977). The sequence of a region of bacteriophage ØX174 coding for parts of genes A and B. J. Mol. Biol. *116*: 1–28.

Brutlag, D., Atkinson, M.R., Setlow, P. and Kornberg, A. (1969). An active fragment of DNA polymerase produced by proteolytic cleavage. Biochem. Biophys. Res. Commun. *37*: 982–989.

Buchardt, O., Egholm, M., Berg, R.H. and Nielsen, P.E. (1993). Peptide nucleic acids and their potential applications in biotechnology. Tibtech *11*: 384–386.

Chamberlin, M.J. (1965). Cooperative properties of DNA, RNA and hybrid homopolymer pairs. Fed. Proc. *24*: 1446–1457.

Chambers, R.W., Moffatt, J.G. and Khorana, H.G. (1957). A new synthesis of guanosine-5'-phosphate. J. Am. Chem. Soc. *79*: 3747–3752.

Conner, B.J., Reyes, A.A., Morin, C., Itakura, K., Teplitz, R.L. and Wallace, R.B. (1983). Detection of sickle-cell β^S-globin allele by hybridization with synthetic oligonucleotides. Proc. Natl. Acad. Sci. USA *80*: 278–282.

Crooke, S.T. (1993). Oligonucleotide therapy. Curr. Opin. Biotechnol. *3*: 656–661.

Doel, M.T. and Smith, M. (1973). The chemical synthesis of deoxyribo-oligonucleotides complementary to a portion of the lysozyme gene of phage T4 and their hybridization to phage-specific RNA and phage DNA. FEBS Letts. *34*: 99–102.

Fischer, E. (1917). Source unknown.

Gilham, P.T. (1962). Complex formation in polynucleotides and its application to the separation of polynucleotides. J. Am. Chem. Soc. *84*: 1311–1312.

Gilham, P.T. and Robinson, W.E. (1964). The use of polynucleotide-celluloses in sequence studies of nucleic acids. J. Am. Chem. Soc. *86*: 4885–4989.

Gillam, S. and Smith, M. (1972). Enzymatic synthesis of deoxyribo-oligonucleotides of defined sequence. Nature New Biol. *238*: 233–234.

Gillam, S. and Smith, M. (1974). Enzymatic synthesis of deoxyribo-oligonucleotides of defined sequence. Properties of the enzyme. Nucleic Acids Res. *1*: 1631–1648.

Gillam, S., Waterman, K. and Smith, M. (1975). The base-pairing specificity of cellulose-pdT$_9$. Nucleic Acids Res. *2*: 625–634.

Gillam, S., Rottman, F., Jahnke, P. and Smith, M. (1977). Enzymatic synthesis of oligonucleotides of defined sequence: synthesis of a segment of yeast iso-1-cytochrome *c* gene. Proc. Natl. Acad. Sci. USA *74*: 96–100.

Gillam, S., Jahnke, P. and Smith, P. (1978). Enzymatic synthesis of oligo-deoxyribonucleotides of defined sequence. J. Biol. Chem. *253*: 2532–2539.

Gillam, S. and Smith, M. (1979a). Site-specific mutagenesis using synthetic oligodeoxyribonucleotide primers: optimum conditions and minimum oligodeoxyribonucleotide length. Gene *8*: 81–97.

Gillam, S. and Smith, M. (1979b). Site-specific mutagenesis using synthetic oligodeoxyribonucleotide primers: *in vitro* selection of mutant DNA. Gene *8*: 99–106.

Gillam, S. and Smith, M. (1980). Use of *E. coli* polynucleotide phosphorylase for the synthesis of oligodeoxyribonucleotides of defined sequence. Methods Enzymol. *65*: 687–701.

Gillam, S., Atkinson, T., Markham, A. and Smith, M. (1985). Gene K of bacteriophage ∅X174 codes for a protein which affects the burst size of phage production. J. Virol. *53*: 708–709.

Goulian, M., Kornberg, A. and Sinsheimer, R.L. (1967). Synthesis of infectious phage ∅X174 DNA. Proc. Natl. Acad. Sci. USA *58*: 2321–2328.

Goulian, M. (1968a). Incorporation of oligodeoxynucleotides into DNA. Proc. Natl. Acade. Sci. USA *61*: 284–291.

Goulian, M. (1968b). Initiation of the replication of single-stranded DNA by *Echerichia coli* DNA polymerase. Cold Spring Harbor Symp. Quant. Biol. *33*: 11–20.

Goulian, M., Goulian, S.H., Codd, E.E. and Blumenfield, A.Z. (1973). Properties of oligodeoxynucleotides that determine priming activity with *Escherichia coli* deoxyribonucleic acid polymerase I. Biochemistry *12*: 2893–2901.

Hutchison, C.A. III and Edgell, M.H. (1971). Genetic assay for small fragments of bacteriophage ∅X174 deoxyribonucleic acid. J. Virol. *8*: 181–189.

Hutchison, C.A. III, Phillips, S., Edgell, M.H., Gillam, S., Jahnke, P. and Smith, M. (1978). Mutagenesis at a specific position in a DNA sequence. J. Biol. Chem. *253*: 6551–6560.

Ingles, C.J., Trevithick, J.R., Smith, M. and Dixon, G.H. (1966). Biosynthesis of protamine during spermatogenesis in salmonid fish. Biochem. Biophys. Res. Comm. *22*: 627–634.

Khorana, H.G., Tener, G.M. Moffatt, J.G. and Pol, E.H. (1956) A new approach to the synthesis of polynucleotides. Chem. and Ind. London, 1523.

Khorana, H.G. (1961). Some Recent Developments in the Chemistry of Phosphate Esters of Biological Interest. J. Wiley and Sons. N.Y.

Khorana, H.G. (1969). Nucleic acid synthethis in the study of the genetic code. In: Les Prix Nobel en 1968. P.A. Norstedt and Sons, Stockholm, pp. 196–220.

Khorana, H.G. (1979). Total synthesis of a gene. Science *203*: 614–625.

Klenow, H. and Henningsen, I. (1970). Selected elimination of the exonuclease activity of the deoxyribonucleic acid polymerase from *Escherichia coli* B by limited proteolysis. Proc. Natl. Acad. Sci. USA *65*: 168–175.

Klenow, H., Overgaard-Hansen, K. and Patkar, S.A. (1971). Proteolytic cleavage of native DNA polymerase into two different catalytic fragments. Influence of assay conditions on the change of exonuclease activity and polymerase activity. Eur. J. Biochem. *22*: 371–381.

Klett, R.P., Cerami, A. and Reich, E. (1968). Exonuclease VI, a new nuclease activity associated with *E. coli* DNA polymerase. Proc. Natl. Acad. Sci. USA *60*: 943–950.

Kornberg, A. (1980). DNA Replication. W.H. Freeman and Company. San Francisco, CA.

Koski, R.A., Clarkson, S.G., Kurjan, J., Hall, B.D. and Smith, M. (1980). Mutations at the yeast *SUP4* tRNATyr locus; transcription of the mutant genes *in vitro*. Cell *22*: 415 – 425.

Kunkel, T.A. (1985). Rapid and efficient site-specific mutagenesis without phenotypic selection. Proc. Natl. Acad. Sci. USA *82*: 488 – 492.

Kurjan, J., Hall, B.D., Gillam, S. and Smith, M. (1980). Mutations at the yeast *SUP4* tRNA^{Tyr} locus: DNA sequence changes in mutants lacking suppresor activity. Cell *20*: 701 – 709.

Lederberg, J. (1959). A view of genetics. In: Les Prix Nobel en 1958. Norstedt and Sons, Stockholm. pp. 170 – 189.

Maxam, A.M. and Gilbert, W. (1977). A new method for sequencing DNA. Proc. Natl. Acad. Sci. USA *74*: 560 – 564.

McBride, L.J. and Caruthers, M.H. (1983). An investigation of several deoxyribonucleoside phosphoramidates useful for synthesizing deoxyoligonucleotides. Tetrahedron Lett. *24*: 245 – 248.

McConaughy, B.L. and McCarthy, B.J. (1967). The interaction of oligodeoxyribonucleotides with denatured DNA. Biochim. Biophys. Acta *149*: 180 – 189.

McNeil, J.B. and Smith, M. (1985). *Saccharomyces cerevisiae CYC1* mRNA 5'-end positioning: analysis by *in vitro* mutagenesis, using synthetic duplexes with random mismatch base pairs. Mol. Cell Biol. *5*: 3545 – 3551.

Michelson, A.M. and Monny, C. (1967). Oligonucleotides and their association with polynucleotides. Biochim. Biophys. Acta *149*: 107 – 126.

Montgomery, D.L., Hall, B.D., Gillam, S. and Smith, M. (1978). Identification and isolation of the yeast cytochrome *c* gene. Cell *14*: 673 – 680.

Mullis, K. (1994). Les Prix Nobel en 1993. Norstedt and Sons, Stockholm. pp. 107 – 117.

Niyogi, S.K. and Thomas, C.A., Jr. (1968). The stability of oligoadenylate-polyuridylate complexes as measured by thermal chromatography. J. Biol. Chem. *243*: 1220 – 1223.

Niyogi, S.K. (1969). The influence of chain length and base composition on the specific association of oligoribonucleotides with denatured deoxyribonucleic acid. J. Biol. Chem. *244*: 1576 – 1581.

Razin, A, Hirose, T., Itakura, K. and Riggs, A.D. (1978). Efficient correction of a mutation by use of chemically synthesized DNA. Proc. Natl. Acad. Sci. USA *75*: 4268 – 4270.

Sambrook, J. Fritsch, E.F. and Maniatis, T. (1989). Molecular Cloning. A Laboratory Manual, 2nd Edition (3 volumes). Cold Spring Harbor Laboratory Press, Cold Spring Harbor, N.Y.

Sanger, F. and Coulson, A.R. (1975). A rapid method for determining sequences in DNA by primed synthesis with DNA polymerase. J. Mol. Biol. *94*: 441 – 448.

Sanger, F., Coulson, A.R., Friedmann, T., Air, G.M., Barrell, B.G., Brown, N.L., Fiddes, J.C., Hutchison, C.A. III, Slocombe, P.M. and Smith, M. (1978). The nucleotide sequence of bacteriophage ØX174. J. Mol. Biol. *125*: 225 – 246.

Sanger, F. (1981). Determination of nucleotide sequences in DNA. In: Les Prix Nobel en 1980. Norstedt and Sons, Stockholm. pp. 143 – 159.

Sanger, F. (1988). Sequences, sequences and sequences. Ann Rev. Biochem. *57*: 1 – 28.

Sanger, F., Nicklen, S. and Coulson, A.R. (1977). DNA sequencing with chain terminating inhibitors. Proc. Natl. Acad. Sci. USA *74*: 5463 – 5467.

Sasavage, N.L., Smith, M., Gillam, S., Astell, C., Nilson, J.H. and Rottman, F. (1980). Use of oligodeoxynucleotide primers to determine poly(adenylic acid) adjacent sequences in messenger ribonucleic acid. Biochemistry *19*: 1737 – 1743.

Sasavage, N.L., Smith, M., Gillam, S., Woychik, R.P. and Rottman, F.M. (1982). Variation in the polyadenylation site of prolactin messenger RNA. Proc. Natl. Acad. Sci. USA *79*: 223 – 227.

Smith, M. and Khorana, H.G. (1958). An improved and general method for the synthesis of ribo- and deoxyribo-nucleoside-5' triphosphates. J. Am. Chem. Soc. *80*: 1141−1145.

Smith, M. and Khorana, H.G. (1959). Specific synthesis of the C5'-C3' inter-ribonucleotide linkage: the synthesis of uridylyl-(3'«5')-uridine. J. Am. Chem. Soc. *81*: 2911.

Smith, M. Drummond, G.I. and Khorana, H.G. (1961). The synthesis and properties of ribonucleoside-3',5' cyclic phosphates. J. Am. Chem. Soc. *83*: 698−706.

Smith, M., Rammler, D.H., Goldberg, I.H. and Khorana, H.G.. (1961). Synthesis of uridylyl-(3'« 5')-Uridine and uridylyl-(3'«5')-adenosine. J. Am. Chem. Soc. *84*: 430−440.

Smith, M. (1964). Synthesis of deoxyribonucleoside-3'5' cyclic phosphates by base-catalysed transesterification. J. Am. Chem. Soc. *86*: 3586.

Smith, M., Leung, D.W., Gillam, S., Astell, C.R., Montgomery, D.L. and Hall, B.D. (1979). Sequence of the gene for iso-1-cytochrome *c* in *Saccharomyces cerevisiae*. Cell *16*: 753−761.

Smith, M. and Gillam, S. (1981). Constructed mutants using synthetic oligodeoxyribonucleotides as site-specific mutagens. In: Genetic Engineering. Principles and Methods (J.K. Setlow and A. Hollaender, eds.). Volume 3. Plenum Press, New York, N.Y. pp. 1−52.

Smith, M. (1983). Synthetic oligodeoxyribo nucleotides as probes for nucleic acids and as primers in sequence determination. In: Methods of DNA and RNA Sequencing (S.M. Weissman, ed.). Praeger Publishers, New York, N.Y. pp. 23−68.

Smith, M. (1985). *In vitro* mutagenesis. Ann. Rev. Genet. *19*: 423−462.

Stewart, J.W. and Sherman, F. (1974). Yeast frameshift mutations identified by sequence changes in iso-1-cytochrome *c*. In: Molecular and Environmental Aspects of Mutagenesis (L. Prakash, F. Sherman, M.W. Miller, C.W. Lawrence and Taber, H.W., eds.). Charles C. Thomas, Springfield, IL pp. 102−107.

Suggs, S.V., Wallace, R.B., Hirose, T., Kawashima, E.H. and Itakura, K. (1981). Use of synthetic oligonucleotides as hybridization probes: isolation of cloned cDNA sequences for human β_2-microglobulin. Proc. Natl. Acad. Sci. USA *78*: 6613−6617.

Taylor, J.W., Ott, J. and Eckstein, F. (1985). The rapid generation of oligonucleotide-directed mutations at high frequency using phosphothioate-modified DNA. Nucleic Acids Res. *13*: 1451−1455.

Tigerstrom, R. von and Smith, M. (1970). Oligodeoxyribo-nucleotides: chemical synthesis in anhydrous base. Science *167*: 1266−1268.

Tigerstrom, R. von, Jahnke, P. and Smith, M. (1975a). The synthesis of the internucleotide bond by a base-catalysed reaction. Nucleic Acids Res. *2*: 1727−1736.

Tigerstrom, R. von, Jahnke, P., Wylie, V. and Smith, M. (1975b). Application of base-catalysed reaction to the synthesis of dinucleotides containing the four common deoxyribonucleosides and of oligodeoxythymdylates. Nucleic Acids Res. *2*: 1737−1743.

Wallace, R.B., Johnson, M.J., Hirose, T., Miyake, T., Kawashima, E.H. and Itakura, K. (1981). Hybridization of oligonucleotides of mixed sequence to rabbit b-globin DNA. Nucleic Acids Res. *9*: 3647−3656.

Watson, J.D., Hopkins, N.H., Roberts, J.W., Steitz, J.A. and Weiner, A.M. (1987). Molecular Biology of the Gene. 4th Edn. The Benjamin/Cummings Publishing Company, Inc. Menlo Park, CA.

Watson, J.D., Gilman, M., Witkawski, J. and Zoller, M. (1993). Recombinant DNA. 2nd Edn. Scientific American Books, New York, N.Y.

Weisbeek, P.J. and van de Pol, J.H. (1970). Biological activity of ∅X174 replicative form DNA fragments. Biochim. Biophys. Acta *224*: 328−338.

Winter, G., Fersht, A.R., Wilkinson, A.J., Zoller, M.J. and Smith, M. (1982). Redesigning enzyme structure by site-directed mutagenesis: tyrosyl t-RNA synthetase and ATP binding . Nature *299*: 756−758.

Zoller, M.J. and Smith, M. (1982). Oligonucleotide-directed mutagenesis using M13-derived vectors: an efficient and general procedure for the production of point mutations in any fragment of DNA. Nucleic Acids Res. *10*: 6487–6500.

Zoller, M.J. and Smith, M. (1983). Oligonucleotide-directed mutagenesis of fragments cloned in M13 vectors. Methods Enzymol. *100*: 468–500.

Zoller, M.J. and Smith, M. (1984). Oligonucleotide-directed mutagenesis: a simple method using two oligonucleotide primers and a single-stranded DNA template DNA *3*: 479–488.

Zoller, M.J. and Smith, M. (1987). Oligonucleotide-directed mutagenesis: a simple method using two oligonucleotide primers and a single-stranded DNA template. Methods Enzymol. *154*: 329–350.

Zoller, M.J. (1991). New molecular biology methods for protein engineering. Curr. Opin. Biotechnol. *2*: 526–531.

Zoller, M.J. (1992). New recombinant DNA methodology for protein engineering. Curr. Opin. Biotechnol. *3*: 348–354.

Chemistry 1994

GEORGE A. OLAH

for his contributions to carbocation chemistry

THE NOBEL PRIZE IN CHEMISTRY

Speech by Professor Salo Gronowitz of the Royal Swedish Academy of Sciences.
Translation from the Swedish text.

Your Majesties, Your Royal Highnesses, Ladies and Gentlemen,

The preparation of complicated organic compounds from simple and inexpensive starting materials is one of the prerequisites of our civilization, the chemical era in which we live. Organic synthesis has given us efficient methods to obtain drugs, vitamins, textile fibers, plastics, insecticides and herbicides, lacquers, paints and fuels. For chemists it is very important to understand in detail what is going on when the molecules in the starting materials react with each other and create the molecules characteristic of the product. This is the process of determining the mechanism of the reaction. Knowledge about mechanisms makes it possible to develop better and less expensive methods to prepare products of technical importance. My teacher, the late Professor Arne Fredga, said from this podium in 1975: "Not knowing the mechanism is like seeing the first and the last scene of Hamlet." The performance would not be very rewarding, and one would wonder what actually happened!

In many cases the reactions proceed via so-called intermediates, which are products with very short lifetimes. One type of reactive intermediates is called "carbocations." Charged atoms and groups of atoms are common in inorganic chemistry. All of us know about table salt, which consists of positively charged sodium ions, or cations, and negatively charged chloride ions, or anions. The opposite is true of the large number of organic compounds, especially hydrocarbons, which are composed of only two elements, carbon and hydrogen. There were many indications that carbocations were intermediates in two common and frequently used reactions in synthetic organic chemistry. In that case these intermediates had to have an extremely short lifetime, a billionth of a second or less, and due to their high reactivity their concentrations had to be very low. Their existence has been indicated by measurements of reaction rates and observations of the spatial arrangement of the atoms in space. For these purposes a variety of ingenious experiments have been carried out. However, nobody was able to see these carbocations, not even with the most powerful microscopes or by spectroscopic methods. These techniques can be regarded as extensions of human vision. Consequently there was no evidence for the existence of carbocations, in other words whether they were a reality independent of human consciousness, or were only created by human imagination in order to describe the experimental results.

Because it was not possible to detect carbocations with spectroscopic methods, different scientists interpreted their experiments differently, and a scientific feud took place in organic chemistry during the 1960s and 1970s.

Through a series of brilliant experiments Professor George Olah solved the problem. He created methods to prepare long-lived carbocations in high concentrations, which made it possible to study their structure, stability and reactions with spectroscopic methods. He achieved this by using special solvents, which did not react with the cations. He observed that in these solvents at low temperatures, carbocations could be prepared with the aid of superacids, acids eighteen powers of ten stronger than concentrated sulfuric acid. Through Olah's pioneering work he and the scientists who followed in his footsteps could obtain detailed knowledge about the structure and reactivity of carbocations. Olah's discovery resulted in a complete revolution for scientific studies of carbocations, and his contributions occupy a prominent place in all modern textbooks of organic chemistry.

Olah found that there are two groups of carbocations, the trivalent ones called carbenium ions, in which the positive carbon atom is surrounded by three atoms, and those in which the positive carbon is surrounded by five atoms, called carbonium ions. The disputed existence of these penta-coordinated carbocations was the reason for the scientific feud. By providing convincing proof that penta-coordinated carbocations exist, Olah demolished the dogma that carbon in organic compounds could at most be tetra-coordinated, or bind a maximum of four atoms. This had been one of the cornerstones of structural organic chemistry since the days of Kekulé in the 1860s.

Olah found that the superacids were so strong that they could donate a proton to simple saturated hydrocarbons, and that these penta-coordinated carbonium ions could undergo further reactions. This fact has contributed to a better understanding of the most important reactions in petrochemistry. His discoveries have led to the development of methods for the isomerization of straight chain alkanes, which have low octane numbers when used in combustion engines, to produce branched alkanes with high octane numbers. Furthermore, these branched alkanes are important as starting materials in industrial syntheses. Olah has also shown that with the aid of superacids it is possible to prepare larger hydrocarbons with methane as the building block. With superacid catalysis it is also possible to crack heavy oils and liquefy coal under surprisingly mild conditions.

Professor Olah,

I have in these few minutes tried to explain your immense impact on physical organic chemistry through your fundamental investigations of the structure, stability and reactions of carbocations. In recognition of your important contribution, the Royal Swedish Academy of Sciences has decided to confer upon you this year's Nobel Prize for Chemistry. It is an honour and pleasure for me to extend to you the congratulations of the Royal Swedish Academy of Sciences and to ask you to receive your Prize from the hands of his Majesty the King.

George Olah

GEORGE A. OLAH

I was born in Budapest, Hungary, on May 22, 1927 the son of Julius Olah and Magda Krasznai. My father was a lawyer and to my best knowledge nobody in my family before had interest in science. I grew up between the two world wars and received a rather solid general education, the kind middle class children enjoyed in a country whose educational system had its roots dating back to the Austro-Hungarian Monarchy. I attended a Gymnasium (a combination of junior and senior high school) at one of the best schools in Budapest run by the Piarist Fathers a Roman Catholic order. A strict and demanding curriculum heavily emphasizing the humanities included 8 years of Latin, with German and French as other obligatory languages. Although we had an outstanding science teacher who later became a professor of physics in the University of Budapest I can not recollect any particular interest in chemistry during my school years. My main interest was in the humanities, particularly history, literature, etc. I was (and still am) and avid reader and believe that getting attached too early to a specific field frequently short-changes a balanced broad education. Although reading the classics in Latin in school may be not as fulfilling as it would be at a more mature age, few scientists can afford the time for such diversion later in life.

After graduating from high school and having survived the ravages of war in Budapest and realizing the difficulties facing life in a small and war torn country, I started to study chemistry upon entering university, being attracted by the wide diversity it offered.

Classes at the Technical University of Budapest were relatively small. We probably started with a class of 70 of 80, whose numbers were rapidly pared down during the first year to maybe half by rather demanding "do or die" oral examinations, where the ones who failed could not connntinue. This was a rather cruel process, because laboratory facilities were so limited that only few could be accommodated. At the same time the laboratory training was thorough. For example, in the organic laboratory we did some 40 Gatterman preparations. It certainly gave a solid foundation.

Organic chemistry particularly intrigued me and I was ffortunate later to become a research assistant to Professor Geza Zemplen, the senior professor of organic chemistry in Hungary, who himself was a student of Emil Fischer in Berlin. He established in Hungary a reputable school in organic chemistry. As Fischer, he too expected his students to pay their own way and even paying for the privilege to work in his laboratory. Becoming an assistant to him although meant no remuneration but also no fee. Zemplen had a formidable reputation, and working for him was quite an experience. He also liked partying and these remarkable events in neighboring pubs lasted frequently for days. Certainly one's stamina developed through these experiences.

Zemplen was a carbohydrate chemist, much interested in glycosides. Early in our association it became clear that my ideas and interst were not always closely matching his. When I suggested that fluorine containing carbohydrates may be of interest in coupling reactions, his reaction was not unexpectedly very negative. To try to pursue fluorine chemistry in post-war Hungary was indeed far fetched. Eventually, however, he gave in. Even basic chemicals needed for the work, such as HF, FSO_3H or BF_3 were non existent and I made them myself, with enthusiastic help by some of my early associates (A. Pavlath, S. Kuhn). Laboratory space, particularly hoods (the kind exhausted only by draft caused by a gas burner causing warm air to raise and take some of the abnooxious fumes through a chimney) was very scarce and even by the time I became an assistant professor it was not welcome to "pollute" more important conventional work. However, the Institute which was on the second floor of the chemistry building, had in the back an open balcony, used to store chemicals. In one of his unexpected gestures Zemplen agreed that I can have the use of this balcony. With some effort we enclosed it, installed two old hoods and were soon in business in what was referred to as the "balcony laboratory". I am not sure that Zemplen even set foot in it. We enjoyed, however, our new quaters and the implicit understanding that our flluorine chemistry and related study of Friedel-Crafts reactions and their intermediates was now officially tolerated.

Some of my publications in the early 50's from Hungary caught the eye of Hans Meerwein. It is still a mystry to me how he came to read them in a Hungrian journal, although there also was a foreign language edition of the Hungarian Chimica Acta. Anyhow, I received an encouraging letter from him and we followed up correspondence (not easy at a time in completely isolated Hungary). He must have sympathized with my difficulties because one day through his efforts I received cyinder of boron trifluoride. What a precious gift it was!

The Hungarian educational system after the Communist takeover was realigned according to the Soviet example. University research was deemphasized and research institutes were established under the auspices of the Academy of Sciences. I was invited to join the newly established Central Chemical Research Institute of the Hungarian Academy of Sciences in 1954 and was able to establish a small research group in organic chemistry, housed in temporary laboratories of an industrial research institute. With my group, which by now also included my wife, we were able to expand our work and made the best of our possibilities. In October 1956 Hungary revolted against the Soviet rule, but the uprising was soon put down by drastic measures and much loss of life. Budapest was again devastated and the future looked rather dim. In November-December 1956 some 200,000 Hungarians, mostly of the younger generation fled their country. With my family and much of my research group we also decided to follow this path and look for a new life in the West.

I married in 1949 Judith Lengyel, the best thing ever to happen to me in my life. We knew each other from our early youth and are happily married now for more than 45 years. Judy worked initially as a technical secretary at the Technical University. After we were married she enrolled to study chemistry. She probably rightly recalls that I was entirely responsible for this step and she only agreed to get along with her single minded husband who seemed to believe that there is little in life outside chemistry. From my point of view for husband and wife to closely understand each other's work and may even work together was most desirable. Our older son George John was born in Budapest in 1954. After we fled Hungary in early December of 1956, we reached late in December London where my wife had relatives. We subsequently moved on in the spring of 1957 to Canada, where my mother-in-law lived in Montreal after the war. During our stay in London for the first time I was able to establish personal contact with some of the organic chemists, whose work I knew and admired from the literature. I found them most gracious and helpful. In particular Christopher Ingold and Alexander Todd extended efforts on behalf of a young, practically unknown Hungarian refugee chemist in a way which I never forget and for which I am always grateful.

Dow Chemical, with its home base at Midland, Michigan was establishing at the time a small exploratory research laboratory 100 miles across the border in Sarnia, Ontario where its Canadian Subsidaries major operations were located. I was offered a position to join this new laboratory and they also hired two of my original Hungarian Collaborators, including Steven Kuhn. We moved to Sarnia in late May of 1957. As our moving expenses where paid we checked in two cardboard boxes containing all of our worldly possessions unto the train from Montreal and started our new life. Our younger son Ronald Peter was born in Sarnia in 1959. There was no possibility for Judy to continue her career at the time. Sacrificing her own career she devoted herself to bring up our children. She rejoined in our research only a decade later in Cleveland after I returned to academic life.

The Sarnia years at Dow were productive. It was during this period in the late 50's that my initial work on stable carbocations was started. Dow was and is a major user of carbocationic chemistry, such as the Friedel-Crafts type manufacture of etylbenzene for styrene production. My work thus also had practical significance and helped to improve some industrial processes. In return I was treated well and given substantial freedom to pursue my own ideas. Eventually I was promoted to company Scientist, the highest research position without administrative responsibility.

In the spring of '64 I transferred to Dow's Eastern Research Laboratories in Framingham, Massachusetts established under Fred McLarrerty's directorship. The laboratory was subsequen6tly moved to Wayland, just outside Boston.

In the summer of 1965 I was invited to join Western Reserve University in Cleveland, Ohio and returned to academic life as professor with the added responsibility of becoming also Department Chairman.

My Cleveland years were both scientifically and personally most rewarding. My wife Judy was able to rejoin me in our research and my research group grew rapidly. The chemistry departments of Western Reserve University and neighboring Case Institute of Technology were practically adjacent, separated only by a parking lot. It became obvious that it would make sense to join the two into a single, stronger department. We achieved this by 1967 with surprisingly little friction and I was asked to serve as the Chair of the joint department till things settled down. It was in 1969 that I was able to give up my administrative responsibility. As I worked hard my research never suffered during this period and as a matter of fact these were probably some of my most productive years.

After 12 years in Cleveland it was time again to move on. Our older son George was approaching the end of his college years and our younger son Ron who was finishing high school set his mind to go to Stanford. He convinced us that it sould be nice for the whole family to resettle in California. Coincidentally, in the fall of 1976 Sid Benson, an old friend called me to find out whether I would be interested to join him at the University of Southern California in Los Angeles. After some visits to LA the challenge of trying to build up chemistry in a dynamic university and the attractiveness of life in Southern California convinced us to move. We fell in love with California and we still are. As USC had limited chemistry facilities, it was offered to establish a research institute in the broad area of hydrocarbon research and provide it with its own building and facilities. We moved in May of 1977. Some 15 members of my research group joined the move West. By arrangements worked out we were able to take with us most of the laboratory equipment, chemicals, etc. Two weeks after our arrival with some large moving vans we were back doing chemistry in temporary quarters, while our research institute was constructed. The Institute was established at USC with generous support by Mr. & Mrs. D.P. Loker, friends and great supporters of the University. The Institute was subsequently named after them. Don Loker passed away some years ago, but Katherine still chairs the Institute's board. Through her and other friends' generosity a wonderful new addition to our Institute is just completed doubling our space.

As rewarding as the Nobel Prize is personally to any scientist, I feel it is also recognition of all my past and present students and associates (by now numbering close to 200), who contributed over the years so much through their dedicated hard work to our joint effort. It also recognizes fundamental contributions by many colleagues and friends from around the world to a field of chemistry, which is not frequently highlighted or recognized.

Awards and Fellowships:

American Chemical Society Award in Petroleum Chemistry, 1964
Leo Hendrick Baekeland Award, 1967
Morley Medal, 1970
Fellow of the J.S. Guggenheim Foundation, 1972 and 1988
Fellow of the Society for the Promotion of Science, Japan, 1974
Member of the U.S. National Academy of Sciences, 1976
Centenary Lectureship, British Chemical Society, 1977
American Chemical Society Award for Creative Work in Synthetic Organic Chemistry, 1979
Alexander von Humbolt-Stiftung Award for Senior U.S. Scientists, 1979
Foreign Member of the Italian National Academy dei Lincei, 1982
Michelson-Morley Award of Case Western Reserve University, 1987
Honorary Member Italian Chemical Society, 1988
California Scientist of the Year Award, 1989
American Chemical Society Roger Adams Award in Organic Chemistry, 1989
Member of the European Academy of Arts, Sciences and Humanities, 1989
Honorary Member Hungarian Academy of Sciences, 1990
Richard C. Tolman Award, American Chemical Society, Southern California Section, 1992
Chemical Pioneers Award, The American Institute of Chemists, Inc., 1993
Nobel Prize in Chemistry, 1994.

Honorary Degrees:

D.Sc. Honoris Causa, University of Durham, England, 1988; Technical University of Budapest, 1989; University of Munich, 1990; University of Crete, 1994; University of Szeged, Hungary, 1995; University of Veszprem, Hungary, 1995; University of Southern California, Los Angeles, 1995; Case Western Reserve University, Cleveland, 1995.

Publications and Patents:

Over 1000 published scientific papers. 100 patents. Books include: "Introduction to Theoretical Organic Chemistry," (in German) Akademie Verlag, Berlin, 1960; "Friedel-Crafts and Related Reactions," Vols. I – IV, Wiley-Interscience, 1963 – 65; "Carbonium Ions" with Paul v. R. Schleyer, Vols. I-IV, Wiley Interscience, 1968 – 73; "Carbocations and Electrophilic Reactions," (German and English versions) Verlag Chemie-Wiley-Interscience, 1973;; "Friedel – Crafts Chemistry," Wiley-Interscience, 1973; "Halonium Ions," Wiley-Interscience, 1975; "Superacids" with G.K. Surya Prakash and J. Sommer, Wiley-Interscience, New York, 1985;

"Hypercarbon Chemistry" with G. K. Surya Prakash, K. Wade, L.D. Field, and R. E. Williams, Wiley-Interscience, New York, 1987; "Nitration: Methods and Mechanism" with R. Malhotra and S.C. Narang, VCH Publishers, New York, 1989; "Cage Hydrocarbons," G. A. Olah (Ed.), Wiley-Interscience, New York, 1990; "Electron Deficient Baron and Carbon Clusters," with R. E. Williams and K. Wade (Eds.), Wiley-Interscience, New York, 1991; "Chemistry of Energetic Materials" with D. R. Squire (Eds.), Academic Press, New York, 1991; "Synthetic Fluorine Chemistry," with R.D. Chambers and G. K. S. Prakash (Eds.), Wiley-Interscience, New York, 1992; "Hydrocarbon Chemistry" with A. Molnar, Wiley-Interscience, New York, 1994.

MY SEARCH FOR CARBOCATIONS AND THEIR ROLE IN CHEMISTRY

Nobel Lecture, December 8, 1994

by

GEORGE A. OLAH

Loker Hydrocarbon Research Institute and Department of Chemistry, University of Southern California, Los Angeles, CA 90089-1661, USA

> *"Every generation of scientific men (i.e. scientists) starts where the previous generation left off, and the most advanced discoveries of one age constitute elementary axioms of the next. – – –*
>
> Aldous Huxley

INTRODUCTION

Hydrocarbons are compounds of the elements carbon and hydrogen. They make up natural gas and oil and thus are essential for our modern life. Burning of hydrocarbons is used to generate energy in our power plants and heat our homes. Derived gasoline and diesel oil propel our cars, trucks, airplanes. Hydrocarbons are also the feed-stock for practically every man-made material from plastics to pharmaceuticals. What nature is giving us needs, however, to be processed and modified. We will eventually also need to make hydrocarbons ourselves, as our natural resources are depleted. Many of the used processes are acid catalyzed involving chemical reactions proceeding through positive ion intermediates. Consequently, the knowledge of these intermediates and their chemistry is of substantial significance both as fundamental, as well as practical science.

Carbocations are the positive ions of carbon compounds. It was in 1901 that Norris[1a] and Kehrman[1b] independently discovered that colorless triphenylmethyl alcohol gave deep yellow solutions in concentrated sulfuric acid. Triphenylmethyl chloride similarly formed orange complexes with aluminum and tin chlorides. von Baeyer (Nobel Prize, 1905) should be credited for having. recognized in 1902 the salt like character of the compounds formed (equation 1).[1c]

$$Ar_3C\text{-}X \; \rightleftharpoons \; Ar_3C^+ + X^- \tag{1}$$

He then proceeded to suggest a correlation between the appearance of color and formation of salt – the so called "halochromy". Gomberg[1d] (who had just shortly before discovered the related stable triphenylmethyl radical) as well as Walden[1e] contributed to the evolving understanding of the structure of related dyes, such as malachite green 1.

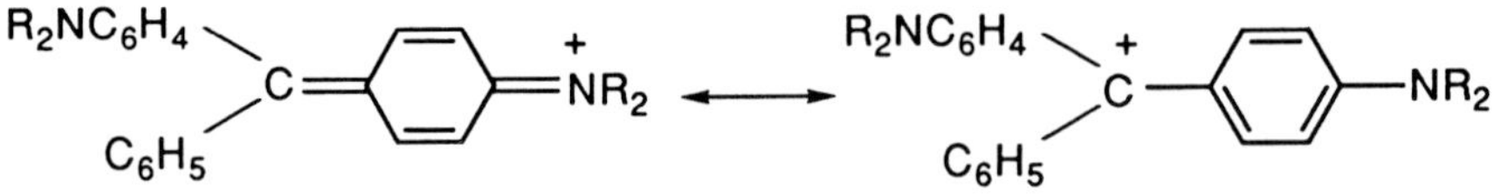

$$R_2NC_6H_4-\overset{C_6H_5}{\underset{}{C}}=\underset{}{\bigcirc}=NR_2 \longleftrightarrow R_2NC_6H_4-\overset{+}{\underset{C_6H_5}{C}}-\bigcirc-NR_2$$

Stable carbocationic dyes were soon found to be even present in nature. The color of red wine as well as of many flowers, fruits, leaves, etc. is due in part to flavylium and anthocyanin compounds formed upon cleavage of their respective glycosides.

The constitution of flavylium and anthocyanin compounds envolved based on Robinson's and Willstatter's pioneering studies. Werner formulated[1f] the parent benzopyrilium or xanthylium salts 2 and 3 as oxonium salts, while Baeyer[1g] emphasized their great similarity to triarylmethylium salts and considered them as carbenium salts. Time has indeed justified both point of view with the realization of the significance of the contribution of both oxonium and carbenium ion resonance forms.

2

3

Whereas the existence of ionic triarylmethyl and related dyes was established at around the turn of the 20th century, the more general significance of carbocations in chemistry was for long unrecognized. Triarylmethyl cations were considered as an isolated curiosity of chemistry, not unlike Gomberg's triarylmethyl radicals. Simple hydrocarbon cations in general were believed not only to be of unstable nature but their fleeting existence was even doubted.

One of the most original and significant ideas in organic chemistry was the suggestion that carbocations (as we now call all the positive ions of carbon compounds) might be intermediates in the course of reactions that start from nonionic reactants and lead to nonionic covalent products.

It was Hans Meerwein[2] who in 1922, while studying the Wagner rearrangement of camphene hydrochloride 4 to isobornyl chloride 5, found that the rate of the reaction increased with the dielectric constant of the solvent. Further, he found that certain Lewis acid chlorides – such as $SbCl_5$, $SnCl_4$,

FeCl$_3$, AlCl$_3$, and SbCl$_3$ (but not BCl$_3$ or SiC$_4$), as well as dry HCl (which promote the ionization of triphenylmethyl chloride by formation of carbocationic complexes) considerably accelerated the rearrangement of camphene hydrochloride to isobornyl chloride. Meerwein concluded that the isomerization actually does not proceed by way of migration of the chlorine atom but by a rearrangement of a cationic intermediate, which he formulated as

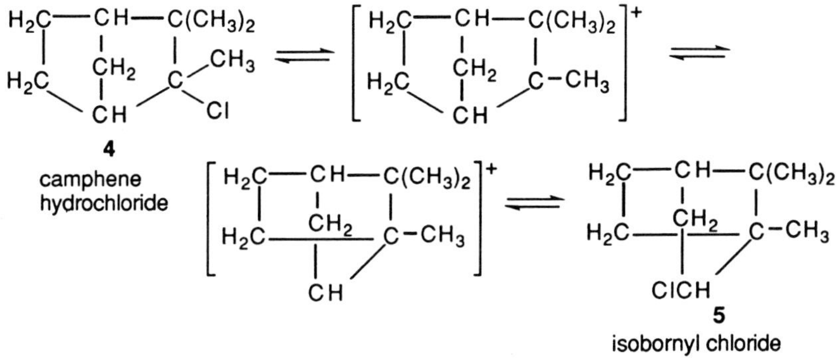

4
camphene
hydrochloride

5
isobornyl chloride

Hence, the modern concept of carbocationic intermediates was born. Meerwein's views were, however, greeted with much skepticism by his contemporary peers in Germany, discouraging him to follow up on these studies.

C. K. Ingold, E. H. Hughes, and their collaborators in England, starting in the late 1920s carried out detailed kinetic and stereochemical investigations on what became known as nucleophilic substitution at saturated carbon and polar elimination reactions.[3] Their work relating to unimolecular nucleophilic substitution and elimination called S$_N$1 and El reactions (equations 2 and 3) laid the foundation for the role of electron deficient carbocationis intermediates in organic reactions.

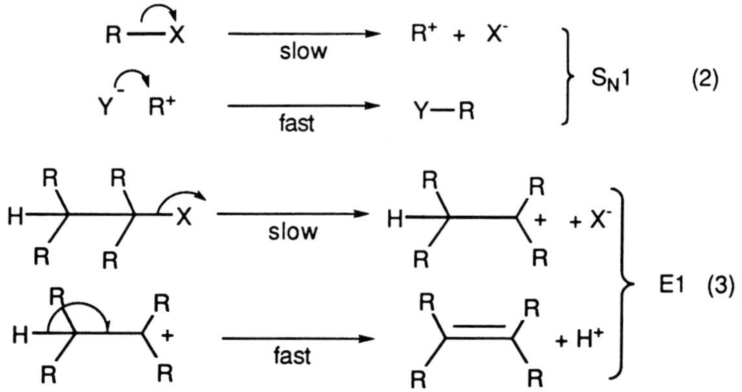

F. Whitmore[4] in the US in a series of papers in the thirties generalized these concepts to many other organic reactions. Carbocations, however, were generally considered to be unstable and transient (short lived) as they could not be directly observed. Many leading chemists, including Roger Adams, determinedly doubted their existence as real intermediates and strongly opposed even mentioning them. Whitmore consequently never was able in any of his publications in the prestigeous Journal of the American Chemical Society use their name or the notation of ionic R_3C^+. The concept of carbocations, however, slowly grew to maturity through kinetic, stereochemical and product studies of a wide variety of reactions. Leading investigators such as P. D. Bartlett, C. D. Nenitzescu, S. Winstein, D. J. Cram, M. J. S. Dewar, J. D. Roberts, P. v. R. Schleyer, and others have contributed fundamentally to the development of modern carbocation chemistry. The role of carbocations as one of the basic concepts of modern chemistry was firmly established and is well reviewed.[5-7] With the advancement of mass spectrometry the existence of gaseous cations was proven, but this could give no indication of their structure or allow extrapolation to solution chemistry. Direct observation and study of stable, long-lived carbocations, such as of alkyl cations in the condensed state remained an elusive goal.

My involvement with the study of carbocations dates back to the fifties and resulted in the first direct observation of alkyl cations and subsequently the whole spectrum of carbocations as long lived species in highly acidic (superacidic) solutions.[5-8-9] The low nucleophilicity of the involved counter anions (SbF_6^-, $Sb_2F_{11}^-$, etc.) greatly contributed to the stability of the carbocations, which were in some instances even possible to isolate as crystalline salts. The developed superacidic, "stable ion" methods also gained wide application in the preparation of other ionic intermediates (nitronium, halonium, oxonium ions, etc.). At the same time the preparation and study of an ever increasing number of carbocations allowed the evolvement of a general concept of carbocations, which I suggested in a 1972 paper, as well as naming the cations of carbon compounds as "carbocations".[10] "Carbocations" is now the IUPAC approved[11] generic name for all cationic carbon compounds, similarly to the anionic compounds being called "carbanions".

FROM ACYL CATIONS TO ALKYL CATIONS

The transient nature of carbocations in their reactions arises from their high reactivity towards reactive nucleophiles present in the system. The use of relatively low nucleophilicity counter-ions, particularly tetrafluoroborate (BF_4^-) enabled Meerwein in the forties to prepare a series of oxonium and carboxonium ion salts, i. e., $R_3O^+BF_4^-$ and $HC(OR)_2^+$, respectively.[12] These Meerwein salts are effective alkylating agents, and transfer alkyl groups in S_N2 type reactions. However, no acyl ($RCO^+BF_4^-$) or alkyl cation salts ($R^+BF_4^-$) were obtained in Meerwein's studies.

Acetic acid or anhydride with Lewis acids such as boron trifluoride was shown to form complexes. The behavior of acetic acid and anhydride in strong protic acids (sulfuric acid, oleum, perchloric acid, etc.) was also extensively studied. None of these resulted, however, in the isolation or unequivocal characterization of the acetyl cation (or other related homologous acyl cations). F. Seel prepared for the first time in 1943 acetylium tetrafluoroborate[13] by reacting acetyl fluoride with boron trifluoride (equation 4).

$$CH_3\overset{\cdot}{C}OF \ + \ BF_3 \ \rightleftharpoons \ CH_3CO^+BF_4^- \qquad (4)$$

In the early fifties while working at the Organic Chemical Institute of the Technical University in Budapest, led by the late Professor G. Zemplen, a noted carbohydrate chemist and former student of Emil Fischer (Nobel Prize, 1902) whose "scientific grandson" I thus can consider myself, I became interested in organic fluorine compounds. Zemplen was not much impressed by this, as he thought attempts to study fluorine compounds necessitating "outrageous" reagents such as hydrogen fluoride to be extremely foolish. Eventually, however, I prevailed and was allowed to convert an open balcony at the rear of the top floor of the chemistry building into a small laboratory where together with some of my early dedicated associates (A. Pavlath, S. Kuhn) we started up the study of organic fluorides as reagents. Seel's previous work particularly fascinated me. As Zemplen's interest was in glycoside synthesis and related carbohydrate chemistry, I thought that selective α- or β-glycoside synthesis could be achieved by reacting either acetofluoroglucose (as well as other fluorinated carbohydrates) or their deacetylated relatively stable fluorohydrins with the corresponding aglucons. In the course of the project -COF compounds were needed. As Seel did not seem to have followed up his earlier study, I got interested in exploring in general acylation with acyl fluorides. The work was subsequently extended to alkylation with alkyl fluorides using boron trifluoride as catalyst to achieve Friedel-Crafts type reactions. These studies also arose my interest in the mechanistic aspects of the reactions, including the complexes of RCOF and RF with BF_3 and subsequently with other Lewis acid fluorides (equations 5 and 6). Thus, my long fascination with the chemistry of carbocationic complexes began.

$$ArH \ + \ RCOF \ \xrightarrow{\ BF_3\ } \ ArCOR \ + \ HF \qquad (5)$$

$$ArH \ + \ RF \ \xrightarrow{\ BF_3\ } \ ArR \ + \ HF \qquad (6)$$

Carrying out such research in post-war Hungary was not easy. There was no access to such chemicals as anhydrous HF, FSO_3H or BF_3 and we needed to prepare them ourselves. After we prepared HF from fluorspar (CaF_2) and sulfuric acid, its reaction with SO_3 (generated from oleum) gave FSO_3H. By reacting boric acid with fluorosulfuric acid we made BF_3. Handling these reagents and carrying out related chemistry in a laboratory equipped with the barest of necessities was indeed a challenge. It was only around 1955 through Meerwein's kindness, who read some of my early publications, started up a correspondence and offered his help, that we received a cylinder of BF_3 gas. What a precious gift it was![14]

My early work with acyl fluorides also involved formyl fluoride, HCOF,[15] which was first prepared by Nesmejanov in the thirties,[16] but did not pursue its use in synthesis. We also prepared a series of higher homologous acyl fluorides and studied their chemistry.[15]

In Friedel-Crafts chemistry it was known that when pivaloyl chloride was reacted with aromatics in the presence of aluminum chloride besides the expected ketones *tert*-butylated products were also obtained (equation 7).[17]

Formation of the latter was assumed to involve decarbonylation of the intermediate pivaloyl complex or cation. In the late fifties, now working at the Dow Chemical Company laboratory in Sarnia, Ontario (Canada), I was able to return to my studies and extend them by using IR and NMR spectroscopy in the investigation of isolable acyl fluoride – Lewis acid fluoride complexes, including those with higher valency Lewis acid fluorides such as SbF_5, AsF_5, PF_5. Consequently, it was not unexpected that when the $(CH_3)_3CCOF-SbF_5$ complex was obtained it showed substantial tendency for decarbonylation.[18] What turned out to be exciting was that it was possible to follow this process by NMR spectroscopy and to observe what turned out to be the first stable, long-lived alkyl cation salt, i.e., *tert*-butyl hexafluoroantimonate.[18 – 20]

$$(CH_3)_3CCOF + SbF_5 \longrightarrow (CH_3)_3CCO^+SbF_6 \xrightarrow{-CO} (CH_3)_3C^+SbF_6^- \quad (8)$$

This breakthrough was first reported in my 1962 papers,[18 – 19] and was followed by further studies which led to methods to obtain long lived alkyl cations in solution.[20] Before recollecting some of this exciting development, however, a brief review of the long quest for these long elusive alkyl cations is in order.

EARLIER UNSUCCESSFUL ATTEMPTS TO OBSERVE ALKYL CATIONS IN SOLUTIO N

Alkyl cations were considered until the early sixties only as transient species. Their existence had been indirectly inferred from kinetic and stereochemical studies.[3] No reliable spectroscopic or other physical measurements of simple alkyl cations in solution or in the solid state were reported despite decades of extensive studies (including conductivity and cryoscopic measurements). Gaseous alkyl cations under electron bombardment of alkanes, haloalkanes, and other precursors have been investigated from the fifties in mass spectrometric studies, but these studies of course did not provide structural information.[21]

The existence of Friedel-Crafts alkyl halide-Lewis acid halide complexes had been established from observations, such as Brown's study of the vapor pressure depression of CH_3Cl and C_2H_5Cl in the presence of gallium chloride.[22] Conductivity measurements were carried out of aluminium chloride in alkyl chlorides[23] and of alkyl fluorides in boron trifluoride[24] and the effect of ethyl bromide on the dipole moment of aluminium bromide was studied.[25] However, in no case could well-defined, stable alkyl cation complexes been established or obtained even at very low temperatures.

Electronic spectra. of alcohols and olefins in strong protic acids such as sulfuric acid were obtained by Rosenbaum and Symons.[26] They observed for a number of simple aliphatic alcohols and olefins to give an absorption maximum around 290 nm and ascribed this characteristic absorption to the corresponding alkyl cations. Finch and Symons,[27a] on reinvestigation, however, showed that condensation products formed with acetic acid (used as solvent for the precursor alcohols and olefins) were responsible for the spectra, not the simple alkyl cations. Moreover, protonated mesityl oxide was also identified as the absorbing species in the isobutylene-acetic acid sulfuric acid system.

Deno and his coworkers[27b] carried out an extensive study of the fate of alkyl cations formed from alcohols or olefins in undiluted H_2SO_4 and oleum and showed the formation of equal amounts of saturated hydrocarbon mixture (C_4 to C_{18}) insoluble in H_2SO_4 and a mixture of cyclopentenyl cations (Cg to C_{20}) in the H_2SO_4 layer. These cations exhibit strong ultraviolet absorption around 300 nm.

Olah, Pittman and Symons subsequently reviewed and clarified the question of electronic spectra of various carbocationic systems and the fate of various precursors in different acids.[27c]

At this stage it was clear that all earlier attempts to prove the existence of long-lived, well-defined alkyl cations were unsuccessful in acids such as sulfuric acid, perchloric acid, etc. and at best inconclusive in case of the interaction of alkyl halides with Lewis acid halides. Proton elimination from any intermediately formed alkyl cation giving olefin which then react further can lead to complex systems affecting conductivity, as well as other chemical and physical studies.

It was not realized till the breakthrough in superacid chemistry (*vide infra*) that in order to depress the deprotonation of alkyl cations to olefins (equation 9) (with subsequent formation of very complex systems via reactions such as alkylation, oligomerizatiorl, polymerization, cyclization, etc. of olefins with alkyl cations) acids much stronger than those known and used at the time were needed.

$$(CH_3)_3C^+ \rightleftharpoons H^+ + (CH_3)_2C=CH_2 \qquad (9)$$

Finding such acids (called "superacids") turned out to be the key for being able to finally obtain stable, long lived alkyl cations and in general carbocations. If any deprotonation would still take place, in however, limited equilibrium, the alkyl cation (a strong acid) would immediately react with the olefin (a good π-base) leading to the multitude of mentioned reactions.

LONG LIVED ALKYL CATIONS FROM ALKYL FLUORIDES IN ANTIMONY PENTAFLUORIDE AND RELATED CONJUGATE SUPERACIDS

The idea that ionization of alkyl fluorides to stable alkyl cations could be possible with excess of strong Lewis acid fluorides serving themselves as solvents first came to me while still working in Hungary in the early fifties and studying the boron trifluoride catalyzed alkylation of aromatics with alkyl fluorides. In the course of these studies I attempted to isolate $RF:BF_3$ complexes. Realizing the difficulty to find suitable solvents which would allow ionization but at the same would not react with developing potentially highly reactive alkyl cations, neat alkyl fluorides were condensed with boron trifluoride at low temperatures. At the time I had no access to spectroscopic methods such as IR or NMR (which were still in their infancy). I remember a visit by Costin Nenitzescu (an outstanding but never fully recognized Rumanian chemist, who carried out much pioneering research on acid catalyzed reactions). We commiserated on our lack of access to even an IR spectrometer: (The story of Nenitzescu's cyclobutadiene-Ag^+ complex travelling on the Orient Express to a colleague in Vienna for IR studies, but decomposing en route was recalled by him later.) All we could carry out at the time on our RF-BF_3 systems were conductivity measurements. The results showed that methyl fluoride and ethyl fluoride gave only low conductivity complexes, whereas the isopropyl fluoride and tertiary butyl fluoride complexes were highly conducting (equation 10). The latter systems, however, also showed some polymerization (from deprotonation of the involved carbocations giving olefins). Thus, the conductivity data must have been affected by acid formation.[24]

$$R\!-\!F + BF_3 \rightleftharpoons \overset{\delta+}{R}\!-\!\overset{\delta-}{F}\!\longrightarrow\!BF_3 \rightleftharpoons R^+BF_4^- \qquad (10)$$

After the defeat of the 1956 Hungarian revolution I escaped with my family and after spending some months in London we moved to Canada where I was able to continue my research at the Dow Chemical Company Research Laboratory in Sarnia, Ontario. After a prolonged, comprehensive search of many Lewis acid halides I finally hit on antimony pentafluoride.[18 – 20] It turned out to be an extremely strong Lewis acid and for the first time allowed ionization of alkyl fluorides in solution to stable, long lived alkyl cations. Neat SbF_5 solutions are viscous, but diluted with liquid sulfur dioxide the solutions could be cooled and studied down to − 78°C (subsequently, I also introduced even lower nucleophilicity solvents such as SO_2ClF or SO_2F_2 which allowed studies at much lower temperatures). Following up the mentioned observation of the decarbonylation of the pivaloyl cation which gave the first spectral evidence for the tertiary butyl cation, *tert*-butyl fluoride was ionized in excess antimony pentafluoride. The solution of the *tert*-butyl cation turned out to be remarkably stable allowing chemical and spectroscopic studies alike.[28 – 29]

In the late fifties the research director of the Canadian Dow laboratories was not yet convinced about the usefulness of NMR spectroscopy. Consequently we had no such instrumentation of our own. Fortunately, the Dow laboratories in Midland (Michigan) just 100 miles across the border had excellent facilities run by E. B. Baker, a pioneer of NMR spectroscopy, who offered his help. To probe whether our SbF_5 solutions of alkyl fluorides indeed contained alkyl cations we routinely drove with our samples in the early morning to Midland and watched Ned Baker obtain their NMR spectra. *tert*-Butyl fluoride itself shows in its ^1H NMR spectrum due to the fluorine-hydrogen coupling of J_{HF} 20 Hz a characteristic doublet. In SbF_5 solution the doublet disappeared and the methyl protons became significantly deshielded from about δ 1.5 to δ 4.3. This was very encouraging but not yet entirely conclusive to prove the presence of the *tert*-butyl cation. If one assumes that *tert*-butyl fluoride forms with SbF_5 only a polarized donor-acceptor complex which undergoes fast fluorine exchange (on the NMR time scale) then the fluorine-hydrogen coupling would be "washed out", while still a significant deshielding of the methyl protons would be expected. The differentiation of a polarized, rapidly exchanging donor-acceptor complex from the long sought-after real ionic $\underline{t} – C_4H_9^+SbF_6^-$, thus, became a major challenge (equation 11).

$$(CH_3)_3C{-}F \ + \ SbF_5 \ \rightleftharpoons \ (CH_3)_3C{-}\underset{F}{\overset{F}{\underset{|}{\overset{|}{Sb}}}}{<}^F_F \quad \text{or} \quad (CH_3)_3C^+ \ SbF_6^- \qquad (11)$$

Ned Baker despite being himself a physicist showed great interest in our chemical problem. In order to solve it he devised a means to obtain the carbon-13 spectra of our dilute solutions, an extremely difficult task before the advent of Fourier transform NMR techniques. Carbon-13 labeling was generally possible at the time only up to ~ 50% level (from available $Ba^{13}CO_3$).

When we prepared 50% C-13 labeled *tert*-butyl fluoride, we could obtain at best only 5% solutions in SbF_5. Thus, the C-13 content of the solution was highly diluted. Baker, however, undaunted devised an INDOR (inter nuclear double resonance) method. Using the high sensitivity of the proton signal he was able by the double resonance technique observe the C-13 shifts of our dilute solutions (a remarkable achievement around 1960!). Understandably to our great joy the tertiary carbon shift ($\delta^{13}C$ 335.2) in $(CH_3)_3CF$-SbF_5 turned out to be more than 300 ppm deshielded from that of the covalent starting material ! Such very large chemical deshielding (the most deshielded C-13 signal at the time) could not be reconciled with only a donor-acceptor complex. It indicated rehybridization from sp^3 to sp^2 and at the same time showed the effect of significant positive charge on the carbocationic carbon. For simplicity I am not discussing here the nature of the counter ion (which can be dimeric Sb_2F_{11}- or even oligomeric) or questions of ion pairing and separation (concepts developed by Winstein).

Besides the *tert*-butyl cation we also succeeded in preparing and studying the related isopropyl and the *tert*-amyl cations (equations 12 and 13).

$$(CH_3)_2CHF \ + \ SbF_5 \ \longrightarrow \ (CH_3)_2CH^+ \ SbF_6^- \qquad (12)$$

$$(CH_3)_2CFCH_2CH_3 \ + \ SbF_5 \ \longrightarrow \ (CH_3)_2\overset{+}{C}CH_2CH_3 \ SbF_6^- \qquad (13)$$

The isopropyl cation was of particular interest. Whereas in the *tert*-butyl cation the methyl protons are attached to carbons which are only adjacent to the carbocationic center, in the isopropyl cation a proton is directly attached to it. When we obtained the proton NMR spectrum of the iC_3H_7F-SbF_5 system the CH proton showed up as an exceedingly deshielded septet at δ 13.5, ruling out the possibility of a polarized donor-acceptor complex and indicating the formation of $(CH_3)_2CH^+$ ion. The C-13 spectrum was also conclusive showing a very highly deshielded (by >300 ppm) +C chemical shift (δ ^{13}C 320.6). In case of the *tert*-amyl cation an additional interesting feature was the observation of strong long range H-H coupling of the methyl protons adjacent to the carbocationic center with the methylene protons. If only the donor-acceptor complex would be involved such long range coupling through an sp3 carbon would be small(1 – 2 Hz). Instead the observed significant coupling (J_{H-H} 10 Hz) indicated that the species studied indeed had an sp^2 center through which the long range H-H coupling became effective. Fig. 1 reproduces the ^1H NMR spectra of the *tert*-butyl, *tert*-amyl and isopropyl cations. These original spectra are framed and hang in my office as a momento, as does the ESCA spectrum of the norbornyl cation (*vide infra*).

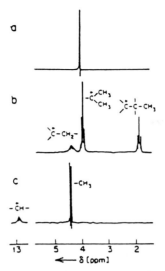

Fig. 1. ¹H NMR spectra of: a) the *tert*-butyl cation [trimethylcarbenium ion, $(CH_3)_3C^+$]; b) the *tert*-amyl cation [dimethylethylcarbenium ion, $(CH_3)_3C+C_2H_5$]; c) the isopropyl cation [dimethylcarbenium ion, $(CH_3)_2C^+H$]. (60 MHz, in $SbF_5:SO_2ClF$ solution, -60°C).

Our studies also included IR spectroscopic investigation of the observed ions (Fig. 2).²⁹ John Evans, at the time a spectroscopist at the Midland Dow laboratories, offered his cooperation and was able to obtain and analyze the vibrational spectra of our alkyl cations. It is rewarding to see that some 30 years later FT-IR spectra obtained by Denis Sunko and his colleagues in Zagreb using low temperature matrix deposition techniques and Schleyer's calculations of the spectra, show good agreement with our early work, even considering that our work was carried out in neat SbF_5 at room temperature long before the advent of the FT-IR methods.³⁰'

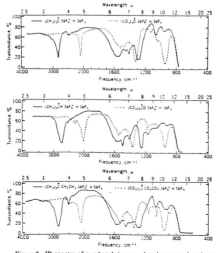

Figure 2. IR spectra of *tert*-butyl, isopropyl and *tert*-amyl cations

Fig. 2. IR spectra of *tert*-butyl, isopropyl and *tert*-amyl cations

Subsequently in 1968 – 70 with Jack DeMember and August Commeyras[31] we were able to carry out more detailed IR and laser Raman spectroscopic studies of alkyl cations. Comparison of the data of light and deuterium labelled tert-butyl cations 6 with those of isoelectronic trimethylborons 7 proved the planar carbocation center of the former (Table I).

This was also an early example of the realization that for nearly all carbocations there exist neutral isoelectronic isostructural boron analogues, which later in the hands of R. E. Williams and others proved itself so useful.

Table 1. Raman and IR Frequency of the *tert*-Butyl Cation and [D9]-*tert*-Butyl Cation and Their ColTelation with Those of $(CH_3)_3B$ and $(CD_3)_3B$

	Frequency of Vibration [cm⁻¹]											
Species	v_1 v_{12} v_7 v_{19}	v_2 v_{13}	v_{21}	v_{14}	v_{15}	v_{17}	v_5	V_{16}	V_6	v_9	v_{10}	v_{18}
$(CH_3)_3C^+$	2947	2850		1450		1295			667		347	306
$(CH_3)_3B$	2975	2875	1060	1440	1300	1150	906	866	675	973(486?)	336[a]	320
$(CD_3)_3C^+$	2187	2090		1075		980			720		347	300
$(CD_3)_3B$	2230	2185		1033	1018	1205			620	870	(289)[b]	(276)[b]

a IR frequency.

b Calculated.

When in the summer of 1962, I was able for the first time to present in public our work at the Brookhaven Organic Reaction Mechanism Conference[28a] and subsequently in a number of other presentations and publications,[28b – e,29] I had convincing evidence in hand to substantiate that after a long and frequently frustrating search stable, long-lived alkyl cations were finally obtained in superacidic solutions.[29] Stable, long lived carbocation chemistry, as it was to be known, began and its progress became fast and wide spread. Working in an industrial laboratory, publication of research is not always easy. I would therefore like to thank again the Dow Chemical Company for allowing me not only to carry out the work but also to publish the results.

After successful preparation of long-lived, stable carbocations in antimony pentafluoride solution the work was extended to a variety of other superacids. Protic superacids such as FSO_3H (fluorosulfuric acid) and CF_3SO_3H (triflic acid) as well as conjugate acids such as $HF-SbF_5$, FSO_3HSbF_5 (Magic

Acid), $CF_3SO_3H\text{-}SbF_5$, $CF_3SO_3H\text{-}B(O_3SCF_3)_3$, etc. were extensively used in ionizing varied precursors, including alcohols. Superacids based on fluorides such as AsF_5, TaF_5, NbF_5 and other strong Lewis acids such as $B(O_3SCF_3)_3$ were also successfully introduced. The name Magic Acid for the $FSO_3H\text{-}SbF_5$ system was given by J. Lukas, a German post-doctoral fellow working with me in Cleveland in the sixties who after a laboratory party put remainders of a Christmas candle into the acid. The candle dissolved and the resulting solution gave an excellent NMR spectrum of the tert-butyl cation. This observation understandably evoked much interest and hence he named the acid "Magic". The name stuck in our laboratory. I think it was Ned Arnett who subsequently introduced the name into the literature where it increasingly became used. When a former graduate student of mine, J. Svoboda, started a small company (Cationics) to commercially make some of our ionic reagents he obtained trade name protection for Magic Acid® and it is marketed as such since.

Many contributed to the study of long lived carbocations. The field rapidly expanded and allowed successful study of practically any carbocationic system. Time does not allow to credit here to all of my former associates and the many researchers around the world who contributed so much to the development of the field. Their work can be found in the extensive. literature. I would like, however, to specifically mention the pioneering work of D. M. Brouwer and H. Hogeveen, as well as their colleagues at the Shell Laboratories in Amsterdam in the sixties and seventies. They contributed fundamentally to the study of long lived carbocations and related superacidic hydrocarbon chemistry. The first publication from the Shell laboratories on alkyl cations appeared in *Chemical Communications* in 1964,[32] following closely my initial reports of 1962 – 64. Similarly, I would like to emphasize the fundamental contributions of R. J. Gillespie to strong acid (superacid) chemistry[33] and also his generous help while I was working at the Dow Laboratories in Canada. I reestablished during this time contact with him after we first met in the winter of 1956 at University College in London where he worked with C. K. Ingold. Subsequently, he moved to McMaster University in Hamilton, Ontario. In the late fifties he had there an NMR spectrometer and in our study of SbF_5 containing highly acidic systems of carbocations we were gratified by his help allowing us to run some of our spectra on his instrument. His long standing interest in fluorosulfuric acid and our studies in SbF_5 containing systems found common ground in studies of $FSO_3H\text{-}SbF_5$ systems.[34] It was also Gillespie who suggested to call protic acids stronger than 100% sulfuric acid as superacids.[33] This arbitrary, but most useful definition is now generally used. It should be, however, pointed out that the name "superacid" goes back to J. B. Conant of Harvard who in 1927 used it to denote acids, such as perchloric acid, which he found stronger than conventional mineral acids and capable of protonating even such weak bases as carbonyl compounds.[35] Our book "Superacids" published in 1985 with Surya Prakash and Jean Sommer[33c] was appropriately dedicated to the memory of

Conant, although few of today's chemists are aware of his contributions to this field.

My memories of the already mentioned 1962 Brookhaven Mechanism Conference[28a] where I first reported on long lived carbocations in public are still clear in my mind. The scheduled "main event" of the meeting was the continuing debate between Saul Winstein and Herbert C. Brown (the pioneer of hydroboration chemistry, Nobel Prize, 1979) on the classical or nonclassical nature of some carbocations (or carbonium ions as they were still called at the time)[36] It must have come to them and others in the audience as quite a surprise that a young chemist from an unknown industrial laboratory was invited to give a major lecture and claimed to have obtained and studied stable, long lived carbonium ions (i.e. carbocations) by the simple new method of using highly acidic (superacidic) systems. I remember to be called aside separately by both Winstein and Brown during the conference and cautioned that a young chemist should be exceedingly careful making such claims. Each pointed out that most probably I must be wrong and could not have obtained longlived carbonium ions. Just in case, however, that my method would turn out to be real, I certainly should obtain evidence for the "nonclassical" or "classical" nature, respectively, of the much disputed 2-norbornyl cation 8. Their much heralded controversy[36b,c] centered around the question whether experimentally observed significant rate enhancement of the hydrolysis of 2-*exo* over 2-*endo*-norbornyl esters and high *exo* selectivity in the systems were caused, as suggested by Winstein, by ~participation of the C_1-C_6 single bond with delocalization to a bridged "nonclassical" ion or only, as stated by Browns by steric hindrance in case of the *endo*-system and involving equilibrating "classical" trivalent ions. Non-classical ions, a term first used by J. D. Roberts,[37] were suggested by P. D. Bartlett to contain too few electrons to allow a pair for each "bond" i.e. must contain ground state delocalized σ-electrons.[36a] As my method allowed us to prepare carbocations as long lived species, clearly the opportunity was given to experimentally decide the question through direct observation of the ion. At the time I had obtained only the proton spectrum of 2-norbornyl fluoride in SbF_5 at room temperature, which displayed a single broad peak indicating complete equilibration through hydride shifts and WagnerMeerwein rearrangement (well known in solvolysis reactions and related transformations of 2-norbornyl systems). However, my curiosity was aroused and subsequently when in 1964 I transferred to Dow's Eastern Research Laboratory (established under Fred McLafferty as laboratory director first in Framingham, MA. and then moved to Weyland, MA.) the work was further pursued in cooperation with Paul Schleyer from Princeton and Marty Saunders from Yale.[38a] Paul, who became a life-long friend, even at that time had a knack to bring together cooperative efforts acting as the catalyst. Using SO_2 as solvent, we were able to lower the temperature of our solution to −78°C. We also prepared the ion by ionization of cyclopentenylethyl fluoride or by protonation of nortricyclene in FSO_3H;SbF_5/SO_2ClF (Scheme 1).

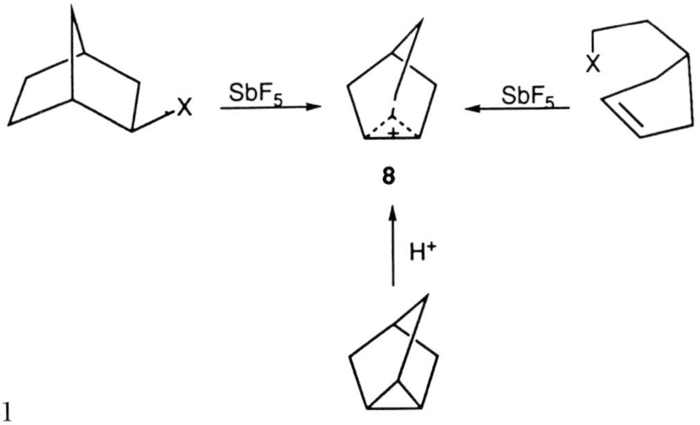

Scheme-1

We still did not have access to suitable low temperature instrumentation of our own to carry out needed NMR studies, but Marty Saunders did. Thus, our samples now traveled the Massachusetts turnpike to New Haven where Saunders was able to study solutions of the norbornyl cation at increasingly low temperatures using his own home-built NMR instrumentation housed in the basement of the Yale chemistry building. We were able to obtain NMR spectra of the ion at − 70°C where the 3,2hydride shift was frozen out. It took, however, till 1969 following my move to Cleveland to Case Western Reserve University to develop efficient low temperature techniques using solvents such as SO_2ClF and SO_2F_2, to be able to obtain high resolution ¹H and ¹³C NMR spectra of the 2-norbornyl cation evetually down to − 159°C in super- cooled solutions.[38b,c] Both 1,2,6 hydride shifts and the Wagner-Meerwein rearrangement could be frozen out at such a low temperature and the static, bridged ion was observed (Fig.s 3a and 3b).[38c]

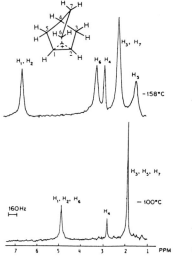

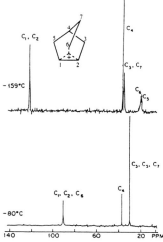

Fig. 3a: 395 Mhz ¹H NMR spectra of 2-norbornyl cation in sbF₅/SO₂ClF/SO₂F₂ solution.

Fig. 3b: 50 MHz proton decoupled ¹³C NMR spectra of 2.-norbornyl cation (¹³C enriched) in SbF₅/SO₂ClF/SO₂F₂ solution.

The differentiation of bridged non-classical from rapidly equilibrating classical carbocations based on NMR spectroscopy is difficult since NMR is a relatively slow physical method with a limited time scale. We addressed this question in some detail in our work using estimated shifts of the two differing systems in comparison with model systems.[38b,c] Of course these days this task is greatly simplified by highly efficient theoretical methods, such as IGLO, GIAO, etc. to calculate NMR shifts of differing ions and comparing them with the experimental data.[38d] It is rewarding to see that our results and conclusions stood up well in comparison with all the more recently advanced studies.

As mentioned we also carried out IR studies (using fast vibrational spectroscopy) early in our work on carbocations. In our studies of the norbornyl cation we also obtained Raman-spectra[38b] and although at the time it was not possible to theoretically calculate the spectra, comparison with model compounds (2-norbornyl system and notricyclene, respectively) indicated the symmetrical, bridged nature of the ion. Sunko and Schleyer recently were able to obtain the FT-IR spectrum in elegant studies and compare it with theoretical calculations.[30]

Kai Siegbahn's (Nobel Prize in Physics, 1981) core electron spectroscopy (ESCA) was another fast physical method we applied to resolve the question of bridged vs. rapidly equilibrating ions. We were able to study carbocations in the late sixties by this method adapting it to superacidic matrixes. George Mateescu and Louise Riemenschneider in my Cleveland laboratory established ESCA instrumentation and the needed methodology for obtaining the ESCA spectra of a number of carbocations, including the *tert*-butyl and the 2-norbornyl cation in SbF_5 based superacidic matrixes (Fig. 4).[39] These studies again convincingly showed the non-classical nature of the 2-norbornyl cation as no trivalent carbenium ion center was observed in the ESCA spectrum characteristic of a "classical" ion, such as is the case for the *tert*-butyl

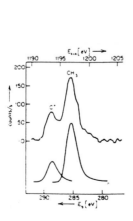

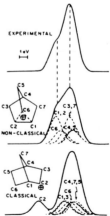

Fig. 4a: Carbon 1s photoelectron spectrum of *tert*-butyl cation.

Fig. 4b: 1s core-hole-state spectra for the 2-norbornyl cation and Clark's simulated spectra for the classical and nonclassical ions.

cation. Although some criticism was leveled at our work by proponents of the equilibrating classical ion concept subsequent studies by Dave Clark fully justified our results and conclusions.[39C] So did theoretical calculations in comparison with the experimental data.

It is proper to mention here some significant some additionel further studies. Saunders' studies showed the absence of deuterium isotopic perturbation of a possible classical equilibrating system.[40] Myhre and Yannoni[41] at very low (5°K!) temperatures were able to obtain solid state ^{13}C NMR spectra which showed no indication of freezing out any equilibrating classical ions; the barriers at this temperature should be as low as 200 cal/mol (the energy of a vibrational transition). Laube was able to carry out single crystal X-ray structural studies on substituted 2norbornyl cations.[42] Schleyer's theoretical studies[38d] including IGLO and related calculation of NMR shifts and their comparison with experimental data contributed further to the understanding of σ-bridged carbonium ion nature of the 2-norbornyl cation. (The classical 2-norbornyl cation was not even found to be a high lying intermediate!) So did Arnett's calorimetric studies.[43] In a 1983 paper entitled "Conclusion of the Norbornyl Ion Controversy" with Prakash and Saunders we were able to state[44] that "all these studies unequivocally ended the so called non-classical ion controversy". Winstein's original views were fully justified by the extensive structural studies made possible through my "stable ion" chemistry .

Although many believe that too much effort was expended on this problem, in my view, the norbornyl ion controversy had significant positive consequences to chemistry. It not only helped to extend the limits of available techniques for structural studies and theoretical calculations but also laid the foundation of developing the electrophilic chemistry of C-H and C-C single bonds and thus of saturated hydrocarbons (*vide infra*).

Intensive, critical studies of a controversial topic always help to eliminate the possibility of any errors. One of my favorite quotation is that by George von Bekessy (Nobel Prize in Medicine, 1961).[45] *"[One] way of dealing with errors is to have friends who are willing to spend the time necessary to carry out a critical examination of the experimental design beforehand and the results after the experiments have been completed. An even better way is to have an enemy. An enemy is willing to devote a vast amount of time and brain power to ferreting out errors both large and small, and this without any compensation. The trouble is that really capable enemies are scarce; most of them are only ordinary. Another trouble with eneznies is that they sometimes develop into friends and lose a good deal of their zeal. It was in this way the writer lost his three best enemies. Everyone, not just scientists, need a few good enemies!"*

Clearly there was no lack of devoted adversaries (perhaps a more proper term than enemies) on both sides of the norbornyl controversy. It is to their credit that we know probably more today about the structure of carbocations, such as the norbornyl cation, than of any other chemical species. Their efforts resulted in the most rigorous studies and development or improvement of many techniques.

To me the most significant consequence of the norbornyl cation studies was the realization of the ability of C-H and C-C single bonds to act as two electron σ-donors not only in intramolecular but also in intermolecular transformations and electrophilic reactions. Two electron three center (2e – 3c) bonding (familiar in boron and organometallic chemistry) is the key for these reactions. Much new chemistry rapidly evolved and allowed the recognition of the broad scope and significance of hypercoordinated carbon compounds (in short hypercarbon) chemistry.[46]

THE GENERAL CONCEPT OF CARBOCATIONS

Based on the study of carbocations by direct observation of longlived species and related superacid chemistry, it became apparent that the carbocation concept is wider than previously thought and needed a more general definition which I offered in a 1972 paper.[10] The definition takes into account the existence of two major limiting classes of carbocations with a continuum of varied degree of delocalization bridging them.

a) Trivalent ["classical"] carbenium ions contain an sp²-hybridized electron deficient carbon atom, which tends to be planar in the absence of constraining skeletal rigidity or steric interference. (It should be noted that sp-hybridized, linear oxocarbonium ions and particularly vinyl cations also show substantial electron deficiency on carbon). The carbenium carbon contains six valence electrons, thus is highly electron deficient. The structure of trivalent carbocations can always be adequately described by using only two-electron two-center bonds (Lewis valence bond structures). CH_3^+ is the parent for trivalent ions.

b) Penta- (or higher) coordinate ["nonclassical"] carbonium ions, which contain five or (higher) coordinated carbon atoms. They cannot be described by two-electron two-center single bonds alone, but also necessitate the use of two electron three (or multi) center bonding. The carbocation center is always surrounded by eight electrons, but overall the carbonium ions are electron deficient due to sharing of two electrons between three (or more) atoms. CH_5^+ can be considered the parent for carbonium ions.

Brown and Schleyer subsequently in 1977 offered a related definition:[36c] "a nonclassical carbonium ion is a positively charged species which cannot be represented adequately by a single Lewis structure. Such a cation contains one or more carbon or hydrogen bridges joining the two electron-deficient centers. The bridging atoms have coordination numbers higher than usual, typically five or more for carbon and two or more for hydrogen. Such ions contain two electron-three (or multiple) center bonds including a carbon or hydrogen bridge."

Lewis' concept that a covalent chemical bond consists of a pair of electrons shared between the two atoms became a cornerstone of structural chemistry. Chemists tend to brand compounds as anomalous whose structures cannot be depicted in terms of such bonds alone. Carbocations with too few electrons to allow a pair for each "bond", came to be referred to as "nonc-lassical", a name first used by J. D. Roberts[37] for the cyclopropylcarbinyl cation and adapted by Winstein to the norbornyl cation.[47] The name is still used even though it is now recognized that like other compounds, they adopt the structures appropriate for the number of electrons they contain with two electron-two or three (even multi) center bonding (not unlike the bonding principles established by Lipscomb, Nobel Prize, 1976, for boron com-pounds). The prefixes "classical" and "nonclassical" are expected, however, to gradually fade away as the general nature of bonding will be recognized.

Whereas the differentiation of trivalent carbenium, and pentacoordinated carbonium ions serves a useful purpose to define them as limiting ions, it should be clear that in carbocationic systems there always exist varying degrees of delocalization. This can involve participation by neighboring *n*-donor atoms, π-donor groups or σ-donor CH or C-C bonds.

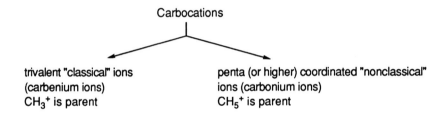

Trivalent carbenium ions are the key intermediates in electrophilic reactions of π-donor unsaturated hydrocarbons. At the same time pentacoordinated carbonium ions are the key to electrophilic reactions of σ-donor saturated hydrocarbons (C-H or C-C single bonds). The ability of single bonds to act .as electron donors lies in their ability to form carbonium ions via two elec-tron three-center (2e – 3c) bond formation.

Expansion of the carbon octet via 3d- orbital participation does not seem possible; there can be only eight valence electrons in the outer shell of car-bon, a small first row element.[46] The valency of carbon cannot exceed four. Kekule's concept of the tetravalency of carbon in bonding terms represents attachment of four atoms (or groups) involving 2e – 2c Lewis type bonding. There is, however, nothing that prevents carbon to also participate in multicen-ter bonding involving 2e – 3c (or multicenter) bonds. Penta- (or higher) – coordination of carbon implies five (or more) atoms or ligands simultane-ously attached to it within reasonable bonding distance.[46]

Neighboring group participation with the vacant *p*-orbital of a carbenium ion center contributes to its stabilization via delocalization which can invol-ve atoms with unshared electron pairs (n-donors), π-electron systems (direct conjugative or allylic stabilization), bent σ-bonds (as in cyclopropylcarbinyl cations), and C-H or C-C σ-bond hyperconjugation. Trivalent carbenium ions with the exception of the parent CH_3^+ therefore always show varying degrees of delocalization without becoming pentacoordinated carbonium

ions. The limiting cases define the extremes of a continuous spectrum of carbocations.

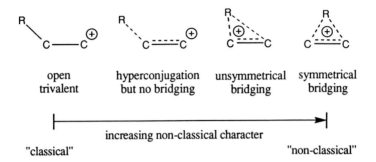

| open
trivalent | hyperconjugation
but no bridging | unsymmetrical
bridging | symmetrical
bridging |

increasing non-classical character

"classical"   "non-classical"

The Role of Carbocations in Electrophilic Reactions

Acid catalyzed electrophilic reactions and transformations such as isomerization, alkylation, substitution, addition, elimination, rearrangements, etc.
involve carbocationic intermediates. Many of these reactions also gained significance in industrial applications. Aromatic hydrocarbon chemistry and
that based on acetylene laid the foundation for organic industries a century
ago. Subsequently olefin based chemistry took on great significance. In all
this chemistry reactive π-bonded systems are the electron donor substrates.
In electrophilic reactions they readily form trivalent carbocationic intermediates (equations 14 and 15).

$$RCH{=}CH_2 + E^+X^- \rightleftharpoons \overset{+}{R}CHCH_2E \xrightarrow[\text{X}^-]{} \overset{\nearrow \text{RCHXCH}_2\text{E}}{\underset{-H^+}{\searrow}} RCH{=}CHE \qquad (14)$$

$$E^+ = H^+, R^+, NO_2^+, Hal^+ \text{ etc.}$$

$$ArH + E^+ \rightleftharpoons Ar\overset{\nearrow H^+}{\underset{\searrow E}{}} \xrightarrow{-H^+} ArE \qquad (15)$$

The discovery of pentacoordinate carbonium ions discussed precedingly led
to the realization that they play an important role not only in understanding
the structure of nonclassical ions, but more importantly represent the key to
electrophilic reactions at single bonds, e.g. of saturated aliphatic hydrocarbons (alkanes and cycloalkanes). Such reactions include not only acid-cata-
lyzed hydrocarbon isomerizations, fragmentations, cyclizations, but also varied substitutions and related electrophilic reactions and transformations.

In ionization of β-phenylethyl systems neighboring π-participation into the
carbocationic center occurs, which can be considered as intramolecular π-
alkylation giving Cram's phenonium ions. The corresponding ionization of
2-norbornyl systems involves participation of a properly oriented C-C single
bond (i.e. intramolecular σ-alkylation) giving the bridged ion.

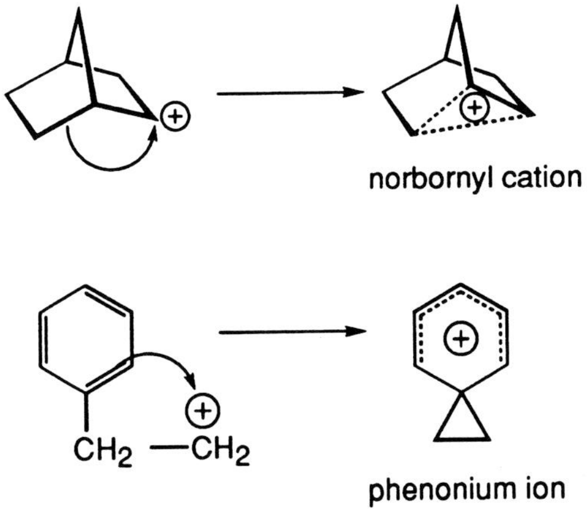

norbornyl cation

$CH_2 - CH_2$

phenonium ion

Alkylation of π-systems in Friedel-Crafts type reactions (either by an inter- or intramolecular way) is for long well known and studied. Extending these relationships it was logical to ask why intermolecular alkylation (and other electrophilic reactions) of σ-donor hydrocarbons could not be affected?

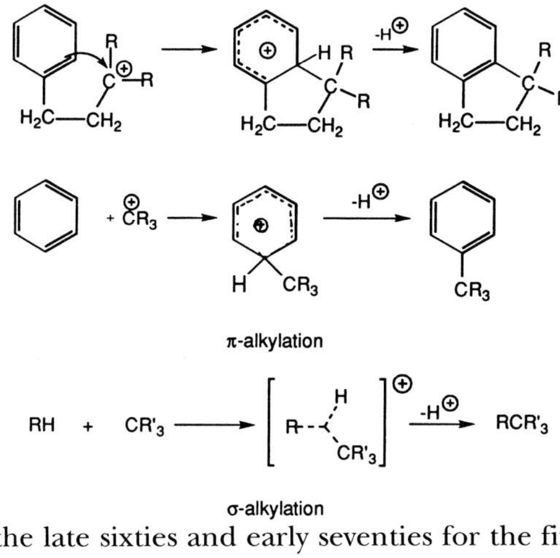

π-alkylation

σ-alkylation

Our studies in the late sixties and early seventies for the first time provided evidence for the general reactivity of covalent C-H and C-C single bonds of alkanes and cycloalkanes in various protolytic processes as well as in hydrogen-deuterium exchange, alkylation, nitration, halogenation, etc. (equations 16 and 17) This reactivity is due to the σ-donor ability (sigma basicity) of single bonds allowing bonded electron pairs to share their electron pairs with an electron deficient reagent via two-electron, three-center bond formation. The reactivity of single bonds thus stems from their ability to participate in the formation of pentacoordinated carbonium ions. Subsequent cleavage of the three-center bond in case of C-H reaction results in formation of substitution products, C-C reaction results in bond cleavage and formation of a fragment carbenium ion, which then can react further.

$$R_3C\text{-}H + E^+ \rightleftharpoons \left[R_3C \cdots \overset{\cdot H}{\underset{E}{\cdot\!<}} \right]^+ \begin{array}{l} \nearrow R_3C^+ + EH \\ \searrow R_3CE + H^+ \end{array} \qquad (16)$$

$$R_3C\text{-}CR_3 + E^+ \rightleftharpoons \left[\begin{array}{c} R_3C \diagdown \quad \diagup CR_3 \\ \vdots \\ E \end{array} \right]^+ \rightleftharpoons R_3CE + R_3C^+ \qquad (17)$$

$$E^+ = D^+, H^+, R^+, NO_2^+, Hal^+ \text{ etc.}$$

As bond-to-bond shifts can readily take place within five coordinate carbonium ions through low barriers, the involved intermediates can be more complex but always involving interconverting carbonium ions.

Superacidic hydrocarbon chemistry involving conditions favoring carbocationic intermediates is also gaining in significance in practical applications. Relatively low temperature effective isomerization of alkanes, much improved and environmentally adaptable alkylation, new approaches to the functionalization of methane and possibilities in its utilization as a building block for higher hydrocarbons and their derivatives, as well as moderate conditions for coal liquefaction are just a few examples to be mentioned here.[48]

PROTOSOLVOLYTIC ACTIVATION OF CARBOCATIONIC SYSTEMS

Carbocations are electrophiles, i.e. electron deficient compounds. In electrophilic reactions of unsaturated, π-donor hydrocarbons and their derivatives (such as acetylenes, olefins, aromatics) the reaction with the electrophilic reagents is facilitated by the nucleophilic assistance of the substrates. In reactions with increasingly weaker (deactivated) π-donors and even more so with only weakly electron donating saturated hydrocarbons (σ-donors) the electrophile itself must provide the driving force for the reactions. Hence the need for very strongly electron demanding electrophiles and comparable low nucleophilicity reaction media (such as superacidic systems).

It was only more recently realized[49] that electrophiles capable of further interaction (coordination, solvation) with strong Bronsted or Lewis acids can be greatly activated. The resulting enhancement of reactivity can be very significant compared to that of the parent electrophiles under conventional conditions and indicates superelectrophile formation i.e. electrophiles with greatly enhanced electron deficiency (frequently of dipositive nature). I have reviewed[49] elsewhere the superelectrophilic activation of varied electrophiles. and will not discuss it here with the exception of some superelectrophilic activation which can also affect the reactivity of carbocations.

In carboxonium ions, originally studied by Meerwein, alkyl groups of an alkyl cation, such as the *tert*-butyl cation, are replaced by alkoxy, such as methoxy groups. The methoxy groups delocalize charge (by neighboring oxygen participation) and thus make these ions increasingly more stable.

$$(CH_3)_3C^+ \; < \; (CH_3)_2\overset{+}{C}OCH_3 \; < \; CH_3\overset{+}{C}(OCH_3)_2 \; < \; \overset{+}{C}(OCH_3)_3$$

At the same time their reactivity as carbon electrophiles decreases. For example they do not alkylate aromatics or other hydrocarbons. Strong oxygen participation thus greatly diminishes the carbocationic nature.

Neighboring oxygen participation, however, can be decreased if a strong acid protosolvates (or protonates) the non-bonded oxygen electron pairs (equation 18). Consequently, carboxonium ions (and related ions such as acyl cations) in superacidic media show greatly enhanced carbon electrohilic reactivity indicative of dicationic nature.

$$\underset{H_3C}{\overset{H_3C}{>}}C \cdots \overset{+}{O}CH_3 \quad \xrightarrow{\; H^+ \;} \quad \underset{H_3C}{\overset{H_3C}{>}}\overset{+}{C}\text{---}\underset{H}{\overset{+}{O}}CH_3 \quad (18)$$

Similarly halogen substituted carbocations, such as the trichloromethyl cation Cl_3C^{+},[50] are strongly stabilized by n-p back donation (not unlike BCl_3). They can also be greatly activated by superacidic media, which protosolvate (protonate) the halogen non-bonded electron pairs diminishing neighboring halogen participation (equation 19).

$$\underset{Cl}{\overset{Cl\cdots\overset{+}{C}\cdots Cl}{|}} \quad \xrightarrow{\; H^+ \;} \quad \underset{Cl}{\overset{Cl}{>}}\overset{+}{C}\text{---}\overset{+}{C}lH \quad (19)$$

This explains for example why carbon tetrachloride highly enhances the reactivity of protic superacids for alkane transformations. Lewis acids have similar activating effect.[49]

Alkyl cations themselves, in which only hyperconjugative C-H or C-C single bond interactions stabilize the electron deficient center, are also activated by superacidic solvation. Theoretical calculations and hydrogen-deuterium exchange of long lived alkyl cations in deuterated superacids, under conditions where no deprotonation-reprotonation can take place substantiate the existence of these protoalkyl dications as real intermediates (equation 20).[51]

$$\underset{CH_3}{\overset{CH_3}{>}}\overset{+}{C}\text{---}CH_3 \quad \xrightarrow{\; H^+ \;} \quad \underset{CH_3}{\overset{CH_3}{>}}\overset{+}{C}\text{---}CH_2\cdots\overset{+}{<}\overset{H}{H} \quad (20)$$

The well recognized high reactivity of alkanes for isomerization alkylation reactions in strongly acidic media is very probably assisted by protosolvation of the intermediate alkyl cations. Similar activation can be involved in other acid catalyzed hydrocarbon transformations, which are preferentially carried out in excess acid solutions.

ACTIVATION BY SOLID SUPERACIDS AND POSSIBLE RELEVANCE TO ENZYMATIC SYSTEMS

The chemistry of carbocations and their activation was discussed so far in superacidic solutions. However, superacidic systems are not limited to solution chemistry. Solid superacids, possessing both Bronsted and Lewis acid sites, are of increasing significance. They range from supported or intercalated systems, to highly acidic perfluorinated resinsulfonic acids (such as Nafion-H and its analogues), to certain zeolites (such as H-ZSM-5).

In order to explain why their remarkable activity, for example in catalytic transformations of alkanes (even methane), an appraisal of the *de facto* activity at the acid sites of such solid acids is called for.[49]

Nafion-H is known to contain acidic SO$_3$H groups clustered together (Fig. 5).

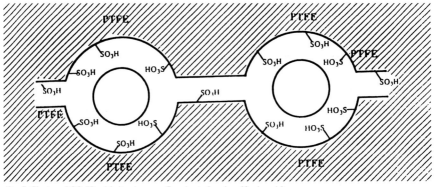

Fig. 5. Clustered SO$_3$H acid sites in a perfluorinated resinsulfonic acid.

H-ZSM-5, which also displays superacidic activity, was found by Haag et al.[52] to isomerize and alkylate alkanes readily (H$_2$ was observed as the protolytic by-product in stoichiometric amounts). In this zeolite, active Bronsted and Lewis acid sites are again in close proximity, approximately 2.5 Å apart (Fig. 6).

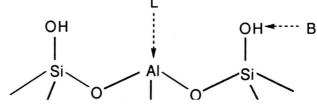

Fig. 6. Bronsted (B) and Lewis acid sites (L) in zeolites.

In comparison with solution chemistry it is reasonable that in these (and other) solid superacid catalyst systems, bi- or multidentate interactions forming highly reactive intermediates are possible. This amounts to the solid-state equivalent of protosolvation (protonation).[49]

Nature is able to perform its own transformations in ways which chemists have only begun to understand and can not yet come close to duplicate. At enzymatic sites many significant transformations take place which are acid catalyzed (including electron deficient metal-ion-catalyzed processes). Because of the unique geometry involved at enzymatic sites bi- and multidentate interactions must be possible and the concepts discussed previously may also have relevance to our understanding of some enzymatic processes.[49]

CONCLUSIONS

The chemistry of long-lived carbocations became a very active and fast developing field with contributions by researchers from all around the world. It is with understandable satisfaction that I look back at the progress achieved and the possibilities ahead. What started out as obtaining one of the most significant class of chemical reaction intermediates (i.e. carbocations) as long lived species and study of their structure, led subsequently to the development of the general concept of the electrophilic reactivity of single bonds, such as C-H and C-C bonds and related superacidic hydrocarbon chemistry.

Despite of all the progress that has been made I believe that most exciting chemistry in the field still lies ahead for future generations to explore. I wish them as much excitement and satisfaction in their work as I had.

The concept of the tetravalency of carbon stated by Kekule well over a century ago remains an essential foundation of organic chemistry. Carbon as a small first new element can not extend it valance shell and the octet rule allows formation of only four two electron two center (2e – 2c) Lewis type bonds (or their equivalent multiple bonds). It is, however, possible for one (or more) electron pair of carbon to be involved in two electron-three center (2e – 3c) bonding.[46] This allows carbon to simultaneously bond five (or even six) atoms or groups. This is the case in carbonium ions which contain hypercarbon (hypercoordinate carbon). It also provides the key to understand the rapidly emerging electrophilic chemistry of saturated hydrocarbons, including parent methane and that of C-H and C-C bonds in general. Whereas hypercoordinate carbocations are 8e carbon systems which do not violate the octet rule, carbanionic S_N2 transition states $[Y–CR_3–X]^-$ are 10e systems and thus can not be intermediates .

I wrote more than twenty years ago[9] "The realization of the electron donor ability of shared (bonded) electron pairs (single bonds) could one day rank equal in importance with G. N. Lewis' realization of the electron donor unshared (non-bonded) electron pairs (or for this reason I could add the electron pairs involved in multiple[4] bonding). We can now not only explain the reactivity of saturated hydrocarbons and single bonds in general electrophilic reactions, but indeed use this understanding to explore new areas and reactions of carbocation chemistry." It is with some satisfaction that I feel this promise is being fulfilled.

ACKNOWLEDGMENTS

I was fortunate to be able to build on the foundations laid by many. I would like to acknowledge particularly the fundamental contributions of Hans Meerwein (1879 – 1965) and Christopher Kelk Ingold (1893 – 1870) who recognized the role of carbocations in some chemical reactions and Frank Whitmore (1887 – 1947) who generalized it to many others. I am also greatly indebted to all my former students and associates, whose dedication, hard work and major scientific contributions made our joint effort possible.

My wife Judy till her retirement was not only an integral part of our scientific effort but all those who ever worked in the Olah group appreciated greatly her warmth, caring and concern for our "scientific family". G. K. Surya Prakash in our 20 years of association from brilliant graduate student to trusted friend and colleague made invaluable contributions. The Loker Hydrocarbon Research Institute of the University of Southern California provided a wonderful home and support for our work in the last 15 years. The board of the Institute, particularly Mrs. Katherine Loker, Harold Moulton and Carl Franklin are thanked for their support and friendship over the years. I also cherish close association with such wonderful colleagues and friends as Ned Arnett, Joseph Casanova, Paul Schleyer, Jean Sommer, Peter Stang, Ken Wade and Robert Williams who are senior distinguished fellows of the Loker Institute. The National Institutes of Health and the National Science Foundation gave support over the years to our studies of carbocations and the Loker Institute mainly supported the work on hydrocarbon chemistry.

REFERENCES

(1) a) J. F. Norris, AmChem.J. 1901, 25,117; b) F. Kehrmann, F. Wentzel, Chem. Ber.1901,34,3815; c) A. Baeyer, V. Villiger, *Chem.Ber.* **1902**, 35, 1189; 3013; d) M. Gomberg, *Chem.Ber.* **1902**, *35, 2397*; e) P. Walden, *Chem.Ber.* **1902**, *35,2018*; f) A. Werner, *Chem.Ber.* **1901**, *34, 3300*; g) A. Baeyer, *Chem.Ber.* **1905**, *38, 569.*

(2) H. Meerwein, K. van Emster, *Chem.Ber.* **1922**, *55, 2500.*

(3) C. K. Ingold "Structure and Mechanism in Organic Chemistry", Cornell University Press, 1953, Ithaca, New York and references therein, 2nd ed. (1969).

(4) F. C. Whitmore, J.*Am.Chem.Soc.* 1932, 54, 3274, 3276; *Ann.Rep.Progr.Chem.* (Chem. Soc. London) 1933, 177; *Chem.Eng.News*, 1948,26, 668.

(5) G. A. Olah, P.v.R. Schleyer (Eds.) "Carbonium Ions", Vols. I-V, WileyInterscience, New York, NY, 1968-76 and reviews therein.

(6) D. Bethell, V. Gold "Carbonium Ions", Academic Press, London-New York, 1967.

(7) P. Vogel "Carbocation Chemistry", Elsevier, Amsterdam, 1985.

(8) G. A. Olah, Chem.Eng.News, **1967**, *45*, 76; Science, **1970**, *168, 1798.*

(9) G. A. Olah, *Angew.Chem.Int.Ed.* **1973**, *12*, 173; *Angew.Chem.* **1973**, 85, 183; G. A. Olah "Carbocations and Electrophilic Reactions", Verlag Chemie (Weinheim)-Wiley (New York) 1974.

(10) G. A. Olah, J.*Am.Chem.Soc.* **1972**, *94*, 808.

(11) Compendium of Chemical Terminology: IUPAC Recommendations, Blackwell Scientific Publication: Oxford, 1987.

(12) H. Meerwein in "Methoden der Organischen Chemie (Houben-Weyl), Ed. E. Muller, 4th ed. Vol. VI/3, Thieme, Stuttgart 1965 and references therein.

(13) F. Seel, *Z.Anorg.Allgem.Chem.* **1943**, *250*, 331; **1943**, *252*, 24.

(14) For my recollection see: *Topics in Current Chemistry*, Vol. 80 "In memory of H. L. Meerwein" 1979, Springer, Heidelberg, p.21.

(15) G. A. Olah, S. Kuhn, *Acta Chim.Acad.Sci.Hung.*, **1956**, *10*, 233; *Chem.Ber.* **1956**, *89*, 866; *J.Am.Chem.Soc.* **1960**, *82*, 2380.

(16) A. N. Nesmejanov, E. J. Kahn, *Ber.* **1934**, *67*, 370.

(17) D. E. Pearson, *J.Am.Chem.Soc.* **1950**, *72*, 4169.

(18) G. A. Olah, S. J. Kuhn, W. S. Tolgyesi, E. B. Baker, *J.Am.Chem.Soc.* **1962**, *84*, 2733.

(19) G. A. Olah, *Rev.Chim.* (Buchrest), **1962**, *7*, 1139 (Nenitzescu issue).

(20) G. A. Olah, W. S. Tolgyesi, S. J. Kuhn, M. E. Moffatt, I. J. Bastien, E. B. Baker, *J.Am.Chem.Soc.* **1963**, *85*, 1328.

(21) F. W. McLafferty (Ed.), Mass Spectrometry of Organic Ions, New York, Academic Press, 1963.

(22) H. C. Brown, H. Pearsall, L. P. Eddy, *J.Am.Chem.Soc.* **1950**, *72*, 5347.

(23) E. Wertyporoch, T. Firla, *Ann.Chim.* **1933**, *500*, 287.

(24) G. A. Olah, S. J. Kuhn, J. A. Olah, *J.Chem.Soc.* **1957**, 2174.

(25) F. Fairbrother, *J.Chem.Soc.* **1945**, 503.

(26) J. Rosenbaum, M. C. R. Symons, *Proc.Chem.Soc.(London)*, **1959**, *92*; J. Rosenbaum, M. Rosenbaum, M.C. R. Symons, *Mol.Phys.* **1960**, *3*, 205; J. Rosenbaum, M. C. R. Symons, J.Chem.Soc. (London), 1961, 1.

(27) a) A. C. M. Finch, M. C. R. Symons, *J.Chem.Soc. (London)*, **1965**, 378; b) For a summary, see M. C. Deno, *Progr.Phys.Org.Chem.* **1964**, 2, 129; c) G. A. Olah, C. U. Pittman, Jr., M.C.R. Symons in "Carbonium Ions" G. A. Olah, P.v.R. Schleyer, Eds., Vol. I, p.153, Wiley Interscience Publishers, New York, 1968.

(28) a) G. A. Olah, Conference Lecture at 9th Reaction Mechanism Conference, Brookhaven, New York, August 1962; b) G. A. Olah, Abstr. 142nd National Meeting of the American Chemical Society, Atlantic City, N.J., September, 1962, p.45; c) G. A. Olah, W. S. Tolgyesi, J. S. MacIntyre, I. J. Bastien, M. W. Meyer, E. B. Baker, Abstracts A, XIX International Congress of Pure and Applied Chemistry, London, June, 1963, p.l21; d) G. A. Olah *Angew.Chem.* **1963**, *75*, 800; e) G. A. Olah in C. D. Nenitzescu's 60th Birthday Issue, *Revue de Chimie*, **1962**, *7*, 1139; f) G. A. Olah, American Chemical Society 1964 Petroleum Award Lecture, Reprints, Division of Petroleum Chemistry, American Chemical Society, Vol.9, No.7, C31 (1964); g) G. A. Olah,

"Intermediate Complexes and Their Role in Electrophilic Aromatic Substitutions", Conference Lecture at Organic Reaction Mechanism Conference, Cork, Ireland, June,1964 in Special Publication No.19, the Chemical Society, London,1965; h) G. A. Olah, C. U. Pittman, Jr., in "Advances in Physical Organic Chemistry" (Ed. V. Gold), Vol. 4, pp.305, Academic Press, London-New York,1966.

(29) G. A. Olah, E. B. Baker, J. C. Evans, W. S. Tolgyesi, J. S. McIntyre, I. J. Bastien, *J.Am.Chem.Soc.* **1964**, *86*, 1360.

(30) (a) H. Vancik, D. E. Sunko, *J.Am.Chem.Soc.* **1989**, *111*, 3742.(b) St. Sibert, P. Buzer, P.v. R. Schleyer w. Koch, J. W. de M. Carneiro *J.Am.Chem. Soc* **1993**, *115*, 259

(31) G. A. Olah, J. R. DeMember, A. Commeyras, J. L. Bribes, *J.Am.Chem.Soc.* **1971**, *93*, 459 and references therein.

(32) D. M. Brouwer, E. L. Mackor, *Proc.Chem.Soc.* **1964**, 147.

(33) a) R. J. Gillespie, *Acc.Chem.Res.* **1968**, *1*, 202; b) R. J. Gillespie, T. E. Peel, *Adv.Phys.Org.Chem.* **1972**, *9*, 1; idem, *J.Am.Chem.Soc.* **1973**, *95*, 5173; c) G. A. Olah, G.K.S. Prakash, J. Sommer, "Superacids", Wiley, New York, 1985; d) R. J. Gillespie, *Can.Chem.* News, May 1991, p.20.

(34) a) J. Bacon, P. A. W. Dean, R. J. Gillespie, *Can.J.Chem.* **1969**, *47*, 1655; b) G. A. Olah, M. Calin, *J.Am.Chem.Soc.* **1968**, *90*, 938.

(35) N. F. Hall, J. B. Conant, *J.Am.Chem.Soc.* **1927**, *49*, 3047.

(36) a) P. D. Bartlett "Nonclassical Ions", W. A. Benjamin, New York, 1965; b) S. Winstein, *Quart. Rev. (London),* **1969**, 23, 1411; c) H. C. Brown (with commentary by P. v. R. Schleyer) "The Nonclassical Ion Problem", Plenum Press, New York, 1977.

(37) J. D. Roberts, R. H. Mazur, *J.Am.Chem.Soc.* **1951**, *73*, 3542.

(38) a) M. Saunders, P.v.R. Schleyer, G. A. Olah, *J.Am.Chem.Soc.* **1964**, *86*, 5680; b) G. A. Olah, A. M. White, J. R. DeMember, A. Commeyras, C. Y. Lui, *J.Am.Chem.Soc.* **1970**, *92*, 4627; c) G. A. Olah, G. K. S. Prakash, M. Arvanaghi, F. A. L. Anet, *J.Am.Chem.Soc.* **1982**, *104*, 7105; d) P. v. R. Schleyer, S. Sieber, *Angew.Chem.* **1993**, *105*, 1676; *Angew.Chem.Int.Ed.Engl.* **1993**, *32*, 1606 and references cited therein.

(39) a) G. A. Olah, G. D. Mateescu, L. A. Wilson, M. H. Gross, *J.Am.Chem.Soc.* **1970**, *92*, 7231; b) G. A. Olah, G. D. Mateescu, J. L. Riemenschneider, *J.Am.Chem.Soc.* **1972**, *94*, 2529; c) S. A. Johnson, D. T. Clark, *J.Am.Chem.Soc.* **1988**, *110*, 4112.

(40 M. Saunders, M. R. Kates, *J.Am.Chem.Soc.* **1980**, *102*, 6867.

(41) C. S. Yannoni, V. Macho, P. C. Myhre, *J.Am.Chem.Soc.* **1982**, *104*, 7380.

(42) T. Laube, *Angew.Chem.* **1987**, *99*, 580; *Angew.Chem.lnt.Ed.Engl.* **1987**, *26*, 560.

(43) E. M. Arnett, N. Pienta, C. Petro, *J.Am.Chem.Soc.* **1980**, *102*, 398.

(44) G. A. Olah, G. K. S. Prakash, M. Saunders, *Acc.Chem.Res.* **1983**, *16*, 440.

(45) G. V. Bekesy, "Experiments in Hearing", McGraw Hill, New York, 1960, p.8.

(46) G. A. Olah, G. K. S. Prakash, R. E. Williams, L. D. Field, K. Wade "Hypercarbon Chemistry", Wiley, New York, 1987.

(47) S. Winstein, D. Trifan, *J.Am.Chem.Soc.* **1952**, 74, 1154.

(49) a) G. A. Olah, A. Molnar "Hydrocarbon Chemistry", Wiley, New York, 1995 and references therein; b) I. Bucsi, G. A. Olah, *Cat.Letters,* **1992**, *16*, 27; c) G. A. Olah, *Acc.Chem..Res.* **1987**,*20*, 422; d) G. A. Olah, M. Bruce, E. H. Edelson, *Fuel,* **1984**, *63*, 1130.

(50) G. A. Olah, *Angew.Chem.* **1993**, *105*, 805; *Angew.Chem.Int.Ed.Engl.* **1993**, *32*, 767 and references therein.

(51) G. A. Olah, L. Heiliger, G. K. S. Prakash, *J.Am.Chem.Soc.* **1989**, *111*, 8020.

(52) a) G. A. Olah, N. Hartz, G. Rasul, G. K. S. Prakash, *J.Am.Chem.Soc.* **1993**, *115*, 6985; b) G. A. Olah, N. Hartz, G. Rasul, G. K. S. Prakash, M. Burkhart, K. Lammertsma, *J.Am.Chem.Soc.* **1994**, *116*, 3187.

(53) W. O. Haag, R. H. Dessau, International Catalysis Congress (West Germany), **1984**, II, 105.

Chemistry 1995

PAUL CRUTZEN
MARIO MOLINA and
F. SHERWOOD ROWLAND

for their work in atmospheric chemistry, particularly concerning the formation and decomposition of ozone

THE NOBEL PRIZE IN CHEMISTRY

Speech by Professor Ingmar Grenthe of the Royal Swedish Academy of Sciences.
Translation of the Swedish text.

Your Majesties, Your Royal Highnesses, Ladies and Gentlemen,

About thirty years ago, for the first time, we humans were able to view our planet from space. We saw white cloud formations, blue oceans, green vegetation and brown soils and mountains. From space, we could view and study the earth as a whole. We have come to understand that we influence and are influenced by our biosphere, our life zone. One of the tasks of science is to describe and explain how this happens. In their research on the chemical reactions occurring in the earth's atmosphere, the 1995 Nobel Laureates in Chemistry–Paul Crutzen, Mario Molina and Sherwood Rowland–have adopted this global perspective.

The sun is the engine of life. Solar radiation is the source of energy for nearly all living organisms. But only some of the sun's rays are beneficial. It also emits ultraviolet radiation that harms living beings. Many of us have painful experience of excessive sunbathing. Life in the forms we are familiar with is the result of photosynthesis in green plants, which transforms the carbon dioxide in the air into biomass and oxygen. It has taken hundreds of millions of years for the biosphere to develop the atmospheric composition we have today. In the upper atmosphere, or stratosphere, solar radiation can transform oxygen into ozone. The highest ozone concentrations are found at an altitude of between 15 and 50 km. This ozone layer absorbs the sun's ultraviolet radiation very effectively, thereby reducing hazardous radiation on the earth's surface. This, in turn, makes efficient photosynthesis possible. Here we see an example of a feedback mechanism between the chemistry of the biosphere and the atmosphere. If it is disrupted, there may be serious consequences for life on our planet.

This year's laureates have made a series of major contributions to our knowledge of atmospheric chemistry. This has included studying how ozone is formed and decomposes and how these processes can be affected by chemical substances in the atmosphere, many of them the result of human activity. In 1970 Paul Crutzen demonstrated that nitrogen oxides, formed during combustion processes, could affect the rate of ozone depletion in the stratosphere. He suggested that dinitrogen monoxide, popularly known as "laughing gas" and formed through microbiological processes in the ground, could have the same effect. He has also studied the formation of ozone in the lower atmosphere. Ozone is one ingredient of "smog," which is formed by the influence of solar radiation on air pollutants, especially exhaust gases from motor vehicles and other combustion systems. Whereas stratospheric ozone is a prerequisite for life, tropospheric ozone is strongly toxic and harmful to most organisms, even in small quantities.

In 1974 Mario Molina and Sherwood Rowland showed that chlorine compounds formed by the photochemical decomposition of chlorofluorocarbons (CFC or "Freon" gases) could decompose the stratospheric ozone. They presented detailed hypotheses on how these complicated processes occurred.

The discoveries of the three researchers have an unusually close connection with the consequences of modern technology. Supersonic aircraft release nitrogen oxides in the stratosphere. Motor vehicles and stationary combustion plants release the same substances into the lower atmosphere. CFC gases from refrigerators and air conditioners, and in the form of aerosol spray propellants–combined with a "throwaway culture"–result in large-scale emissions of chlorine compounds into the atmosphere. The findings presented by this year's laureates in chemistry have had an enormous political and industrial impact. This was because they clearly identified unacceptable environmental hazards in a large, economically important sector. Their models were also subjected to very rigorous examination which eventually confirmed the main features of their original hypotheses. One obvious result is an international agreement known as the Montreal Protocol, which regulates the manufacture and use of CFCs.

Perhaps the most spectacular observation of changes in the stratospheric ozone content was made in 1985 over Antarctica by Joseph Farman and his colleagues. They observed a rapid and dramatic depletion of ozone in the polar region when sunlight returned after the polar night. The ozone content then built up to more normal levels during the subsequent polar summer and winter, after which the process was repeated. This recurring "ozone hole" was completely unexpected. Eventually a scientific explanation was found, mainly through the research of Susan Solomon, with important contributions from this year's laureates in chemistry as well.

Professor Crutzen, Professor Molina, and Professor Rowland,

You have demonstrated the importance of homogeneous and heterogeneous chemical processes in the earth's atmosphere. You have developed models that combine these data with knowledge of the large-scale transport processes in the atmosphere, and how these models can be utilized as a forecasting tool to evaluate the consequences of emissions of anthropogenic substances of various kinds. You have thereby not only created a clearer understanding of fundamental chemical phenomena, but also of the large-scale and often negative consequences of human behavior. In the words of Alfred Nobel's will, your work has been of very great "benefit to mankind." It is a privilege to congratulate you on behalf of the Royal Swedish Academy of Sciences, and I now ask you to receive your Nobel Prizes from the hands of His Majesty the King.

PAUL CRUTZEN

The following is the curriculum vitae of Paul Josef Crutzen, born 3 December, 1933, in Amsterdam, Holland, Director at the Max-Planck-Institute for Chemistry in Mainz, Germany. He is married, with two children.

Education

High School: 1946–1951, Amsterdam, Holland.
Civil Engineering, 1951–1954, Amsterdam, Holland.
Academic Studies and Research Activities 1959–1973 at the University of Stockholm
M.Sc. (Filosofie Kandidat), 1963.
Ph.D. (Filosofie Licentiat), Meteorology, 1968,
Title: *"Determination of parameters appearing in the 'dry' and the 'wet' photochemical theories for ozone in the stratosphere"*, Examiner: Prof. Dr. Bert Bolin, Stockholm.
D.Sc. (Filosofie Doktor), 1973, Stockholm, Sweden.
Title: *"On the photochemistry of ozone in the stratosphere and troposphere and pollution of the stratosphere by high-flying aircraft"*, Promoters: Prof. Dr. John Houghton, FRS, Oxford, and Dr. R.P. Wayne, Oxford.
(Ph.D. and D.Sc. degrees were given with the highest possible distinctions).

Employment

1954–1958:	Bridge Construction Bureau of the City of Amsterdam, The Netherlands.
1956–1958:	Military Service, The Netherlands.
1958–1959:	House Construction Bureau (HKB), Gävle, Sweden.
1959–1974:	Various computer consulting, teaching and research positions at the department of Meteorology of the University of Stockholm, Sweden, Latest positions: Research Associate and Research Professor.
1969–1971:	Post-doctoral fellow of the European Space Research Organization at the Clarendon Laboratory of the University of Oxford, England.

1974–1977: 1) Research Scientist in the Upper Atmosphere Project, National Center for Atmospheric Research (NCAR), Boulder, Colorado, USA.

2) Consultant at the Aeronomy Laboratory, Environmental Research Laboratories, National Oceanic and Atmospheric Administration (NOAA), Boulder, Colorado, USA.

1977–July 1980: Senior Scientist and Director of the Air Quality Division, National Center for Atmospheric Research (NCAR), Boulder, Colorado, USA.

1976–1981: Adjunct professor at the Atmospheric Sciences Department, Colorado State University, Fort Collins, Colorado.

Since 1980: Member of the Max-Planck-Society for the Advancement of Science and Director of the Atmospheric Chemistry Division, Max-Planck-Institute for Chemistry, Mainz, West Germany.

1983–1985: Executive Director, Max-Planck-Institute for Chemistry, Mainz, West Germany.

1987–1991: Professor (part-time) at the Department of Geophysical Sciences, University of Chicago, USA.

Since 1992: Professor (part-time), Scripps Institution of Oceanography, University of California, San Diego, LaJolla, USA.

Since 1993: Honorary Professor at the University of Mainz, Germany.

Other functions

Member of the International Ozone Commission and of the International Commission of the Upper Atmospherc of IAMAP (International Association for Meteorology and Atmospheric Physics) (1974–1984).

Member of NASA Stratospheric Research Advisory Committee (1975–1977).

Member, Atmospheric Sciences Advisory Committee, National Foundation (1977–1979).

Member, Committee of Atmospheric Sciences (CAS), National Academy of Sciences, U.S.A. (1978–1980).

Member, Advisory Committee, High Altitude Pollution Program, Federal Aviation Authority (FAA), U.S.A. (1978–1982).

Member, Commission of Air Chemistry and Global Air Pollution (CACGP) of the International Association of Meteorology and Atmospheric Physics (IAMAP) (1979–1990).

Member of Special Inter-Ministerial Advisory Commission on Forest Damage in the Federal Republic of Germany (1983–1987).

Member, Commission of the Parliament of the F.R.G. for the "Protection of the Earth's Atmosphere" (Enquete-Kommission zum Schutz der Erdatmosphäre) (1987–1990).

Member (present and past) of various research advisory committees of the German Science Foundation (DFG) and Ministry of Research and Technology (BMFT) of the Federal Republic of Germany.

Member of Steering Committees of the SCOPE/ICSU effort to estimate the environmental consequences of a nuclear war (SCOPE/ENUWAR) (1984–1988).

Member of the Kuratorium (Board of Trustees) of the Max-Planck-Institut für Meteorologie, Hamburg (1984–present).

Member, Executive Board of SCOPE (Scientific Committee on Problems of the Environment) of the International Council of Scientific Unions and Chairman of the National SCOPE Committee of the FRG (1986–1989).

Member of Special Committee and Executive Committee, and Chairman of Coordinating Panel I of the International Geosphere–Biosphere Programme (IGBP) (1986–1990).

Chairman of the Steering Committee of the International Global Atmospheric Chemistry (IGAC) Programme, a Core Project of the IGBP (1987–1990); Vice-Chairman (1990–1996).

Chairman of the European IGAC Project Office (1992–1996).

Member of the Kuratorium (Board of Trustees) of the Fraunhofer-Institut für atmosphärische Umweltforschung, Garmisch-Partenkirchen (1987–1994); Chairman (1992–1994).

Editor "Journal of Atmospheric Chemistry".

Member, Editorial Board "Tellus".

Member, Editorial Board "Climate Dynamics".

Member Editorial Advisory Board "Issues in Environmental Science and Technology", Royal Society of Chemistry, Britain.

Member of the Advisory Board of the Institute for Marine and Atmospheric Research, University Utrecht, Netherlands (1993–present).

Reviewing Editor "Science" (1993–present).

Member of the Advisory Council of the Volvo Environment Prize (1993–present).

Member of STAP (Scientific and Technical Advisory Panel) Roster of Experts of the United Nations Environment Programme (1993–present).

Member of the European Environmental Research Organisation (EERO) (1993–present).

Member of the *Prix Lemaitre* Committee (1994–present).

Member of the Steering Committee on Global Environmental Change of International Institute for Applied Systems Analysis (IIASA), Laxenburg, Austria (1994–present).

Member of the *SPINOZA* Prize Committee of Nederlandse Organisatie voor Wetenschappelijk Onderzoek (Dutch Organization for Scientific Research) (1994–1995).

Member of the Scientific Advisory Group of the School of Environmental Sciences, University of East Anglia, Norwich, Britain (1995–)

Member, Executive Board of Gesellschaft Deutscher Naturforscher und Ärzte (GDNÄ, German Society of Natural Scientists and Physicians) (1995–present).

Awards and honours

1969–1971   Visiting Fellow of St. Cross College, Oxford, England.

1976   Outstanding Publication Award, Environmental Research Laboratories, National Oceanic and Atmospheric Administration (NOAA), Boulder, Colorado, U.S.A.

1977   Special Achievement Award, Environmental Research Laboratories, NOAA, Boulder, Colorado, U.S.A.

1984   Rolex-Discover Scientist of the Year.

1985   Recipient of the Leo Szilard Award for "Physics in the Public Interest" of the American Physical Society.

1986   Elected to Fellow of the American Geophysical Union.

1986   Honorary Doctoral Degree, York University, Canada.

1986   Foreign Honorary Member of the American Academy of Arts and Sciences, Cambridge, U.S.A.

1987   Lindsay Memorial Lecturer, Goddard Space Flight Center, National Aeronautics and Space Administration.

1988   Founding Member of Academia Europaea.

1989   Recipient of the Tyler Prize for the Environment.

1990   Tracy and Ruth Scorer Lecturer at the University of California, Davis, U.S.A.

         Corresponding Member of The Royal Netherlands Academy of Science.

1991   Recipient of the Volvo Environmental Prize.

1992   Honorary Doctoral Degree, Université Catholique de Louvain, Belgium.

         Member of the Royal Swedish Academy of Sciences.

         Member of the Royal Swedish Academy of Engineering.

         Member Leopoldina, Halle.

1993   Ida Beam Visiting Professor, The University of Iowa, U.S.A.

1994   Raymond and Beverly Sackler Distinguished Lecturer in
              Geophysics and Planetary Sciences, Tel Aviv University, Israel.

              Recipient of the Deutscher Umweltpreis of the Umweltstiftung
              (German Environmental Prize of the Federal Foundation for
              the Environment).

              Foreign Associate of the U.S. National Academy of Sciences.

              Honorary Doctoral Degree, School of Environmental Sciences,
              University of East Anglia, Norwich, U.K.

              Recipient of the Max-Planck-Forschungspreis (with Dr. M.
              Molina, U.S.A.).

1995   Recipient of the Nobel Prize in Chemistry (with Dr. M. Molina
              and Dr. F.S. Rowland, U.S.A.).

              Recipient of United Nations Environment Ozone Awards for
              Outstanding Contribution for the Protection of the Ozone
              Layer.

Publications

Over 170 refereed and 50 other research publications in specialist publications, 4 books (co-authored), 4 books (edited).

Main research interest

Atmospheric chemistry and its role in biogeochemical cycles and climate.

MY LIFE WITH O_3, NO_x AND OTHER YZO_xs

Nobel Lecture, December 8, 1995

by

PAUL CRUTZEN

Max-Planck-Institute for Chemistry, Department of Atmospheric Chemistry, Mainz, Germany.

> *To the generation of Jamie Paul and our future grandchildren, who will know so much more and who will celebrate the disappearance of the ozone hole. I hope you will not be disappointed with us.*

I was born in Amsterdam on December, 3, 1933, the son of Anna Gurk and Jozef Crutzen. I have one sister who still lives in Amsterdam with her family. My mother's parents moved to the industrial Ruhr region in Germany from East Prussia towards the end of the last century. They were of mixed German and Polish origin. In 1929 at the age of 17, my mother, moved to Amsterdam to work as a housekeeper. There she met my father. He came from Vaals, a little town in the southeastern corner of the Netherlands, Bordering Belgium and Germany and very close to the historical German city of Aachen. He died in 1977. He had relatives in the Netherlands, Germany and Belgium. Thus, from both parents I inherited a cosmopolitan view of the world. My mother, now 84 years old, still lives in Amsterdam, mentally very alert, but since a few months ago, wheelchair-bound. Despite having worked in several countries outside The Netherlands since 1958, I have remained a Dutch citizen.

In May, 1940, The Netherlands were overrun by the German army. In September of the same year I entered elementary school, "de grote school" (the big school), as it was popularly called. My six years of elementary school largely overlapped with the 2nd World War. Our school class had to move between different premises in Amsterdam after the German army had confiscated our original school building. The last months of the war, between the fall of 1944 and Liberation Day on May, 5, 1945, were particularly horrible. During the cold "hongerwinter" (winter of famine) of 1944–1945, there was a severe lack of food and heating fuels. Also water for drinking, cooking and washing was available only in limited quantities for a few hours per day, causing poor hygienic conditions. Many died of hunger and disease, including several of my schoolmates. Some relief came at the beginning of 1945 when the Swedish Red Cross dropped food supplies on parachutes from airplanes. To welcome them we waved our red, white and blue Dutch flags in the streets. I had of course not the slightest idea how important Sweden would become later in my life. We only had a few hours of school each week, but because of special help from one of the teachers, I was allowed together with two other schoolmates to continue to the next and final class of elementary school; unfortunately, all the others lost a year. More or less normal school education only became possible again with the start of the new school year in the fall of 1945.

In 1946, after a successful entrance exam, I entered the "Hogere Burgerschool" (HBS), "Higher Citizen School", a 5 year long middle school, which prepared for University entrance. I finished this school in June, 1951, with natural sciences as my focal subjects. However, we all also had to become proficient in 3 foreign languages: French, English and German. I got considerable help in learning languages from my parents: German from my mother, French from my father. During those years, chemistry definitely was not one of my favourite subjects. They were mathematics and physics, but I also did very well in the three foreign languages. During my school years I spent considerable time with a variety of sports: football, bicycling, and my greatest passion, long distance skating on the Dutch canals and lakes. I also played chess, which in the Netherlands is ranked as a "denksport" (thought sport). I read widely about travels in distant lands, about astronomy, as well as about bridges and tunnels. Unfortunately, because of a heavy fever, my grades in the final exam of the HBS were not good enough to qualify for a university study stipend, which was very hard to obtain at that time, only 6 years after the end of the 2nd world war and a few years after the end of colonial war in Indonesia, which had been a large drain on Dutch resources. As I did not want to be a further financial burden on my parents for another 4 years or more (my father, a waiter, was often unemployed; my mother worked in the kitchen of a hospital), I chose to attend the Middelbare Technische School (MTS), middle technical school, now called the higher technical school (HTS), to train as a civil engineer. Although the MTS took 3 years, the second year was a practical year during which I earned a modest salary, enough to live on for about 2 years. From the summer of 1954 until February, 1958, with a 21-month interruption for compulsory military service in The Netherlands, I worked at the Bridge Construction Bureau of the City of Amsterdam. In the meanwhile, on a vacation trip in Switzerland, I met a sweet girl, Terttu Soininen, a student of Finnish history and literature at the University of Helsinki. A few years later I was able to entice her to marry me. What a great choice I made! She has been the center of a happy family; without her support, I would never have been able to devote so much of my time to studies and science. After our marriage in February, 1958, we settled in Gävle, a little town about 200 km north of Stockholm, where I had found a job in a building construction bureau. In December of that same year our daughter Ilona was born. In March, 1964, she got a little sister, Sylvia. Ilona is a registered nurse, now living in Boulder, Colorado. Her son Jamie Paul is 12 years old. Sylvia is a marketing assistant in München, Germany. All were present in Stockholm, Upsala and Gävle during the Nobel week. We had a happy and unforgettable time.

All this time I had longed for an academic career. One day, at the beginning of 1958, I saw an advertisement in a Swedish newspaper from the Department of Meteorology of Stockholm Högskola (from 1961, Stockholm University) announcing an opening for a computer programmer. Although I had not the slightest experience in this subject, I applied for the job and had the great luck to be chosen from among many candidates. On July 1, 1959, we moved to Stockholm and I started with my second profession. At that time the Meteorology Institute of Stockholm University (MISU) and the associated International Meteorological Institute (IMI) were at the forefront of meteorological research and many top researchers worked in Stockholm for extended periods. Only about a year earlier the founder of the institutes, Prof. Gustav Rossby, one of the greatest meteorologists ever, had died suddenly and was succeeded by Dr. Bert Bolin, another famous meteorologist, now "retired" as director of the Intergovernmental Panel on Climate Change (IPCC). At that time Stockholm University housed the fastest computers in the world (BESK and its successor FACIT).

With the exception of participation in a field campaign in northern Sweden, led by Dr. Georg Witt to measure the properties of noctilucent clouds, which appear during summer at about 85 km altitude in the coldest parts of atmosphere, and some programming work related to this, I was until about 1966 mainly involved in various meteorological projects, especially helping to build and run some of the first numerical (barotropic) weather prediction models. I also programmed a model of a tropical cyclone for a good friend, Hilding Sundquist, now a professor at MISU. At that time programming was a special art. Advanced general computer languages, such as Algol or Fortran, had not been developed, so that all programmes had to be written in specific machine code. One also had to make sure that all operations yielded numbers in the range $-1 \leq x < 1$, which meant that one had to scale all equations to stay within these limits; otherwise the computations would yield wrong results.

The great advantage of being at a university department was that I got the opportunity to follow some of the lecture courses that were offered at the university. By 1963 I could thus fulfill the requirement for the filosofie kandidat (corresponding to a Master of Science) degree, combining the subjects mathematics, mathematical statistics, and meteorology. Unfortunately, I could include neither physics nor chemistry in my formal education, because this would have required my participation in time consuming laboratory excercises. In this way I became a pure theoretician. I have, however, always felt close to experimental work, which I have strongly supported during my later years as director of research at the National Center of Atmospheric Research (NCAR) in Boulder, Colorado (1977–1980) and at the Max-Planck-Institute for Chemistry in Mainz, Germany (since 1980).

Being employed at the meteorological research institute, it was quite natural to take a meteorological topic for my filosofie licentiat thesis (comparable to a Ph.D. thesis). Building on my earlier experience further development of a numerical model of a tropical cyclone had been proposed to me. However, around 1965 I was given the task of helping a scientist from the U.S. to develop a numerical model of the oxygen allotrope distribution in the stratosphere, mesosphere and lower thermosphere. This project got me highly interested in the photochemistry of atmospheric ozone and I started an intensive study of the scientific literature. This gave me an understanding of the status of scientific knowledge about stratospheric chemistry by the latter half of the 1960's, thus setting the "initial conditions" for my scientific career. Instead of the initially proposed research project, I preferred research on stratospheric chemistry, which was generously accepted. At that time the main topics of research at the Meteorological Institute at the University of Stockholm were dynamics, cloud physics, the carbon cycle, studies of the chemical composition of rainwater, and especially the "acid rain" problem which was largely "discovered" at MISU through the work of Svante Odén and Erik Eriksson. Several researchers at MISU, among them Prof. Bolin and my good friend and fellow student Henning Rodhe, now Professor in Chemical Meteorology at MISU, got heavily involved in the issue which drew considerable political interest at the first United Nation Conference on the Environment in Stockholm in 1972 (1). However, I wanted to do pure science related to natural processes and therefore I picked stratospheric ozone as my subject, without the slightest anticipation of what lay ahead. In this choice of research topic I was left totally free. I can not overstate how I value the generosity and confidence which were conveyed to me by my supervisors Prof. Georg Witt, an expert on the aeronomy of the upper atmosphere, and the head of MISU Prof. Bert Bolin. They were always extremely helpful and showed great interest in the progress of my research.

STRATOSPHERIC OZONE CHEMISTRY

As early as 1930 the famous British scientist Sydney Chapman [2] had proposed that the formation of "odd oxygen", $O_x = O + O_3$, is due to photolysis of O_2 by solar radiation at wavelengths shorter than 240 nm

R1: $O_2 + h\nu$ (≤ 240 nm)   →   2 O

Rapid reactions, R2 and R3, next lead to the establishment of a steady state relationship between the concentrations of O and O_3

R2: $O + O_2 + M$ → $O_3 + M$
R3: $O_3 + h\nu$ (≤ 1180 nm)   → $O + O_2$

without affecting the concentration of odd oxygen. Destruction of odd oxygen, counteracting its production by reaction R1, occurs via the reaction

R4: $O + O_3$ → $2 O_2$

Until about the middle of the 1960's it was generally believed that reactions R1–R4 sufficed to explain the ozone concentration distribution in the stratosphere. However, by the mid 1960's, especially following a study by Benson and Axworthy (3), it became clear that reaction R4 is much too slow to balance the production of "odd oxygen" by reaction R1 (see Figure 1). In 1950 David Bates and Marcel Nicolet (4), together with Sydney Chapman the great pioneers of upper atmospheric photochemistry research, proposed

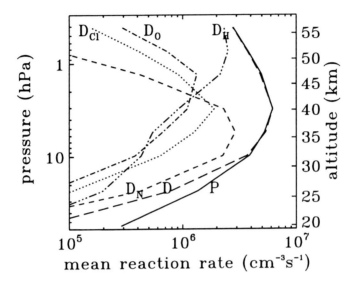

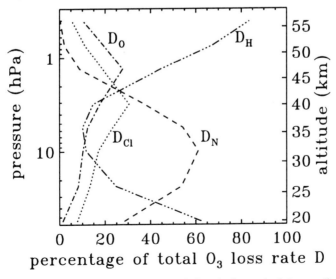

Figure 1: Ozone production and destruction rates, including absolute and relative contributions by the Chapman reaction R4 (D_O), NOx catalysis R11 + R12 (D_N), NO_x catalysis R5 + R6 (D_H) and ClO_x catalysis R21 + R22 (D_{Cl})$_x$. (120). The calculations neglect the heterogeneous halogen activation which become very important below 25 km under cold conditions.

that catalytic reactions involving OH and HO_2 radicals could counterbalance the production of odd oxygen in the mesosphere and thermosphere. Building on their work and on laboratory studies conducted by one of the 1967 Nobel Prize Laureates in Chemistry, Prof. R. Norrish of Cambridge University and his coworkers (5, 6), the ozone destruction reaction pair (R5, R6) involving OH and HO_2 radicals as catalysts were postulated by Hampson (7) and incorporated in an atmospheric chemical model by Hunt (8):

R5: $OH + O_3$ $\rightarrow$ $HO_2 + O_2$
R6: $HO_2 + O_3$ $\rightarrow$ $OH + 2 O_2$
net: $2 O_3$ $\rightarrow$ $3 O_2$

The proposed primary source for the OH radicals was photolysis of O_3 by solar ultraviolet radiation of wavelengths shorter than about 320 nm, leading to electronically excited $O(^1D)$ atoms,

R7: $O_3 + h\nu$ $\rightarrow$ $O(^1D) + O_2$ ($\leq$ 320 nm)

a small fraction of which reacts with water vapour

R8: $O(^1D) + H_2O$ $\rightarrow$ 2 OH.

Most $O(^1D)$ reacts with O_2 and N_2 to rapidly reproduce O_3, leading to a null cycle with no effect on ozone or odd oxygen

R9: $O(^1D) + M$ $\rightarrow$ $O + M$
R2: $O + O_2 + M$ $\rightarrow$ $O_3 + M$

In the absence of laboratory measurements for the rate constants of reactions R5 and R6, and in order for these reactions to counterbalance the production of odd oxygen by reaction R1, Hunt adopted the rate constants

$k_5 = 5 \times 10^{-13}$ cm^3 molec^{-1}s^{-1}
$k_6 = 10^{-14}$ cm^3 molec^{-1}s^{-1}.

In my filosofie licentiat thesis of 1968 I analyzed the proposal by Hampson and Hunt and concluded that the rate constants for reactions R5 and R6 which they had chosen could not explain the vertical distribution of ozone in the photochemically dominated stratosphere above 25 km. Furthermore, I pointed out (9) that the above choice of rate constants would also lead to unrealistically rapid loss of ozone (on a timescale of only a few days) in the troposphere. Anticipating a possible role of OH in tropospheric chemistry, in the same study I also briefly mentioned the potential importance of a reaction between OH with CH_4. We now know that reactions R5 and R6 proceed about 25 and 10 times slower, respectively, than postulated by Hunt and

Hampson and that the CH_4 oxidation cycle plays a very large role in tropospheric chemistry, a topic to which we will return.

Regarding stratospheric ozone chemistry, I discarded the theory of Hampson and Hunt and concluded: "... at least part of the solution of the problem of the ozone distribution might be the introduction of photochemical processes other than those treated here. The influence of nitrogen compounds on the photochemistry of the ozone layer should be investigated ".

Unfortunately no measurements of stratospheric NO_x (NO and NO_2) were available to confirm my thoughts about their potential role in stratospheric chemistry. By the summer of 1969 I had joined the Department of Atmospheric Physics at the Clarendon Laboratory of Oxford University as a postdoctoral fellow of the European Space Research Organization and stayed there for a two year period. The head of the research group, Dr. (now Sir) John Houghton, hearing of my idea on the potential role of NO_x, handed me a solar spectrum, taken on board a balloon by Dr. David Murcray and coworkers of the University of Denver, and indicated to me that it might reveal the presence of HNO_3 (10). After some analysis I could derive the approximate amounts of stratospheric HNO_3, including a rough idea of its vertical distributions. I did not get the opportunity to write up the result, because at about the same time, Rhine et al (11) published a paper, showing a vertical HNO_3 column density of 2.8×10^{-4} atm.cm ($\approx 7.6 \times 10^{15}$ molecules cm^{-2}) above 18.8 km. With this information I knew that NO_x should also be present in the stratosphere as a result of the reactions,

R10 a: OH + NO2 (+ M) $\rightarrow$ HNO_3(+ M),
R10 b: HNO3 + hv $\rightarrow$ OH + NO_2 ($\leq$ 320 nm)

This gave me enough confidence to submit my paper (12) on catalytic ozone destruction by NO and NO_2, based on the simple catalytic set of reactions:

R11: NO + O_3 $\rightarrow$ NO_2 + O_2
R12: NO_2 + O $\rightarrow$ NO + O_2
net   O + O_3 $\rightarrow$ 2 O_2

The net result of reactions R11 and R12 is equivalent to the direct reaction R4. However, the rate of the net reaction can be greatly enhanced by relatively small quantities of NO_x on the order of a few nanomole/mole (or ppbv). I also included a calculation of the vertical distribution of stratospheric HNO_3. As the source of stratospheric NO_x, I initially accepted the proposal by Bates and Hays (13) that about 20% of the photolysis of N_2O would yield N and NO. Subsequent work showed that this reaction does not take place. However, it was soon shown that NO could also be formed to a lesser extent, but still in significant quantities, via the oxidation of nitrous oxide (N_2O)

R13: $N_2O + O(^1D)$ $\rightarrow$ $2\,NO$

(14–16). It was further shown by Davis et al. (17) that reaction R12 proceeds about 3.5 times faster than I had originally assumed based on earlier laboratory work. Later it was also shown that earlier estimates of O_3 production by reactions R1 and R2 had been too large due to overestimations of both the absorption cross sections of molecular oxygen (18) and solar intensities in the ozone producing 200–240 nm wavelength region (19, 20). As a result of these developments it became clear that enough NO is produced via reaction R13 to make reactions R11 and R12 the most important ozone loss reactions in the stratosphere in the altitude region between about 25 and 45 km.

N_2O is a natural product of microbiological processes in soils and waters. A number of anthropogenic activities, such as the application of nitrogen fertilizers in agriculture, also lead to significant N_2O emissions. The rate of increase in atmospheric N_2O concentrations for the past decades has been about 0.3% per year (21). That, however, was not known in 1971. The discovery of the indirect role of a primarily biospheric product on the chemistry of the ozone layer has greatly stimulated interest in bringing biologists and atmospheric scientists together. Other examples of such biosphere–stratosphere interactions are CH_4 and OCS.

MAN'S IMPACT ON STRATOSPHERIC OZONE

In the fall of 1970, still in Oxford, I obtained a preprint of a MIT sponsored Study on Critical Environmental Problems (SCEP) which was held in July of that year (22). This report also considered the potential impact of the introduction of large stratospheric fleets of supersonic aircraft (U.S.: Boeing; Britain/France: Concorde; Soviet Union: Tupolev) and gave me the first quantitative information on the stratospheric inputs of NO_x which would result from these operations. By comparing these with the production of NO_x by reaction R13, I realized immediately that we could be faced with a severe global environmental problem. Although the paper in which I proposed the important catalytic role of NO_x on ozone destruction had already been published in April, 1970, clearly the participants in the study conference had not taken any note of it, since they concluded "The direct role of CO, CO_2, NO, NO_2, SO_2, and hydrocarbons in altering the heat budget is small. It is also unlikely that their involvement in ozone photochemistry is as significant as water vapour". I was quite upset by that statement. Somewhere in the margin of this text I wrote "Idiots".

After it became quite clear to me that I had stumbled on a hot topic, I decided to extend my 1970 study by treating in much more detail the chemistry of the oxides of nitrogen (NO, NO_2, NO_3, N_2O_4, N_2O_5), hydrogen (OH, HO_2), and HNO_3, partially building on a literature review by Nicolet (23). I soon got into big difficulties. In the first place, adopting Nicolet's reaction scheme I calculated high concentrations of N_2O_4, a problem which

I could soon resolve when I realized that this compound is thermally unstable, a fact which was not considered by Nicolet. A greater headache was caused by the supposedly gas phase reactions

R14: $N_2O_5 + H_2O \quad \rightarrow \quad 2\,HNO_3$

and

R15: $O + HNO_3 \quad \rightarrow \quad OH + NO_3$

for which the only laboratory studies available at that time had yielded rather high rate coefficients: $k_{14} = 1.7 \times 10^{-18}$ cm^3 molec^{-1}s^{-1} and $k_{15} = 1.7 - 17 \times 10^{-11}$ cm^3 molec^{-1}s^{-1} at room temperatures. A combination of reactions R14 and R15 with these rate constants would provide a very large source of OH radicals, about a thousand times larger than supplied by reaction R8, leading to prohibitively rapid catalytic ozone loss. This was a terribly nervous period for me. At that time no critical reviews and recommendations of rate coefficients were available. With no formal background in chemistry, I basically had to compile and comprehend much of the needed chemistry by myself from the available publications, although I profited greatly from discussions with colleagues at the University of Oxford, especially Dr. Richard Wayne of the Physical Chemistry Laboratory, a former student of Prof. R. Norrish in Cambridge. I discussed all these difficulties and produced extensive model calculations on the vertical distributions of trace gases in the O_x–NO_x–HO_x–HNO_x system in a paper which was submitted by the end of 1970 to the Journal of Geophysical Research (received January 13, 1971) and which, after revision, was finally published in the October 20 issue of 1971 (15). The publication of this paper was much delayed because of an extended mail strike in Britain. Because of the major problems I had encountered, I did not make any calculations of ozone depletions, but instead drew attention to the potential seriousness of the problem by stating "An artificial increase of the mixing ratio of the oxides of nitrogen in the stratosphere by about 1×10^{-8} may lead to observable changes in the atmospheric ozone level" and further in the text "It is estimated that global nitrogen oxide mixing ratios may increase by almost 10^{-8} from a fleet of 500 SSTs in the stratosphere. Larger increases, up to 7×10^{-8}, are possible in regions of high traffic densities .... Clearly, serious decreases in the total atmospheric ozone level and changes in the vertical distributions of ozone, at least in certain regions, can result from such an activity..."

THE SUPERSONIC TRANSPORT CONTROVERSY IN THE U.S.

Unknown to me, a debate on the potential environmental impact of supersonic stratospheric transport (SST) had erupted in the U.S. Initially the concern was mainly enhanced catalytic ozone destruction by OH and HO$_2$ radicals resulting from the release of H_2O in the engine exhausts (24). By mid-March, 1971, a workshop was organized in Boulder, Colorado, by an Advisory Board of the Department of Commerce, to which Prof. Harold Johnston of the University of California, Berkeley, was invited. As an expert in laboratory kinetics and reaction mechanisms of NO_x compounds (e.g. 25–27), he immediately realized that the role of NO_x in reducing stratospheric ozone had been grossly underestimated. Very quickly (submission 14 April, revision 14 June) on August 6, 1971, his paper appeared in Science (27) with the title "Reduction of Stratospheric Ozone by Nitrogen Oxide Catalysts from Supersonic Transport Exhaust". In the abstract of this paper Johnston stated "... oxides of nitrogen from SST exhaust pose a much greater threat to the ozone layer than does the increase in water. The projected increase in stratospheric oxides of nitrogen could reduce the ozone shield by about a factor of 2, thus permitting the harsh radiation below 300 nanometers to permeate the lower atmosphere". During the summer of 1971, I received a preprint of Johnston's study via a representative of British Aerospace, one of the Concorde manufacturers. This was the first time I had heard of Harold Johnston, for whom I quickly developed a great respect both as a scientist and a human being. Although I had expressed myself rather modestly about the potential impact of stratospheric NOx emissions from SST's, for the reasons given above, I fully agreed with Prof. Johnston on the potential severe consequences for stratospheric ozone and I was really happy to have support for my own ideas from such an eminent scientist. For a thorough resumé of the controversies between scientists and industry, and between meteorologists and chemists, recurring themes also in later years, I refer to Johnston's article "Atmospheric Ozone" (28). It should also be mentioned here that Prof. Johnston's publications in the early 1970's removed several of the major reaction kinetic problems which I had encountered in my 1971 study (15). It was shown, for instance, that neither reaction R14 nor R15 occur to a significant degree in the gas phase, and that the earlier laboratory studies had been strongly influenced by reactions on the walls of the reaction vessels (29), an advice which was earlier also given to me in a private communication by Prof. Sydney Benson of the University of Southern California.

In July, 1971, I returned to the University of Stockholm and devoted myself mainly to studies concerning the impact of NO_x releases from SST's on stratospheric ozone. In May, 1973, I submitted my inaugural dissertation "On the Photochemistry of Ozone in the Stratosphere and Troposphere and Pollution of the Stratosphere by High-Flying Aircraft" to the Faculty of Natural Sciences and was awarded the degree of Doctor of Philosophy with the highest possible distinction, the third time this had ever happened during the history of Stockholm University (and earlier Stockholm "Högskola"). This was one of the last occasions in which the classical and rather solemn "Filosofie Doktor", similar to the Habilitation in Germany and France, was awarded. I had to dress up just like during the Nobel Ceremonies. First and second "opponents" were Dr. John Houghton and Dr.

Richard Wayne of the University of Oxford, who wore their college gowns for the occasion. Dr. Wayne also served as a most capable, not obligatory, third opponent, whose task it was to make a fool of the candidate. Unfortunately, the classical doctoral degree has been abolished (I was one of the last ones to go through the procedure). The modern Swedish Filosofie Doktor degree corresponds more closely to the former Filosofie Licential degree.

In large part as a result of the proposal by Johnston (27) that NO_x emissions from SST's could severely harm the ozone layer, major research programs were started, the Climate Impact Assessment Program (CIAP), organized by the U.S. Department of Transportation (30), and the COVOS/COMESA (31, 32) program, jointly sponsored by France and Great Britain (the producers of the Concorde Aircraft). The aim of these programs was to study the chemical and meteorological processes that determine the abundance and distribution of ozone in the stratosphere, about which so little was known that the stratosphere was sometimes dubbed the "ignorosphere". The outcome of the CIAP study was summarized in a publication by the U.S. National Academy of Sciences in 1975 (33). "We recommend that national and international regulatory authorities be alerted to the existence of potentially serious problems arising from growth of future fleets of stratospheric airlines, both subsonic and supersonic. The most clearly established problem is a potential reduction of ozone in the stratosphere, leading to an increase in biologically harmful ultraviolet light at ground level".

The proposed large fleets of SST's never materialized, largely for economic reasons; only a few Concordes are currently in operation. The CIAP and COVOS/COMESA research program, however, greatly enhanced knowledge about stratospheric chemistry. They confirmed the catalytic role of NOx in stratospheric ozone chemistry. A convincing example of this was provided by a major solar proton event which occurred in August, 1972 and during which, within a few hours, large quantities of NO, comparable to the normal NO_x content, were produced at high geomagnetic latitudes ($> 65°$), as shown in Figure 2. With such a large input of NO, a clear depletion of stratospheric ozone was expected (34), a hypothesis which was confirmed by

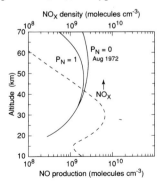

Figure 2: Production of NO at high geomagnetic latitudes during the solar proton event of 1972 for two assumptions about the electronic states of the N atoms formed ($P_N = 0$, or 1). Also shown are the average NO_x concentrations for these locations.

analysis of satellite observations (35). Figure 3 shows results of the calculated and observed ozone depletions, the former obtained with a model which also considered chlorine chemistry (36).

Although I had started my scientific career with the ambition to do basic research related to natural processes, the experiences of the early 1970's had made it utterly clear to me that human activities had grown so much that they could compete and interfere with natural processes. Since then this has been an important factor in my research efforts. Already by the end of 1971 I wrote in an article published in the "The Future of Science Year Book" of the USSR in 1972:

"... the upper atmosphere is an important part of our environment. Let us finish by expressing a sincere hope that in the future environmental dangers

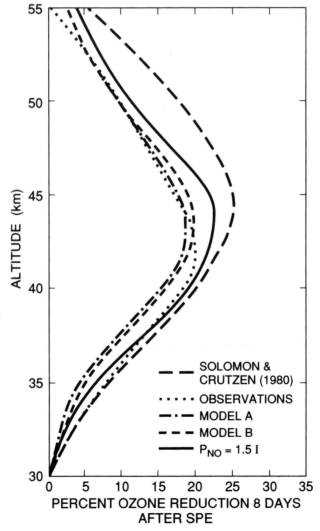

Figure 3: Observed and calculated percentage ozone depletions resulting from the 1972 solar proton event. The various calculated curves correspond to assumed values of parameters that were not well known.

of new technological development will be recognizable at an early stage. The proposed supersonic air transport is an example of a potential threat to the environment by future human activities. Other serious problems will certainly arise in the increasingly complicated world of tomorrow".

Tropospheric Ozone

My first thoughts on tropospheric ozone photochemistry go back to 1968, as discussed briefly above (9). However, in the following 3 years, my research was largely devoted to stratospheric ozone chemistry. Then in 1971 a very important paper with the title "Normal Atmosphere: Large Radical and Formaldehyde Concentrations Predicted" was published by Hiram Levy III, then of the Smithsonian Astrophysical Observatory in Cambridge, Massachusetts (37). Levy proposed that OH radicals could also be produced in the troposphere by the action of solar ultraviolet radiation on ozone (Reaction R8), and that they are responsible for the oxidation of CH_4 and CO, an idea which was also quickly adopted by Jack McConnell, Michael McElroy and Steve Wofsy (38) of Harvard University. The recognition of the important role of OH was a major step forward in our understanding of atmospheric chemistry. Despite very low atmospheric concentrations, currently estimated at 10^6 molecules/cm^3, corresponding to a mean tropospheric volume mixing ratio of 4×10^{-14} (39), it is this ultraminor constituent–and not the 10^{13} times more abundant O_2–which is responsible for the oxidation of almost all compounds that are emitted into the atmosphere by natural processes and anthropogenic activities. The lifetimes of most atmospheric gases are, therefore, largely determined by the concentrations of OH and the corresponding reaction coefficients (40) (see Table 1). Those gases

Table 1: Schematic representation of importance of OH radicals in atmospheric chemistry.

PRIMARY PRODUCTION OF OH RADICALS

$O_3 + h\upsilon \ (\leq 320 \text{ nm}) \quad \rightarrow \quad O\,(^1D) + O_2$
$O\,(^1D) + H_2O \quad \rightarrow \quad 2\,OH$

GLOBAL, 24 HOUR, AVERAGE (OH) $\approx 10^6$ MOLECULES/CM^3
MOLAR MIXING RATIO IN TROPOSPHERE $\approx 4 \times 10^{-14}$

REACTION WITH OH DETERMINES THE LIFETIME OF MOST GASES IN ATMOSPHERE
EXAMPLES:

CH_4:	8 YEARS
C_2H_6:	2 MONTHS
C_3H_8:	10 DAYS
C_5H_8:	HOURS
$(CH_3)_2S$:	2–3 DAYS
CH_3Cl:	≈ 1 YEAR
CH_3CCl_3:	≈ 5 YEARS
NO_2:	≈ 1 DAY

$CFCl_3$, CF_2Cl_2, N_2O do not react with OH. They are broken down in the stratosphere and have a large influence on ozone chemistry.

that do not react with OH have very long atmospheric residence times and are largely destroyed in the stratosphere. Examples of the latter class of compounds are N_2O, and several fully halogenated, industrial organic compounds, such as $CFCl_3$, CF_2Cl_2, and CCl_4. These play a major role in stratospheric ozone chemistry, an issue to which we will return.

Following Levy's paper my attention returned strongly to tropospheric chemistry. Starting with a presentation at the 1972 International Ozone Symposium in Davos, Switzerland, I proposed that *in situ* chemical processes could produce or destroy ozone in quantities larger than the estimated downward flux of ozone from the stratosphere to the troposphere (41, 42). Destruction of ozone occurs via reactions R7 + R8 and R5 + R6. Ozone production takes place in environments containing sufficient NO_x, via

R16: $RO_2 + NO$ $\rightarrow$ $RO + NO_2$
R17: $NO_2 + h\nu$ $\rightarrow$ $NO + O$ ($\lambda \le 405$ nm)
R2: $O + O_2 + M$ $\rightarrow$ $O_3 + M$
net: $RO_2 + O_2$ $\rightarrow$ $RO + O_3$
(with R = H, CH_3, or other organoperoxy radicals)

The catalytic role of NO in atmospheric chemistry is, therefore, twofold. At altitudes above about 25 km, where O atom concentrations are high, ozone destruction by reactions R11 + R12 dominates over ozone production by reactions R16 + R17 + R2. The latter chain of reactions is at the base of all photochemical ozone formation in the troposphere, including that taking place during photochemical smog episodes, originally discovered in southern California, as discussed by Johnston (28). Such reactions can, however, also take place in background air with ubiquitous CO and CH_4 serving as fuels: in the case of CO oxidation

R18: $CO + OH$ $\rightarrow$ $CO_2 + H$
R19: $H + O_2 + M$ $\rightarrow$ $HO_2 + M$
R16: $HO_2 + NO$ $\rightarrow$ $OH + NO_2$
R17: $NO_2 + h\nu$ $\rightarrow$ $NO + O$
R2: $O + O_2 + M$ $\rightarrow$ $O_3 + M$
net: $CO + 2\,O_2$ $\rightarrow$ $CO_2 + O_3$

This reaction chain requires the presence of sufficient concentrations of NO. At low NO volume mixing ratios, below about 10 pmole/mole (p = pico = 10^{-12}), oxidation of CO may lead to ozone destruction since the HO_2 radical then reacts mostly with O_3:

R18: $CO + OH$ $\rightarrow$ $CO_2 + H$
R19: $H + O_2 + M$ $\rightarrow$ $CO_2 + M$
R 6: $HO_2 + O_3$ $\rightarrow$ $OH + 2\,O_2$
net: $CO + O_3$ $\rightarrow$ $CO_2 + O_2$

In a similar way, the oxidation of CH_4 in the presence of sufficient NO_x will lead to tropospheric ozone production.

Besides reacting with NO or O_3, HO_2 can also react with itself

R20: $HO_2 + HO_2 \rightarrow H_2O_2 + O_2$

to produce H_2O_2 which serves as a strong oxidizer of S (IV) compounds in cloud and rain water.

My talk at the International Ozone Symposium was not well received by some members of the scientific establishment of that time. However, in the following years, the idea gradually received increased support. In particular, Bill Chameides and Jim Walker (43), then of Yale University, took it up and went as far as proposing that even the diurnal variation of lower tropospheric ozone could be explained largely by *in situ* photochemical processes. Although I did not agree with their hypothesis (CH_4 and CO oxidation rates are just not rapid enough), it was good to note that my idea was being taken seriously. (I should immediately add that especially Bill Chameides in subsequent years added much to our knowledge of tropospheric ozone). A couple of years later, together with two of my finest students, Jack Fishman and Susan Solomon, we presented observational evidence for a strong *in situ* tropospheric ozone chemistry (44, 45). Laboratory measurements by Howard and Evenson (46) next showed that reaction R16 proceeded about 40 times faster than determined earlier, strongly promoting ozone production and increased OH concentrations with major consequences for tropospheric and stratospheric chemistry (47). A consequence of faster rate of this reaction is a reduction in the estimated ozone depletions by stratospheric aircraft as the ozone production reactions R16 + R17 + R2 are favoured over the destruction reaction R6. Furthermore, a faster reaction R16 leeds to enhanced OH concentrations and thus a faster conversion of reactive NO_x to far less reactive HNO_3. Table 2 summarizes a recent ozone budget calculated with a

Table 2: Tropospheric ozone budgets, globally and for the northern (NH) and southern (SH) hemisphere in 10^{13} mole/year. Only CH_4 and CO oxidation cycles were considered. Calculations were made with the latest version of the global, three-dimensional MOGUNTIA model (48).

	Global	NH	SH
Sources			
$HO_2 + NO$	6.5	4.1	2.4
$CH_3O_2 + NO$	1.7	1.0	0.7
Transport from stratosphere	1.0	0.7	0.3
Sinks			
$O(^1D) + H_2O$	3.8	2.2	1.6
$HO_2 + O_3$ and $OH + O_3$	2.8	1.8	1.0
Deposition on surface	2.7	1.8	0.9
Net chemical source	1.6	1.1	0.5

three-dimensional chemistry transport model of the troposphere. The results clearly show the dominance of *in situ* tropospheric ozone production and destruction. With the same model, estimates were also made of the present and pre-industrial ozone concentration distributions. The calculations, shown in Figures 4 and 5, indicate a clear increase in tropospheric ozone

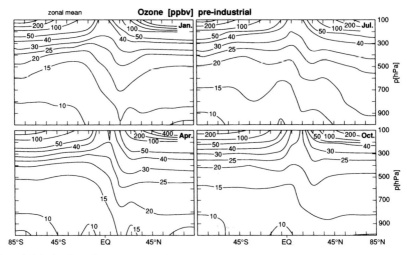

Figure 4: Calculated zonal average ozone volume mixing ratios in units of nanomole/mole (or ppbv) for the pre-industrial era (nano = 10^{-9}) for different months.

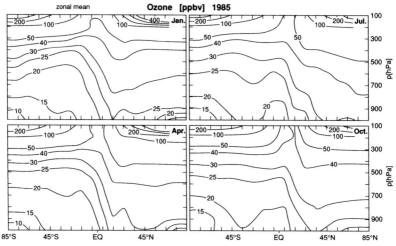

Figure 5: Same as figure 4, but for mid-1980's.

concentrations over the past centuries (48). In Figure 6 we also show the meridional cross sections of zonal average ozone, as compiled by Jack Fishman (unpublished data).

With the same model we have also calculated the OH concentration distributions for pre-industrial and present conditions. Since pre-industrial times, CH_4 volume mixing ratios in the atmosphere has increased (49) from

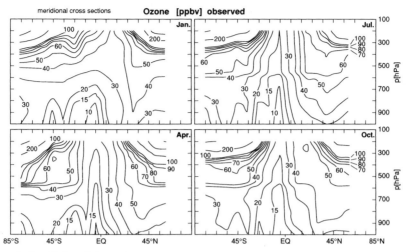

Figure 6: Compilation of observed ozone distributions prior to 1989, compiled by Jack Fishman of NASA Langley Research Center. It hsould be mentioned that the data base is very limited and has not much improved for the tropics and subtropics.

about 0.7 to 1.7 ppmv (1 ppmv = 10^{-6} volume/volume). Because reaction with CH_4 is one of the main sinks for OH, an increase in CH_4 should have led to a decrease in OH concentrations. On the other hand, increased ozone concentrations, leading to enhanced OH production by reactions R7 + R8, and the effect of the reactions,

R6: $HO_2 + O_3$ → $OH + 2\,O_2$

and

R16: $HO_2 + NO$ → $OH + NO_2$

both stimulated by strongly enhanced anthropogenic NO production, should have worked in the opposite direction. Figures 7 and 8 show calculated zonally and diurnally averaged, meridional distributions of the OH con-

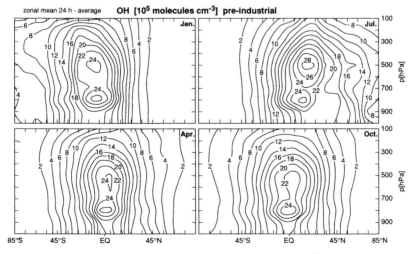

Figure 7: Calculated zonal and 24-hour average OH concentrations in units of 10^5 molecules/cm3 for the pre-industrial period for January, April, July and October.

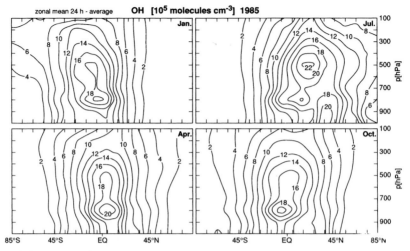

Figure 8: Same as figure 7 for 1985.

centrations, both for the pre-industrial and industrial periods. They indicate:

a) strong maxima of OH concentrations in the tropics, largely due to high intensities of ultraviolet radiation as a consequence of a minimum vertical ozone column. Consequently the atmospheric oxidation efficiency is strongly determined by tropical processes. For instance, most CH_4 and CO is removed from the atmosphere by reaction with OH in the tropics.

b) the possibility of a significant decline in OH concentrations from pre-industrial to industrial conditions.

The results presented in Figure 8 are of great importance, as they allow estimations of the sink of atmospheric CH_4 by reaction with OH. Prior to the discovery of the fundamental role of the OH radical (37), estimates of the sources and sinks of trace gases were largely based on guess work without a sound scientific basis. As shown in Table 3, this recognition has led to very large changes in the budget estimates of CH_4 and CO. "Authoritative" estimates of the CH_4 budget of 1968 (no reference will be given) gave much higher values for CH_4 releases from natural wetlands. With such a dominance of natural sources, it would have been impossible to explain the annual increase in atmospheric CH_4 concentrations by almost 1% per year. Early estimates of CO sources, on the other hand, were much too low.

The dominance of OH concentrations and the high biological productivity in the tropics clearly points at the great importance of the tropics and subtropics in atmospheric chemistry. Despite this fact, research on low latitude chemistry is much neglected, such that we do not even have satisfactory statistics on the ozone distribution in this part of the world. Tropical chemistry is a topic which has played and will continue to play a large role in my research. Contrary to what was commonly believed prior to the early 1980's, the chemical composition of the tropical and subtropical atmosphere is substan-

Table 3: Estimated budgets of important atmospheric trace gases made in 1968 and at present.
DMS denotes dimethylsulfide

	1968		1995	
	CH$_4$ BUDGET (Tg/year)			
Natural wetlands	1180		275	
Anthropogenic	270		265	
	1450		540	
	CO BUDGET (Tg/year)			
Natural	75		860	
Anthropogenic	274		1640	
	350		2500	
	S BUDGET (Tg S/year)			
Pollutants	76		78	
Oceanic emissions	30	(H$_2$S)	25	(DMS)
Land emissions	70	(H$_2$S)	few (various compounds)	
	176		105	
	NO$_x$ BUDGET (Tg N/year)			
Biological	150		10	
Pollution	15		24	
Lightning	–		2–10	
	165		36–44	
	N$_2$O BUDGET (Tg N/year)			
Biological	340		15	
Anthropogenic	–		3.5	
	340		18.5	

tially affected by human activities, in particular biomass burning which takes place during the dry season. The high temporal and spatial variability of ozone in the tropics is shown in figure 9. Highest ozone concentrations are

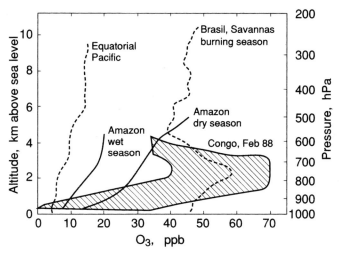

Figure 9: Variability of ozone profiles in the tropics, including contrast between dry and wet season, and continents versus marine soundings.

observed over the polluted regions of the continents during the dry season, lowest values in the clean air over the Pacific. I will return to the topic of tropical tropospheric ozone, but will first review the stormy developments in stratospheric ozone depletion by halogen compounds that started in 1974.

POLLUTION OF THE STRATOSPHERE BY ClO_x

Towards the end of the CIAP programme some researchers had turned their interest to the potential input of reactive chlorine radicals on stratospheric ozone. In the most thorough of these studies, Stolarski and Cicerone (50) calculated significant ozone depletions if inorganic chlorine were to be present in the stratosphere at a volume mixing ratio of 10^{-9} mole/mole of air (1 nanomole/mole). Odd oxygen destruction would take place via the catalytic reaction cycle

R21: $Cl + O_3$ $\rightarrow$ $ClO + O_2$
R22: $O + ClO$ $\rightarrow$ $Cl + O_2$
net: $O + O_3$ $\rightarrow$ $2 O_2$.

This reaction sequence is very similar to the catalytic NO_x cycle R11 + R12 introduced before. The study by Stolarski and Cicerone, first presented at a conference in Kyoto, Japan, in the fall of 1973, mainly considered volcanic injections as a potential source of ClX (their initial interest in chlorine chemistry was, however, concerned with the impact of the exhaust of solid rocket fuels of the space shuttle). Two other conference papers (51, 52), also dealt with ClO_x chemistry. All three papers struggled, however, with the problem of a missing chlorine source in the stratosphere (research over the past 20 years has shown that the volcanic source is rather insignificant).

In the fall of 1973 and early 1974 I spent some time looking for potential anthropogenic sources of chlorine in the stratosphere. Initially my main interest was with DDT and other pesticides. Then by the beginning of 1974 I read a paper by James Lovelock and coworkers (53) who reported atmospheric measurements of $CFCl_3$ (50 picomole/mole) and CCl_4 (71 pmole/mole) over the Atlantic. (Such measurements had been made possible by Lovelock's invention of the electron capture detector for gas chromatographic analysis, a major advance in the environmental sciences). Lovelock's paper gave me the first estimates of the industrial production rates of CF_2Cl_2 and $CFCl_3$. It also stated that these compounds "are unusually stable chemically and only slightly soluble in water and might therefore persist and accumulate in the atmosphere ... The presence of these compounds constitutes no conceivable hazard". This statement had just aroused my curiosity about the fate of these compounds in the atmosphere when a preprint of a paper by M.J. Molina and F.S. Rowland with the title "Stratospheric Sink for Chlorofluoromethanes–Chlorine Atom Catalyzed Destruction of Ozone" was sent to me by the authors. I knew immediately that this was a very important paper and decided to mention it briefly during a presentation on stratospheric ozone to which I had been invited by the Royal Swedish Academy of Sciences in Stockholm . What I did not know was that the press was likewise invited to the lecture. To my great surprise, within a few days, an article appeared in the Swedish newspaper "Svenska Dagbladet". This article quickly attracted wide international attention and soon I was visited by representatives of the German chemical company Hoechst and also by Professor Rowland, who at that time was spending a sabbatical year at the Atomic Energy Agency in Vienna. This was the first time I had heard of Molina or Rowland, which is not surprising as they had not been active in studies on the chemistry of the atmosphere. Needless to say, I remained highly interested in the topic and by September, 1974, about 2 months after the publication of Molina's and Rowland's paper (54), I presented a model analysis of the potential ozone depletion resulting from continued use of the chlorofluorocarbons (CFC's) (55), which indicated the possibility of up to about 40% ozone depletion near 40 km altitude as a result of continued use of these compounds at 1974 rates. Almost simultaneously, Cicerone et al. (56) published a paper in which they predicted that by 1985–1990, continued use of CFCs at early 1970's levels could lead to ClO_x catalyzed ozone destruction of a similar magnitude as the natural sinks of ozone. Following Molina's and Rowland's proposal, research on stratospheric chemistry further intensified, now with the emphasis on chlorine compounds.

By the summer of 1974, together with my family, I moved to Boulder, Colorado, where I assumed two halftime positions, one as a consultant at the Aeronomy Laboratory of the National Oceanic and Atmospheric Administration (NOAA), the other at the Upper Atmosphere Project of the National Center for Atmospheric Research (NCAR). The NOAA group, which under the able direction of Dr. Eldon Ferguson had become the world leading group in the area of laboratory studies of ion-molecule reactions, had just decided to direct their considerable experimental skills to studies of stratospheric chemistry. My task was to guide them in that direction. I still feel proud to have been part of a most remarkable transformation. Together with Eldon Ferguson, scientists like Dan Albritton, Art Schmeltekopf, Fred Fehsenfeld, Paul Goldan, Carl Howard, George Reid, John Noxon, and Dieter Kley rapidly made major contributions to stratospheric research, including such activities as air sampling with balloon borne evacuated cans, so-called "salad bowls" for later gas-chromatographic analysis, optical measurements of the vertical abundances and distributions of NO_2 and NO_3 (later expanded by Susan Solomon to BrO and OClO), the design and operation of an instrument to measure extremely low water vapour mixing ratios, and laboratory simulations of important, but previously poorly known rate coefficients of important reactions. In later years the NOAA group also devoted itself to studies of tropospheric chemistry, reaching a prominent position in this research area as well. At NCAR the emphasis was more on infrared spectrographic measurements by John Gille and Bill Mankin, work which also developed into satellite-borne experiments. Another prominent activity was the analysis of the vertical distributions of less reactive gases, such as CH_4, H_2O, N_2O, and the CFC's, employing the cryogenic sampling technique which had been pioneered by Ed Martell and Dieter Ehhalt.

In 1977, I took up the directorship of the Air Quality Division of NCAR, my first partially administrative position. I continued, however, my scientific work, something which many thought would be impossible. Fortunately, in Nelder Medrud I had a highly competent administrative officer. In my position as director I promoted work on both stratospheric and tropospheric chemistry. My own research was mostly devoted to the development of photochemical models, conducted mostly with my students Jack Fishman, Susan Solomon and Bob Chatfield. Together with Pat Zimmerman we started studies on atmosphere-biosphere interactions, especially the release of hydrocarbons from vegetation and pollutant emissions due to biomass burning in the tropics. I also tried to strengthen interactions between atmospheric chemists and meteorologists to improve the interpretation of the chemical measurements obtained during various field campaigns. To get this interdisciplinary research going was a challenge, particularly in those days.

During this period, as part of various US and international activities, much of my research remained centered on the issue of anthropogenic, chlorine-catalyzed ozone destruction. However, because I am sure that this topic will be covered extensively by my two fellow recipients of this year's Nobel Prize, I would like to make a jump to the year 1985, when Joe Farman and his colleagues (57) of the British Antarctic Survey published their remarkable set of October total ozone column measurements from the Halley Bay station, showing a rapid depletion on the average by more than 3% per year, starting from the latter half of the 1970's. Although their explanation (ClO_x/NO_x interactions) was wrong, Farman et al. (57) correctly suspected a connection with the continued increase in stratospheric chlorine (nowadays more than 5 times higher than natural levels). Their display of the downward trend of ozone, matching the upward trend of the chlorfluorocarbons (with the appropriate scaling) was indeed highly suggestive.

The discovery of the ozone hole came during a period in which I was heavily involved in various international studies on the potential environmental impacts of a major nuclear war between the NATO and Warsaw Pact nations, an issue to which I will briefly return in one of the following chapters. Because so many researchers became quickly involved in the "ozone hole" research, initially I stayed out of it. Then, in early 1986 I attended a scientific workshop in Boulder, Colorado, which brought me up-to-date with the various theories which had been proposed to explain the ozone hole phenomenon. Although it turned out that some of the hypotheses had elements of the truth, in particular the idea put forward by Solomon et al. (58) of chlorine activation on the surface of stratospheric ice particles, via the reaction,

R23: $HCl + ClONO_2 \rightarrow Cl_2 + HNO_3$

followed by rapid photolysis of Cl_2 and production of highly reactive Cl atoms

R24: $Cl_2 + h\upsilon$ →   2 Cl

I felt dissatisfied with the treatment of the heterogeneous chemistry. On my flight back to Germany (I hardly sleep on trans-Atlantic flights), I had good time to think it over and suddenly realized that if HNO_3 and NO_x were removed from the gas phase into the particulate phase, then an important defense against the attack of ClO_x on O_3 would be removed. The thought goes as follows. Under normal stratospheric conditions, there are strong interactions between the NO_x and ClO_x radicals which lead to protection of ozone from otherwise much more severe destruction. Important examples of these are the reactions,

R25: $ClO + NO_2 + M$ → $ClONO_2 + M$

and the pair of reactions,

R26: $ClO + NO$ → $Cl + NO_2$
R27: $Cl + CH_4$ → $HCl + CH_3$

producing HCl and $ClONO_2$, which do not react with O or O_3. Due to these reactions, under normal stratospheric conditions most of the inorganic chlorine is present as HCl and $ClONO_2$. Like two mafia families, the ClO_x and NO_x thus fight each other, to the advantage of ozone. As shown in Figure 10, there are plenty of complex interactions between the OX, HX,

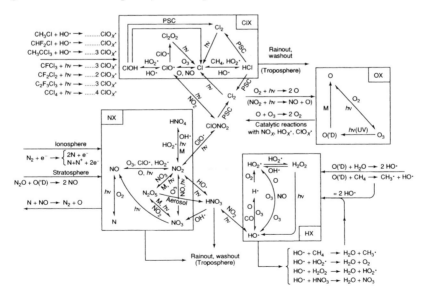

Figure 10: Schematic presentation of the chemical interactions in the stratosphere. At the start of my scientif carreer only the OX and some of the HX reactions had been taken into account. Note that OX stands for the odd oxygen compounds, HX for H, OH, HO_2 and H_2O_2; NX for N, NO, NO_2, NO_3, N_2O_5, HNO_3 and HNO_4; and ClX for all inorganic chlorine compounds, Cl, ClO, Cl_2O_2, $ClONO_2$, HCl, OClO and Cl_2. Not included are the bromine compounds which likewise play a significant role in stratospheric ozone depletion.

NX and ClX families. (We should even have included BrX). Now, if the NX compounds were removed from the gas phase, then reactions R25–R27 would not occur and most inorganic chlorine would be available in the activated, ozone destroying forms. A possible scenario would involve conversion of NO_x + HNO_3, via the nighttime reactions

R11:	$NO + O_3$	$\rightarrow$	$NO_2 + O_2$
R28:	$NO_2 + O_3$	$\rightarrow$	$NO_3 + O_2$
R29:	$NO_3 + NO_2(+M)$	$\rightarrow$	$N_2O_5(+M)$
R14:	$N_2O_5 + H_2O$ (surface)	$\rightarrow$	$2\ HNO_3$ (gas),

followed by uptake of HNO_3 in the aerosol phase

R30:	HNO_3(gas)	$\rightarrow$	HNO_3 (particles)

As noted before, reaction R14 does not occur in the gas phase, but it readily occurs on wetted particulate surfaces. These are always present in the lower stratosphere in the form of sulfate particles, a fact which was first discovered by Christian Junge, a pioneer in atmospheric chemistry and my predecessor as director at the Max-Planck-Institute of Chemistry in Mainz (59). The sulfate particles are formed by nucleation of gas phase H_2SO_4 which is formed from SO_2, following attack by OH (60, 61)

R31:	$SO_2 + OH + M$	$\rightarrow$	$HSO_3 + M$
R32:	$HSO_3 + O_2$	$\rightarrow$	$SO_3 + HO_2$
R33:	$SO_3 + H_2O$	$\rightarrow$	H_2SO_4

The sources of stratospheric SO_2 are either direct injections by volcanic explosions (59) or oxidation of OCS, produced at the earth's surface (62), via

R34:	$OCS + h\nu$	$\rightarrow$	$S + CO$
R35:	$S + O_2$	$\rightarrow$	$SO + O$
R36:	$SO + O_2$	$\rightarrow$	$SO_2 + O$

The possibility of HNO_3 formation via heterogeneous reactions on sulfate particles was already considered in a 1975 paper which I co-authored with Richard Cadle and Dieter Ehhalt (63). Based on laboratory experiments, this reaction was, however, for a long while thought to be unimportant, until it was discovered that the original laboratory measurements were grossly incorrect and that reaction R14 readily occurs on H_2O containing surfaces (64–66). (Earlier tropospheric measurements had already indicated this (67)). The introduction of reaction R14 leads to a significant conversion of reactive NO_x to much less reactive HNO_3, thus diminishing the role of NO_x in ozone chemistry, especially in the lower stratosphere. By including reaction R14, better agreement is obtained between theory and observations

(68). The experience with reaction R14 and earlier discussed reactions R6, R14 and R15, emphasizes the importance of high quality measurements.

As soon as I had returned to Mainz, I contacted Dr. Frank Arnold of the Max-Planck-Institute for Nuclear Physics in Heidelberg to explain to him my idea about NO_x removal from the gas phase. After about a week he had shown that under stratospheric conditions, solid nitric acid trihydrate (NAT) particles could be formed at temperatures below about 200K, that is a temperature about 10 K higher than that needed for water ice particle formation. The paper about our findings was published in Nature at the end of 1986 (69). Independently, the idea had also been developed by Brian Toon, Rich Turco and coworkers (70). Subsequent laboratory investigations, notably by David Hanson and Konrad Mauersberger (71), then of the University of Minnesota, provided accurate information on the thermodynamic properties of NAT. It was next also shown that the NAT particles could provide excellent surfaces to catalyze the production of ClO_x by reactions R23 and R24 (72, 73). Finally, Molina and Molina (74) proposed a powerful catalytic reaction cycle, involving ClO-dimer formation,

R21:	$Cl + O_3$	$\rightarrow$	$ClO + O_2$ (2 x)		
R37:	$ClO + ClO + M$	$\rightarrow$	$Cl_2O_2 + M$		
R38:	$Cl_2O_2 + h\nu$	$\rightarrow$	$Cl + ClO_2$	$\rightarrow$	$2 Cl + O_2$
net:	$2 O_3$	$\rightarrow$	$3 O_2$		

completing the chain of events causing rapid ozone depletion under cold, sun-lit stratospheric conditions. Note that reaction R37 implies an ozone depletion response which is proportional to the square of the ClO concentrations. Furthermore, as chlorine activation by reaction R23 is also non-linearly dependent on the stratospheric chlorine content, a powerful non-linear, positive feedback system is created, which is responsible for the accelerating loss of ozone under "ozone hole" conditions. The "ozone hole" is a drastic example of a man-made chemical instability, which developed at a location most remote from the industrial releases of the chemicals responsible for the effect.

The general validity of the chain of events leading to chlorine activation has been confirmed by both ground based (75, 76) and airborne in-situ (77) radical observations. Especially the latter, performed by James Anderson and his students of Harvard University, have been very illuminating, showing large enhancements in ClO concentrations in the cold, polar region of the lower stratosphere, coincident with a rapid decline in ozone concentrations. Together with other observations this confirms the correctnesss of the ozone depletion theory as outlined above. In the meanwhile the seriousness of this global problem has been recognized by all nations of the world and international agreements have been signed to halt the production of CFC's and halons from this year on. Although the cause–effect relationship is very clear, for the layperson as well, it is depressing to see that it is, nevertheless, not accepted by a small group of very vocal critics without any record of achievements in this area of research. Some of these have recently even succeeded in becoming members of the U.S. Congress.

AND THINGS COULD HAVE BEEN MUCH WORSE

Gradually, over a period of a century or so, stratospheric ozone should recover. However, it was a close call. Had Joe Farman and his colleagues from the British Antarctic Survey not persevered in making their measurements in the harsh Antarctic environment for all those years since the International Geophysical Year 1958/1959, the discovery of the ozone hole may have been substantially delayed and there may have been far less urgency to reach international agreement on the phasing out of CFC production. There might thus have been a substantial risk that an ozone hole could also have developed in the higher latitudes of the northern hemisphere.

Furthermore, while the establishment of an instability in the O_x–ClO_x system requires chlorine activation by heterogeneous reactions on solid or supercooled liquid particles, this is not required for inorganic bromine, which is normally largely present in its activated forms due to gas phase photochemical reactions. This makes bromine on an atom to atom basis almost a hundred times more dangerous for ozone than chlorine (78, 52). This brings up the nightmarish thought that if the chemical industry had developed organobromine compounds instead of the CFCs–or alternatively, if chlorine chemistry would have run more like that of bromine–then without any preparedness, we would have been faced with a catastrophic ozone hole everywhere and at all seasons during the 1970s, probably before the atmospheric chemists had developed the necessary knowledge to identify the problem and the appropriate techniques for the necessary critical measurements. Noting that nobody had given any thought to the atmospheric consequences of the release of Cl or Br before 1974, I can only conclude that mankind has been been extremely lucky, that Cl activation can only occur under very special circumstances. This shows that we should always be on our guard for the potential consequences of the release of new products into the environment. Continued surveillance of the composition of the stratosphere, therefore, remains a matter of high priority for many years ahead.

In the meanwhile, we know that freezing of H_2SO_4–HNO_3–H_2O mixtures to give NAT particle formation does not always occur and that supercooled liquid droplets can exist in the stratosphere substantially below NAT nucleation temperatures, down to the ice freezing temperatures (79). This can have great significance for chlorine activation (80, 81). This issue, and its implications for heterogeneous processes, have been under intensive investigation at a number of laboratories, especially in the U.S., notably by the groups headed by A.R. Ravishankara at the Aeronomy Laboratory of NOAA, Margaret Tolbert at the University of Colorado, Mario Molina at MIT, Doug Worsnop and Chuck Kolb at Aerodyne, Boston, and Dave Golden at Stanford Research Institute in Palo Alto. I am very happy that a team of young colleagues at the Max-Planck-Institute for Chemistry under the leadership of Dr. Thomas Peter is likewise very successfully involved in experimental and theoretical studies of the physical and chemical properties of stratospheric particles at low temperatures. A highly exciting new finding from this work was that freezing of supercooled ternary $H_2SO_4/HNO_3/H_2O$ mixtures may actually start in the small aerosol size ranges when air parcels go through orographically induced cooling events. Under these conditions the smaller particles, originally mostly consisting of a mixture of H_2SO_4/H_2O, will most rapidly be diluted with HNO_3 and H_2O and attain a chemical composition resembling that of a NAT aerosol, which, according to laboratory investigations, can readily freeze (82, 83).

TROPICAL BIOMASS BURNING

By the end of the 1970's considerable attention was given to the possibility of a large net source of atmospheric CO_2 due to tropical deforestation (84). Biomass burning is, however, not only a source of CO_2, but also of a great number of photochemically and radiatively active trace gases, such as NO_x, CO, CH_4, reactive hydrocarbons, H_2, N_2O, OCS, CH_3Cl, etc. Furthermore, tropical biomass burning is not restricted to forest conversion, but is also a common activity related to agriculture, involving the burning of savanna grasses, wood and agricultural wastes. In the summer of 1978, on our way back to Boulder from measurements of the emissions of OCS and N_2O from feedlots in Northeastern Colorado, we saw a big forest fire high up in the Rocky Mountain National Forest, which provided us with the opportunity to collect air samples from a major forest fire plume. After chemical analysis in the NCAR Laboratories by Leroy Heidt, Walt Pollock and Rich Lueb the emission ratios of the above gases relative to CO_2 could be established. Multiplying these ratios with estimates of the global extent of CO_2 production by biomass burning, estimated to be of the order of $2–4 \times 10^{15}$ g C/year (85), we next derived the first estimates of the global emissions of H_2, CH_4, CO, N_2O, NO_x, OCS and CH_3Cl, and could show that the emissions of these gases could constitute a significant fraction of their total global emissions. These first measurements stimulated considerable international research efforts. Except for N_2O (for which our first measurements have since proved incorrect) our original findings were largely confirmed, although large uncertainties in the quantification of the various human activities contributing to biomass burning and individual trace gas releases remain (86). Because biomass burning releases substantial quantities of reactive trace gases, such as hydrocarbons, CO, and NO_x, in photochemically very active environments, large quantities of ozone were expected to be formed in the tropics and subtropics during the dry season. Several measurement campaigns in South America and Africa, starting in 1979 and 1980 with NCAR's Quemadas expedition in Brazil, have confirmed this expectation (87–92). The effects of biomass burning are especially noticeable in the industrially

lightly polluted southern hemisphere, as is clearly shown from satellite observations of the tropospheric column amounts of CO and O_3 in figures 11 and
12 (93, 94).

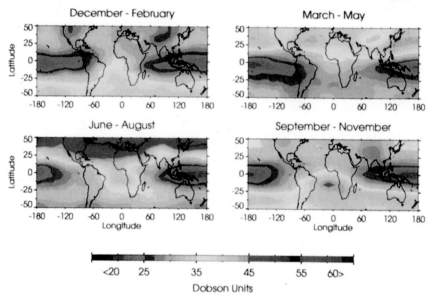

SEASONAL DEPICTIONS OF TROPOSPHERIC OZONE DISTRIBUTION

Panels below depict global climatologies of tropospheric ozone (smog) developed at NASA
Langley.   Note high summertime values in the Northern Hemisphere and enhancements over
South Atlantic Ocean due to widespread biomass burning in Africa September - November.

Figure 11: Observed distributions of vertical column ozone in the troposphere for 4 periods from Fishman
et al. (91, 93). 1 Dobson unit represents a vertical column of $2.62\,10^{16}$ molecules cm^{-2}.

"NUCLEAR WINTER"

My research interests both in the effects of NO_x on stratospheric ozone and
in biomass burning explain my involvement in the "nuclear winter" studies.
When in 1981 I was asked by the editor of Ambio to contribute to a special
issue on the environmental consequences of a major nuclear war, an issue co-
edited by Dr. Joseph Rotblat, this year's Nobel Peace Prize awardee, the initial thought was that I would make an update on predictions of the destruction of ozone by the NOx that would be produced and carried up by the fireballs into the stratosphere (95, 96). Prof. John Birks of the University of
Colorado, Boulder, one of the co-authors of the Johnston study on this topic
(96), who spent a sabbatical in my research division in Mainz, joined me in
this study. Although the ozone depletion effects were significant, it was also
clear to us that these effects could not compete with the direct impacts of the
nuclear explosions. However, we then came to think about the potential climatic effects of the large amounts of sooty smoke from fires in the forests
and in urban and industrial centers and oil storage facilities, which would
reach the middle and higher troposphere. Our conclusion was that the

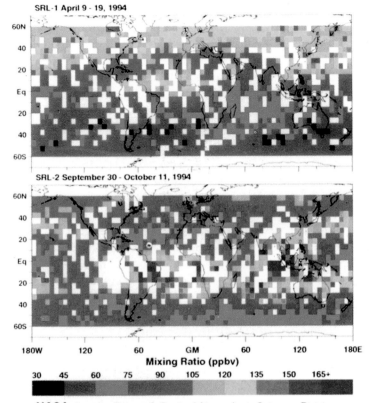

Measurement of Air Pollution from Satellites

Carbon Monoxide Mixing Ratios in Middle Troposphere
during April and October 1994

NASA Langley Research Center / Atmospheric Sciences Division

Figure 12: Observed distributions of vertical column CO in the troposphere for 4 periods, measured on the space shuttle during April and October, 1994. Courtesy of Drs. Vicki Connors, Hank Reichle and the MAPS team. Reference should be made to Connors et al. (94). (1 ppbv is the same as 1 nmole/mole).

absorption of sunlight by the black smoke could lead to darkness and strong cooling at the earth's surface, and a heating of the atmosphere at higher elevations, thus creating atypical meteorological and climatic conditions which would jeopardize agricultural production for a large part of the human population (97). This idea was picked up by others, especially the so-called TTAPS (Turco, Toon, Ackerman, Pollack, Sagan) group (98) who even predicted that subfreezing temperatures could be possible over much of the earth. This was supported by detailed climate modeling (99). A major international study of the issue which was conducted by a group of scientists working under the auspices of SCOPE (Scientific Committee on Problems of the Environment) of ICSU (International Council of Scientific Unions) also supported the initial hypothesis, concluding that far more people could die by the climatic and environmental consequences of a nuclear war than directly by the explosions (100, 101).

Although I do not count the "nuclear winter" idea among my greatest scientific achievements (in fact, the hypothesis can not be tested without performing the "experiment", which it wants to prevent), I am convinced that, from a political point of view, it is by far the most important, because it magnifies and highlights the dangers of a nuclear war and convinces me that in the long run mankind can only escape such horrific consequences if nuclear weapons are totally abolished by international agreement. I thus wholeheartedly agree in this respect with Joseph Rotblat and the Pugwash organization, this year's recipients of the Nobel Prize for Peace.

CURRENT RESEARCH INTERESTS

Realizing the great importance of heterogeneous reactions in stratospheric chemistry, together with my Dutch students Jos Lelieveld (now professor at the University of Utrecht) and Frank Dentener, I have been involved in studies on the effects of reactions taking place in cloud droplets and tropospheric aerosol particles. In general, such reactions result in removal of NO_x and lower concentrations of O_3 and OH (102, 103). Furthermore, even at high enough NO_x concentrations to allow ozone formation by reactions R16 + R17 + R2, such reactions would be much limited within clouds due to the fact that the NO_x molecules, which are only slightly water soluble, stay in the gas phase, while the HO_2 radicals readily dissolve in the cloud droplets, where they can destroy ozone via the reactions,

R39: HO_2 (gas) $\rightarrow$ HO_2 (aqueous) $\rightleftarrows$ $H^+ + O_2^-$

R40: $O_2^- + O_3 (+ H^+)$ $\rightarrow$ $OH + 2 O_2$

The role of rapid transport of reactive compounds from the planetary boundary layer into the upper troposphere is another topic with which I have been involved with some of my students over the past decade. This may have important effects on the chemistry of the upper troposphere (104, 105). My great interest in the role of clouds in atmospheric chemistry has brought me in close contact with a major research group at the University of California, San Diego, headed by my good friend Prof. V. Ramanathan.

A new project in which I am currently much interested is the possibility of Cl and especially Br activation in the marine boundary layer. It is already known that Br activation can explain the near-zero O_3 concentrations which are often found in the high latitude marine boundary layer during springtime (106). In our most recent papers we discuss the possibility that Br activation may also occur in other marine regions and seasons (107, 108).

The ideas outlined above will be tested by field programs and, if confirmed, introduced in advanced photochemical-transport models. The field programs will be mostly carried out by members of my research division at the Max-Planck-Institute for Chemistry, often in collaboration with other experimental groups. The modeling work is conducted within a consortium of researchers from Sweden, The Netherlands, France, Italy and Germany. This effort is funded by the European Union and coordinated by Professors Lennart Bengtsson, Hamburg, Henning Rodhe, Stockholm and Jos Lelieveld, Utrecht.

A LOOK AHEAD

Despite the fundamental progress that has been made over the past decades, much research will be needed to fill major gaps in our knowledge of atmospheric chemistry. In closing I will try to indicate some of those research areas which I consider to be of greatest interest (109).

Tropospheric Ozone Observations: Despite the great importance of tropospheric ozone in atmospheric chemistry, there are still major uncertainties concerning its budget and global concentration distribution. Everywhere, but especially in the tropics and the subtropics, there is a severe lack of data on tropospheric ozone concentrations. Considering the enormous role of tropical ozone in the oxidation efficiency of the atmosphere, the already recognized large anthropogenic impact on ozone through biomass burning, and the expected major agricultural and industrial expansion of human activities in this part of the world, this knowledge gap is very serious. At this stage it is not possible to test photochemical transport models owing to severe scarcity of ozone observations, especially in the tropics and subtropics. Of critical importance in the effort to obtain data from the tropics and subtropics will be the *training and long term active participation of scientists from the developing countries.* Besides the ozone measurements at a number of stations and during intensive measurement campaigns, it will be important to also obtain data on reactive hydrocarbons, CO, NO_x, NX and on chemical constituents in precipitation. Unfortunately, it has been frustrating to note how little response there has been from potential funding agencies to support efforts in this direction.

Long-term observations of atmospheric properties: Two major findings have demonstrated the extreme value of long term observations of important atmospheric chemical properties. One example was the discovery of the rapid depletion of stratospheric ozone over Antarctica during the spring months, as discussed before. Another is the recent, unexpected major, temporary break in the trends of CH_4 and CO. Most surprising were the changes in CO, for which Khalil and Rasmussen (110) derived a downward trend in surface concentrations by $1.4\pm0.9\%$/yr in the northern hemisphere and by as much as $5.2\pm0.7\%$/yr in the southern hemisphere between 1987 and 1992. Even larger downward trends, $6.1\pm1\%$/yr in the northern hemisphere and $7\pm0.6\%$/yr in the southern hemisphere, were reported for the period between June, 1990 and June, 1993 by Novelli et al. (111). Although these trends have again reversed (P. Novelli, private communication) into the previous upward trend of + 0.7% per year for CO and almost 1% per year for CH_4 (112, 113), the temporal break is remarkable. The reasons for this surprising behaviour are not known. They may consist of a combination of: (i) variable annual emissions from biomass burning, (ii) higher concentrations of OH radicals, maybe due to loss of stratospheric ozone, triggered by an increase in reactive aerosol surfaces in the stratosphere following the Pinatubo volcanic eruption in June, 1991, (iii) a dynamically forced global redistribution of CO, introducing a bias due to the location of the limited number of measuring sites, (iv) reduced CO formation from the oxidation of natural hydrocarbons emitted by tropical forests due to globally altered precipitation and temperature patterns, or, most likely, a combination of these plus other, yet unknown factors. At this stage we can only conclude that the causes for the surprisingly rapid CO trend changes are not known, the main reason being incomplete global coverage of the CO measurement network. The same applies for CH_4.

Intensive measurement campaigns: Comprehensive field programmes that have been conducted in the past with detailed observations of all factors that influence the photochemistry of the troposphere will also be much needed in the future, especially in various regions of the marine and continental tropics and subtropics, in order to find out whether we understand the major processes that determine the chemistry of ozone and related photochemically active compounds. Applications of comprehensive chemical-transport models should be an important part of these activities. Topics in which greatly improved knowledge is necessary, are especially: improved quantification of the stratospheric influx of ozone; distributions, sources and sinks of CH_4, reactive hydrocarbons, CO, NO_x and NX; quantification of natural NO emissions from lightning and soils.

Cloud transport: The role of clouds as transporters of boundary layer chemical constituents, such as CO, NO_x, reactive hydrocarbons and their oxidation products to the middle and upper troposphere (and possibly into the lower stratosphere) should be better understood and quantified, so that they can be parameterized for inclusion in large scale photochemical models of the atmosphere. Similarly the production of NO by lightning and its vertical redistribution by convective storms should also be much better quantified, both for marine and continental conditions. Current uncertainties of NO production by lightning are at least a factor of four.

Chemical interactions with hydrometeors: The interactions of chemical constituents emanating from the boundary layer with liquid and solid hydrometeors in the clouds will be of special importance. There is for instance the question of why strong ozone formation has not been noticed around the most convective regions in the continental tropics in which large amounts of forest-derived reactive hydrocarbons, such as isoprene (C_5H_8), and their oxidation products are rapidly lifted to the middle and upper troposphere and mixed with lightning-produced NO to provide favourable conditions for photochemical ozone formation. Could it be that the expected ozone formation is prevented by chemical interactions of the hydrocarbon reaction products and NO_x with the hydrometeors? Could significant ozone destruction take place in cloud water and/or on the surface of ice particles which may be partially covered by water (114, 115)? Such questions regarding potential loss of ozone by reactions with hydrometeors may be especially relevant in connection with observations of record low O_3 volume mixing ratios often of less than 10 ppbv in the upper kilometers of the troposphere in March 1993 in an extended, heavily convective region between Fiji and Christmas Island over the Pacific Ocean (116). Although such low ozone volume mixing ratios had been noted on several occasions in the tropical marine boundary layer and can be explained by the ozone-destroying reactions R6–R8 in the lower troposphere, it should be ascertained whether these reactions alone suffice to explain the extremely low ozone concentrations in such a large volume of air.

Photolysis rates in cloudy atmospheres: Regarding the photochemistry taking place in cloudy atmospheric conditions, recent observations of unexpectedly high absorption of solar radiation in cloudy atmospheres (117), point to the possibility that multiple scattering in broken cloud systems may lead to strongly enhanced photolysis rates and photochemical activity, leading e.g. to much higher O_3 destruction and OH production rates by reactions R7 and R8, or ozone production by reactions R16 + R17 + R2, than thought so far. The influence of clouds on the photochemically active UV radiation field is a potentially very important research topic which should be pursued by measurements and the development of appropriate radiative transfer models.

Biogenic sources of hydrocarbons, CO and NO: The continental biosphere is a large source of hydrocarbons. Quantification of these sources in terms of geophysical (e.g. temperature, humidity, light levels) and biogeochemical (soil physical and chemical properties, land use) parameters are urgently needed for inclusion in atmospheric models. The hydrocarbon oxidation mechanisms in the atmosphere should also be better understood, so that formation of ozone, carbon monoxide, partially oxidized gaseous hydrocarbons, and organic aerosol can be better quantified. The formation of organic aerosol from hydrocarbon precursors and their capability to serve as cloud condensation nuclei are related, potentially important, subjects which have not been studied in any depth so far.

Potential role of halogen radicals in ozone destruction: There are strong observational indications that tropospheric ozone can be destroyed by reactions in addition to those discussed so far. Surface ozone observations during polar sunrise in the Arctic have frequently shown the occurrence of unmeasurably low ozone concentrations, coinciding with high "filterable Br" (106). Further measurements (118) identified BrO as one of the active Br compounds, which, as is well known from stratospheric measurements, may rapidly attack ozone by a series of catalytic reactions, such as

$$2x \ (Br + O_3 \rightarrow BrO + O_2) + (BrO + BrO \rightarrow 2Br + O_2) = (2O_3 \rightarrow 3O_2)$$

or

$$(Br + O_3 \rightarrow BrO + O_2) + (BrO + HO_2 \rightarrow HOBr + O_2) + (HOBr + h\nu \rightarrow$$
$$OH + Br) + (OH + CO + O_2 \rightarrow HO_2 + CO_2) = (CO + O_3 \rightarrow CO_2 + O_2).$$

It should be explored whether halogen activation reactions may also occur under different circumstances than indicated above (106–108).

Heterogeneous reactions on aerosol particles: The issue of interactions between gases and atmospheric aerosol is largely unexplored and very little considered in tropospheric chemistry models. Examples are interactions of dimethyl sulfide-derived sulphur compounds with seasalt in the marine boundary layer and reactions of SO_2, H_2SO_4, NO_x, N_2O_5 and HNO_3 on soil dust particles which remove these compounds from the gas phase. In the case of industrial SO_2, the neglect of such heterogeneous reactions may well have led to overestimations of the climatic cooling effects of anthropogenic aerosol, as any incorporation of sulphur in soil dust or sea salt will prevent the nucleation of new sunlight backscattering sulfate particles.

Ozone/climate feedbacks in the stratosphere: Ozone is a significant greenhouse gas with an infrared absorption band in the atmospheric window region, centered at 9.6 lm. Although the tropospheric ozone amount is only about 10% of that of the stratosphere, the effective longwave optical depth of tropospheric ozone is larger. Of greatest importance would be any changes that might take place in the ozone concentrations in the tropopause regions as a result of human activities, such as H_2O, NO, SO_2 and particulate emissions from expanding fleets of civil aircraft flying in the stratosphere and upper troposphere. On the one hand this may lead to increasing temperatures in the lower stratosphere. However, increased HNO_3 and H_2O concentrations in the lower stratosphere may increase the likelihood of polar stratospheric particle formation and ozone destruction. Such a course of events is also promoted by cooling of the stratosphere by increasing concentrations of CO_2. (This cooling effect increases with height in the stratosphere and mesosphere. The implications of this for the future dynamics of the stratosphere, mesosphere and lower thermosphere is likewise a topic, deserving attention). Changes in chemical and radiative conditions in the lower stratosphere may, therefore, create feedbacks which we need to understand well, including understanding their potential impact on tropopause heights and temperatures, stratospheric water vapor, lower stratospheric cloud characteristics and the tropospheric hydrological cycle. Recent observations of increasing trends of water vapour concentrations in the lower stratosphere over Boulder emphasize this point (119). All these factors should be taken into account before decisions are taken on vast expansions of aircraft operations in the stratosphere.

ACKNOWLEDGEMENTS

Firstly, I have to thank my parents, my wife and my family for their love and support, and for creating the personal environment without which nothing will go.

I thank my secretaries Anja Wienhold and Bettina Krüger for their joyful attitude and hard work in sometimes chaotic times, especially in connection with the "Nobelrummel". Many thanks to Geoff Harris, Mark Lawrence and Jens-Uwe Grooß for a final reading of the manuscript.

I thank my current and former students, post-docs and coworkers at the Max-Planck-Institute for Chemistry for their enthusiastic research efforts. Several of them are now professors at universities or directors of major research activities. With most of them I keep a close contact and we are friends for life.

I also thank the Max-Planck-Society and the various organizations with which I am or have been associated during my scientific career. All of them have been very important in trusting me with long-term funding and giving me excellent opportunities to do research without major interferences. I am particularly happy to be a member of the NSF Center on the role of Clouds in atmospheric Chemistry and Climate at Scripps Institution of Oceanography of the University of California, San Diego, I am learning a lot about clouds, one of the most important elements in the climate system. I thank my good friend, Prof. V. Ramanathan, and SIO for this great opportunity to rejuvenate myself during a few months of the year.

I thank my director colleagues and personel of the Max-Planck-Institute for Chemistry, the University of Mainz, the City of Mainz, and the "Sonderforschungsbereich" for a wonderful welcome and party after my return to Mainz as a fresh Nobel Laureate following a brief vacation in Spain. I will also never forget my "homecoming" at MISU, Stockholm and the welcome (with illegal fireworks) by my Dutch students and "grandstudents" in Wageningen, The Netherlands.

Finally, I have to thank the many colleagues, all around the world, who have congratulated me on the Nobel Prize award. Many of them have themselves contributed greatly to the remarkable progress in our research field over the past quarter of a century; only a few of these I could recognize in this Nobel Lecture. As most of them have written, this is an award to the entire atmospheric chemistry and environmental field. I totally agree and thank you all.

And last, but not least, a great Thank You to the Nobel Committee of the Royal Swedish Academy of Sciences. Your decision is an enormous boost for environmental research.

REFERENCES

1. Sweden's Case Study for the United Nations Conference on the Human Environment 1972: Air Pollution Across National Boundaries. The Impact on the Environment of Sulfur in Air and Precipitation, Stockholm, 1972.

2. Chapman, S., A theory of upper atmospheric ozone, Mem. Roy. Soc., 3, 103–125, 1930.

3. Benson, S.W. and A.E. Axworthy, Reconsiderations of the rate constants from the thermal decomposition of ozone, J. Chem. Phys., 42, 2614, 1965.

4. Bates, D.R. and M. Nicolet, The photochemistry of atmospheric water vapour, J. Geophys. Res., 55, 301, 1950.

5. McGrath, W.D. and R.G.W. Norrish, Studies of the reaction of excited oxygen atoms and molecules produced in the flash photolysis of ozone, Proc. Roy. Soc, A 254, 317, 1960.

6. Norrish, R.G.W. and R.P. Wayne. The photolysis of ozone by ultraviolet radiation. "The photolysis of ozone mixed with certain hydrogen-containing substances. Proc. Roy. Soc. London, A 288, 361, 1965.

7. Hampson, J., Chemiluminescent emission observed in the stratosphere and mesosphere, in "Les problèmes météorologiques de la stratosphère et de la mésosphère", p. 393. Presses universitaires de France, Paris, 1965.

8. Hunt, B.G., Photochemistry of ozone in a moist atmosphere, J. Geophys. Res., 71, 1385, 1966.

9. Crutzen, P.J., Determination of parameters appearing in the "dry" and "wet" photochemical theories for ozone in the stratosphere, Tellus, 21, 368–388, 1969.

10. Murcray, D.G., T.G. Kyle, F.H. Murcray and W.J. Williams, Nitric acid and nitric oxide in the lower stratosphere, Nature, 218, 78, 1968.

11. Rhine, P.E., L.D. Tubbs and D. Williams, Nitric acid vapor above 19 km in the Earth's atmosphere, Appl. Optics, 8, 1501, 1969.

12. Crutzen, P.J., The influence of nitrogen oxides on the atmospheric ozone content, Q.J.R. Meteorol. Soc., 96, 320–325, 1970.

13. Bates, D.R. and P.B. Hays, Atmospheric nitrous oxide, Planet. Space Sci., 15, 189, 1967.

14. Greenberg, R.I. and J. Heicklen. Reaction of O(1D) with N2O, Int. J. Chem. Kin., 2, 185, 1970.

15. Crutzen, P.J., Ozone production rates in an oxygen-hydrogen-nitrogen oxide atmosphere, J. Geophys. Res., 76, 7311, 1971.

16. McElroy, M.B. and J.C. McConnell. Nitrous oxide. A natural source of stratospheric NO, J. Atmos. Sci., 28, 1085, 1971.

17. Davis, D.D. et al., Recent kinetic measurements on the reactions of O(3p), H and HO2, Second Conference on CIAP, p. 126, DOT-TSC-OST-73-4, 1973.

18. Yoshino, K. et al., Improved absorption cross-sections of oxygen in the wavelength region 205–240 nm of the Herzberg Continuum, Planet. Space Sci., 36, 1469, 1988.

19. Frederick, J.E. and J.E. Mentall, Solar irradiance in the stratosphere: Implications for the Herzberg Continuum Absorption of O2, Geophys. Res. Lett., 9, 461, 1982.

20. Nicolet, M., The solar spectral irridiance and its action in the atmospheric photodissociation processes, Planet. Space Sci., 29, 951, 1981.

21. Weiss, R.F., The temporal and spatial distribution of tropospheric nitrous oxide, J. Geophys. Res., 86, 7185, 1981.

22. SCEP (Study on Critical Environmental Problems). Man's Impact on the Global Environment. Assessment and Recommendations for Action, The MIT Press, Cambridge, MA and London, England, 1970.

23. Nicolet, M., Nitrogen oxides in the chemosphere, J. Geophys. Res., 70, 679, 1965.

24. Harrison, H.S., Stratospheric ozone with added water vapour: influence of high altitude aircraft, Science, 170, 734, 1970.

25. Johnston, H.S. and H.J. Crosby, Kinetics of the fast gas phase reaction between ozone and nitric oxide, J. Chem. Phys., 22, 689, 1954,

26. Johnston, H.S. and D. Garvin, Working Papers for a Survey of Rate Data for Chemical Reactions in the Stratosphere, National Bureau of Standards, Report 10931, 1972.

27. Johnston, H., Reduction of stratospheric ozone by nitrogen oxide catalysts from supersonic transport exhaust, J. Geophys. Res., 173, 517, 1971.

28. Johnston, H.S., Atmospheric Ozone, Annu. Rev. Phys. Chem., 43, 1, 1992.

29. Johnston, H.S. and R.A. Graham, Photochemistry of NOx and HNOx compounds, Can. J. Chem., 52, 1415, 1974.

30. CIAP (Climate Impact Assessment Program). Report of Findings: The Effects of Stratospheric Pollution by Aircraft, DOT-TSC-75-50, U.S. Department of Transportation, Washington, DC, 1974.

31. COMESA. The Report of the Committee on Meteorological Effects of Stratospheric Aircraft, U.K. Meteorological Office, Bracknell, England, 1975.

32. COVOS, Comité d'Etudes sur les Conséquences des Vols Stratosphériques, Société Métérologique de France, Boulogne, France, 1976.

33. NAS (National Academy of Sciences). Environmental Impact of Stratospheric Flight, ISBN 0-309-02346-7, Washington, DC, 1975.

34. Crutzen, P.J., I.S.A. Isaksen and G.C. Reid, Solar proton events: Stratospheric sources of nitric oxide, Science, 189, 457, 1975.

35. Heath, D.F., A.J. Krueger and P.J. Crutzen, Solar proton event: Influence on stratospheric ozone, Science, 197, 886, 1977.

36. Solomon, S. and P.J. Crutzen, Analysis of the August 1972 solar proton event, including chlorine chemistry, J. Geophys. Res., 86, 1140, 1981.

37. Levy, H., III, Normal atmosphere: Large radical and formaldehyde concentrations predicted, Science, 173, 141, 1971.

38. McConnell, J.C., M.B. McElroy and S.C. Wofsy, Natural sources of atmospheric CO, Nature, 233, 187, 1971.

39. Prinn, R.G. et al., Atmospheric trends and lifetime of trichloroethane and global average hydroxyl radical concentrations based on 1978–1994 ALE/GAGE measurements, Science, 269, 187, 1995.

40. Levy, H., III, Photochemistry of the lower troposphere, Planet. Space Sci., 20, 919, 1972.

41. Crutzen, P.J., A discussion of the chemistry of some minor constituents in the stratosphere and troposphere, Pure Appl. Geophys., 106–108, 1385, 1973.

42. Crutzen, P.J., Photochemical reactions initiated by and influencing ozone in unpolluted tropospheric air, Tellus, 26, 47, 1974.

43. Chameides, W.L. and J.C.G. Walker, A photochemical theory of tropospheric ozone, J. Geophys. Res., 78, 8751, 1973.

44. Fishman, J. and P.J. Crutzen, The origin of ozone in the troposphere, Nature, 274, 855, 1978.

45. Fishman, J., S. Solomon and P.J. Crutzen, Observational and theoretical evidence in support of a significant in-situ photochemical source of tropospheric ozone, Tellus, 31, 432, 1979.

46. Howard, C.J. and K.M. Evenson, Kinetics of the reaction of HO_2 radicals with NO, Geophys. Res. Lett., 4, 437, 1977.

47. Crutzen, P.J. and C.J. Howard, The effect of the HO_2 + NO reaction rate constant on one-dimensional model calculations of stratospheric ozone depletions, Pure and Appl. Geophys., 116, 497, 1978.

48. Crutzen, P.J. and P.H. Zimmermann, The changing photochemistry of the troposphere, Tellus, 43 A/B, 136, 1991.

49. Intergovernmental Panel on Climate Change, Climate Change: The IPCC Scientific Assessment (J.T. Houghton et al., Eds.), Cambridge University Press, 365 pp, 1990.

50. Stolarski, R.S. and R.J. Cicerone, Stratospheric chlorine: A possible sink for ozone, Can. J. Chem., 52, 1610, 1974.

51. Wofsy, S.C. and M.B. McElroy, HO_x, NO_x and ClO_x: Their role in atmospheric photochemistry, Can. J. Chem., 52, 1582, 1974.

52. Crutzen, P.J., A review of upper atmospheric photochemistry, Can. J. Chem., 52, 1569, 1974.

53. Lovelock, J.E., R.J. Maggs and R.J. Wade, Halogenated hydrocarbons in and over the Atlantic, Nature, 241, 194, 1973.

54. Molina, M.J. and F.S. Rowland, Stratospheric sink of chlorofluoromethanes: Chlorine atom-catalyzed destruction of ozone, Nature, 249, 810, 1974.
55. Crutzen, P.J., Estimates of possible future ozone reductions from continued use of fluoro-chloro-methanes (CF_2Cl_2, $CFCl_3$), Geophys. Res. Lett., 1, 205, 1974.
56. Cicerone, R.J., R.S. Stolarski and S. Walters, Stratospheric ozone destruction by man-made chlorofluoromethanes, Science, 185, 1165, 1974.
57. Farman, J.C., B.G. Gardiner and J.D. Shanklin, Large losses of total ozone in Antarctica reveal seasonal ClO_x/NO_x interaction, Nature, 315, 201, 1985.
58. Solomon, S., R.R. Garcia, F.S. Rowland, and D.J. Wuebbles, On the depletion of Antarctic ozone, Nature, 321, 755, 1986.
59. Junge, C.E., C.W. Chagnon and J.E. Manson, Stratospheric aerosols, J. Meteorol., 18, 81, 1961.
60. Davis, D.D., A.R. Ravishankara and S. Fischer, SO_2 oxidation via the hydroxyl radical: atmospheric fate of the HSO_x radicals, Geophys. Res. Lett, 6, 113, 1979.
61. Stockwell, W.R. and J.G. Calvert, The mechanism of the $HO-SO_2$ reaction, Atmos. Environ., 17, 2231, 1983.
62. Crutzen, P.J., The possible importance of CSO for the sulfate layer of the stratosphere, Geophys. Res. Lett., 3, 73, 1976.
63. Cadle, R.D., P.J. Crutzen and D.H. Ehhalt, Heterogeneous chemical reactions in the stratosphere, J. Geophys. Res., 80, 3381, 1975.
64. Mozurkewich, M. and J.G. Calvert, Reaction probabilities of N_2O_5 on aqueous aerosols, J. Geophys. Res., 93, 15889, 1988.
65. Hanson, D.R. and A.R. Ravishankara, The reaction probabilities of $ClONO_2$ and N_2O_5 on 40 to 75 percent sulfuric acid solutions, J. Geophys. Res., 96, 17307, 1991.
66. Van Doren, J.M., L.R. Watson, P. Davidovits, D.R. Worsnop, M.S. Zahniser and C.E. Kolb, Temperature dependence of the uptake coefficients of HNO_3, HCl, and N_2O_5 by water droplets, J. Phys. Chem., 94, 3265, 1990.
67. Platt, U., D. Perner, A.M. Winer, G.W. Harris and J.N. Pitts, Jr., Detection of NO3 in the polluted troposphere by differential optical absorption, Geophys. Res. Lett, 7, 89, 1980.
68. Pommereau, J.F. and F. Goutail, Stratospheric O_3 and NO_2 observations at the southern polar circle in summer and fall 1988, Geophys. Res. Lett., 15, 895, 1988.
69. Crutzen, P.J. and F. Arnold. Nitric acid cloud formation in the cold Antarctic stratosphere: a major cause for the springtime "ozone hole", Nature, 324, 651, 1986.
70. Toon, O.B., P. Hamill, R.P. Turco and J. Pinto. Condensation of HNO_3 and HCl in the winter polar stratosphere, Geophys. Res. Lett., 13, 1284, 1986.
71. Hanson, D.R. and K. Mauersberger. Vapor pressures of HNO_3/H_2O solutions at low temperatures, J. Phys. Chem., 92, 6167, 1988.
72. Molina, M.J., T.L. Tso, L.T. Molina and F.C.-Y. Wang. Antarctic stratospheric chemistry of chlorine nitrate, hydrogen chloride and ice, Science, 238, 1253, 1987.
73. Tolbert, M.A., M.J. Rossi, R. Malhotra, and D.M. Golden. Reaction of chlorine nitrate with hydrogen chloride and water at Antarctic stratospheric temperatures, Science, 238, 1258, 1987.
74. Molina, L.T. and M.J. Molina. Production of Cl2O2 from the self-reaction of the ClO radical, J. Phys. Chem., 91, 433, 1987.
75. de Zafra, R.L., M. Jaramillo, A. Parrish, P.M. Solomon, B. Connor, and J. Barrett, High concentration of chlorine monoxide at low altitudes in the Antarctic spring stratosphere, I. Diurnal variation, Nature, 328, 408, 1987.
76. Solomon, S., G.H. Mount, R.W. Sanders, and A.L. Schmeltekopf, Visible spectroscopy at McMurdo Station, Antarctica: Observations of OClO, J. Geophys. Res., 92, 8329, 1987.
77. Anderson, J.G., W.H. Brune, and M.H. Proffitt, Ozone destruction by chlorine radicals within the Antarctic vortex: The spatial and temporal evolution of $ClO-O_3$ anticorrelation based on in situ ER-2 data, J. Geophys. Res., 94, 11465, 1989.
78. Wofsy, S.C., M.B. McElroy and Y.L. Yung, The chemistry of atmospheric bromine, Geophys. Res. Lett., 2, 215, 1975.
79. Dye, J.E., D. Baumgardner, B.W. Gandrud, S.R. Kawa, K.K. Kelly, M. Loewenstein,

G.V. Ferry, K.R. Chan and B.L. Gary, Particle size distribution in Arctic polar stratospheric clouds, growth and freezing of sulfuric acid droplets, and implications for cloud formation, J. Geophys. Res., 97, 8015, 1992.

80. Cox, R.A., A.R. MacKenzie, R. Müller, Th. Peter and P.J. Crutzen, Activation of stratospheric chlorine by reactions in liquid sulphuric acid, Geophys. Res. Lett., 21, 1439, 1994.

81. Hanson, D.R., A.R. Ravishankara and S. Solomon, Heterogeneous reactions in sulfuric acid aerosol: A framework for model calculations, J. Geophys. Res., 99. 3615. 1994.

82. Meilinger, S.K.. T. Koop, B.P. Luo, T. Huthwelker, K.S. Carslaw, P.J. Crutzen and T. Peter, Size-dependent stratospheric droplet composition in lee wave temperature fluctuations and their potential role in PSC freezing, Geophys. Res. Lett, 22, 3031, 1995.

83. Koop, T., B.P. Luo, U.M. Biermann, P.J. Crutzen and T. Peter, Freezing of $HNO_3/H_2SO_4/H_2O$ solutions at stratospheric temperatures: Nucleation statistics and Experiments, J. Phys. Chem. (to be submitted).

84. Woodwell, G.M., R.H. Whittaker, W.A. Reiners, G.E. Likens, C.C. Delwiche and D.B. Botkin, The biota and the world carbon budget, Science, 199, 141, 1978.

85. Seiler, W. and P.J. Crutzen, Estimates of gross and net fluxes of carbon between the biosphere and the atmosphere from biomass burning, Climatic Change, 2, 207, 2980.

86. Crutzen, P.J. and M.O. Andreae, Biomass burning in the Tropics: Impact on atmospheric chemistry and biogeochemical cycles, Science, 250, 1669, 1990.

87. Crutzen, P.J., A.C. Delany, J. Greenberg, P. Haagenson, L. Heidt, R. Lueb, W. Pollock, W. Seiler, A. Wartburg and P. Zimmerman, Tropospheric chemical composition measurements in Brazil during the dry season, J. Atmos. Chem., 2, 233, 1985.

88. Amazon Boundary Layer Experiment (ABLE 2A): Dry season 1985, Collection of 24 papers, J. Geophys. Res., 93 (D2), 1349–1624, 1988.

89. Andreae, M.O. et al., Biomass burning emissions and associated haze layers over Amazonia, J. Geophys. Res., 93, 1509, 1988.

90. Andreae, M.O. et al., Ozone and Aitken nuclei over equatorial Africa: Airborne observations during DECAFE 88, J. Geophys. Res., 97, 6137, 1992.

91. Fishman, J., K. Fakhruzzaman, B. Cros and D. Nyanga, Identification of widespread pollution in the southern hemisphere deduced from satellite analyses, Science, 252, 1693, 1991.

92. FOS/DECAFE 91 Experiment, Collection of 13 papers in J. Atmos. Chem., 22, 1–239, 1995.

93. Fishman, J., Probing planetary pollution from space, Environ. Sci. Technol., 25, 612, 1991.

94. Connors, V., M. Flood, T. Jones, B. Gormsen, S. Nolt and H. Reichle, Global distribution of biomass burning and carbon monoxide in the middle troposphere during early April and October 1994, in Biomass Burning and Global Change (J. Levine, Ed.), MIT Press, 1996 (in press).

95. Foley, H.M. and M.A. Ruderman, Stratospheric NO production from past nuclear explosions, J. Geophys. Res., 78, 4441, 1973.

96. Johnston, H.S., G. Whitten and J.W. Birks, Effects of nuclear explosions on stratospheric nitric oxide and ozone, J. Geophys. Res., 78, 6107, 1973.

97. Crutzen, P.J. and J. Birks, The atmosphere after a nuclear war: Twilight at noon, Ambio, 12, 114, 1982.

98. Turco, R.P., O.B. Toon, R.P. Ackerman, H.B. Pollack and C. Sagan, Nuclear winter: Global consequences of multiple nuclear explosion, Science, 222, 1283, 1983.

99. Thompson, S.L., V. Alexandrov, G.L. Stenchikov, S.H. Schneider, C. Covey and R.M. Chervin, Global climatic consequences of nuclear war: Simulations with three-dimensional models, Ambio, 13, 236, 1984.

100. Pittock, A.B., T.P. Ackerman, P.J. Crutzen, M.C. MacCracken, C.S. Shapiro and R.P. Turco, Environmental Consequences of Nuclear War. Volume I: Physical and Atmospheric Effects, SCOPE 28, Wiley, 1986.

101. Harwell, M.A. and T.C. Hutchinson. Environmental Consequences of Nuclear War. Volume II: Ecological and Agricultural Effects, SCOPE 28, Wiley, 1985.

102. Lelieveld, J. and P.J. Crutzen, Influences of cloud photochemical processes on tropospheric ozone, Nature, 343, 227, 1990.
103. Dentener, F.J. and P.J. Crutzen, Reaction of N_2O_5 on tropospheric aerosols: Impact on the global disstributions of NO_x, O_3, and OH, J. Geophys. Res., 98 (D14), 7149, 1993.
104. Chatfield, R. and P.J. Crutzen, Sulfur dioxide in remote oceanic air: Cloud transport of reactive precursors, J. Geophys. Res., 89 (D5), 711, 1984.
105. Lelieveld, J. and P.J. Crutzen, Role of deep convection in the ozone budget of the troposphere, Science, 264, 1759, 1994.
106. Barrie, L.A., J.W. Bottenheim, R.C. Schnell, P.J. Crutzen and R.A. Rasmussen, Ozone destruction and photochemical reactions at polar sunrise in the lower Arctic atmosphere, Nature, 334, 138, 1988.
107. Sander, R. and P.J. Crutzen, Model study indicating halogen activation and ozone destruction in polluted air masses transported to the sea, J. Geophys. Res. *101*, 9121, 1996.
108. Vogt, R., P.J. Crutzen and R. Sander, A new mechanism for bromine and chlorine release from sea salt aerosol in the unpolluted marine boundary layer, Nature (submitted).
109. Crutzen, P.J., Overview of tropospheric chemistry: Developments during the past quarter century and a look ahead, Faraday Discuss. *100*, 1, 1995.
110. Khalil, M.A.K. and R.A. Rasmussen, Global decrease of atmospheric carbon monoxide, Nature, 370, 639, 1993.
111. Novelli, P.C., K.A. Masario, P.P. Tans and P.M. Lang, Recent changes in atmospheric carbon monoxide, Science, 263 1587, 1994.
112. Zander, R., Ph. Demoulin, D.H. Ehhalt, U. Schmidt and C.P. Rinsland, Secular increases in the total vertical abundances of carbon monoxide above central Europe since 1950, J. Geophys. Res., 94, 11021, 1989.
113. Zander, R., Ph. Demoulin, D.H. Ehhalt and U. Schmidt, Secular increases of the vertical abundance of methane derived from IR solar spectra recorded at the Jungfraujoch station, J. Geophys. Res., 94, 11029, 1989.
114. Crutzen, P.J., Global tropospheric chemistry, in Low-Temperature Chemistry of the Atmosphere (G.K. Moortgat et al., Eds), Springer, Berlin, pp 467–498, 1994.
115. Crutzen, P.J., Ozone in the Troposphere, in Composition, Chemistry and Climate of the Atmosphere (H.B. Singh, Ed.), Von Nostrand Reinhold, New York, 349–393, 1995.
116. Kley, D., H.G.J. Smit, H. Vömel, S. Oltmans, H. Grassl, V. Ramanathan and P.J. Crutzen, Extremely low upper tropospheric ozone observations in the convective regions of the Pacific, Science (submitted).
117. Ramanathan, V., B. Subasilar, G.J. Zhang, W. Conant, R.D. Cess, J.T. Kiehl, H. Grassl and L. Shi, Warm pool heat budget and shortwave cloud forcing: a missing physics? Science, 267, 499, 1995.
118. Hausmann, M. and U. Platt, Spectroscopic measurement of bromine oxide and ozone in the high Arctic during Polar Sunrise Experiment 1992, J. Geophys. Res., 99, 25399, 1994.
119. Oltmans, S.J. and D.J. Hofmann, Increase in lower-stratospheric water vapour at a mid-latitude northern hemisphere site from 1981 to 1994, Nature, 374, 146, 1995.
120. Crutzen, P.J., J.U. Grooß, C. Brühl, R. Müller and J.M. Russell III, A reevaluation of the ozone budget with HALOE UARS data: No evidence for the ozone deficit, Science, 268, 705, 1995.

PUBLICATIONS

Books

Crutzen, P.J. and J. Hahn, 1985: *Schwarzer Himmel.* S. Fischer Verlag, 240 pp.

Pittock, A.B., T.P. Ackerman, P.J. Crutzen, M.C. MacCracken, C.S. Shapiro and R.P. Turco, 1986: *Environmental Consequences of Nuclear War, SCOPE 28,* Volume I: Physical and Atmospheric Effects, Wiley, Chichester, 359 pp; 2nd edition 1989.

Crutzen P.J. and M. Müller, 1989: *Das Ende des blauen Planeten?* C.H. Beck Verlag, 271 pp.

Enquete Commission "Preventive Measures to Protect the Earth's Atmosphere", 1989: Interim Report: *Protecting the Earth's atmosphere: an International Challenge.* Ed. Deutscher Bundestag, Referat Öffentlichkeitsarbeit, Bonn, 592 pp.

Enquete Commission "Preventive Measures to Protect the Earth's Atmosphere", 1990: *Protecting the Tropical Forests: A High-Priority International Task.* Ed. Deutscher Bundestag, Referat Öffentlichkeitsarbeit, Bonn, 968 pp.

Enquete Commission "Preventive Measures to Protect the Earth's Atmosphere", 1991: *Protecting the Earth: A Status Report with Recommendations for a new Energy Policy.* Ed. Deutscher Bundestag, Referat Öffentlichkeitsarbeit, Bonn, 2 Volumes.

Crutzen, P.J., J.-C. Gerard and R. Zander (Eds.), 1989: *Our Changing Atmosphere.* Proceedings of the 28th Liège International Astrophysical Colloquium June 26–30, 1989, Université de Liège, Cointe-Ougree, Belgium, 534 pp.

Graedel, T.E. and P.J. Crutzen, 1993: *Atmospheric Change: An Earth System Perspective.* W.H. Freeman, New York, 446 pp.

Crutzen, P.J. and J.G. Goldammer, 1993: *Fire in the Environment: The Ecological, Atmospheric, and Climatic Importance of Vegetation Fires.* Dahlem Konferenz (15–20 March 1992, Berlin), ES13, Wiley, Chichester, 400 pp.

Graedel, T.E. and P.J. Crutzen, 1994: *Chemie der Atmosphäre. Bedeutung für Klima und Umwelt,* Spektrum-Verlag, Heidelberg, 511 pp.

Graedel, T.E. and P.J. Crutzen, 1995: *Atmosphere, Climate, and Change.* W.H. Freeman, New York, 208 pp.

JOURNAL ARTICLES (REFEREED):

1. Blankenship, J.R. and P.J. Crutzen, 1965: A photochemical model for the space-time variations of the oxygen allotropes in the 20 to 100 km layer. *Tellus,* **18,** 160–175.
2. Crutzen, P.J., 1969: Determination of parameters appearing in the "dry" and the "wet" photochemical theories for ozone in the stratosphere. *Tellus,* **21,** 368–388.
3. Crutzen, P.J., 1969: Determination of parameters appearing in the oxygen-hydrogen atmosphere. *Ann. Géophys.,* **25,** 275–279.
4. Crutzen, P.J., 1970: The influence of nitrogen oxides on the atmospheric ozone content. *Quart. J. Roy. Meteor. Soc.,* **96,** 320–325.
5. Crutzen, P.J., 1970: Comments on "Absorption and emission by carbon dioxide in the mesosphere". *Quart. J. Roy. Meteor. Soc.,* **96,** 767–769.
6. Crutzen, P.J., 1971: Energy conversions and mean vertical motions in the high latitude summer mesosphere and lower thermosphere. *In Mesospheric Models and Related Experiments,* G. Fiocco, (ed.), D. Reidel Publ. Co., Dordrecht, Holland, 78–88.
7. Crutzen, P.J., 1971: Calculation of O_2 ($^1\Delta g$) in the atmosphere using new laboratory data. *J. Geophys. Res.,* **76,** 1490–1497.
8. Crutzen, P.J., 1971: Ozone production rates in an oxygen-hydrogen-nitrogen oxide atmosphere. *J. Geophys. Res.,* **76,** 7311–7327.
9. Crutzen, P.J., 1972: SST's–a threat to the earth's ozone shield. *Ambio,* **1,** 41–51.
10. Crutzen, P.J., 1973: A discussion of the chemistry of some minor constituents in the stratosphere and troposphere. *Pure App. Geophys.,* **106–108,** 1385–1399.
11. Crutzen, P.J., 1973: Gas-phase nitrogen and methane chemistry in the atmosphere. In *Physics and Chemistry of Upper Atmospheres,* B.M. McCormac (ed.), Reidel, Dordrecht, Holland, 110–124.
12. Crutzen, P.J., 1974: A review of upper atmospheric photochemistry. *Can. J. Chem.,* **52,** 1569–1581.

13. Crutzen, P.J., 1974: Estimates of possible future ozone reductions from continued use of fluorochloromethanes (CF_2Cl_2, $CFCl_3$). *Geophys. Res. Lett.*, **1**, 205–208.
14. Crutzen, P.J., 1974: Estimates of possible variations in total ozone due to natural causes and human activities. *Ambio*, **3**, 201–210.
15. Crutzen, P.J., 1974: Photochemical reactions initiated by and influencing ozone in unpolluted tropospheric air. *Tellus*, **26**, 48–57.
16. Cadle, R.D., P.J. Crutzen and D.H. Ehhalt, 1975: Heterogeneous chemical reactions in the stratosphere. *J. Geophys. Res.*, **80**, 3381–3385.
17. Crutzen, P.J., I.S.A. Isaksen and G.C. Reid, 1975: Solar proton events: stratospheric sources of nitric oxide. *Science*, **189**, 457–459.
18. Johnston, H.S., D. Garvin, M.L. Corrin, P.J. Crutzen, R.J. Cvetanovic, D.D. Davis, E.S. Domalski, E.E. Ferguson, R.F. Hampson, R.D. Hudson, L.J. Kieffer, H.I. Schiff, R.L. Taylor, D.D. Wagman and R.T. Watson, 1975: Chemistry in the stratosphere, Chapter 5, *CIAP Monograph 1*. The Natural Stratosphere of 1974, DOT-TST-75-51, U.S. Department of Transportation, Climate Impact Assessment Program.
19. Schmeltekopf, A.L., P.D. Goldan, W.R. Henderson, W.J. Harrop, T.L. Thompson, F.C. Fehsenfeld, H.I. Schiff, P.J. Crutzen, I.S.A. Isaksen and E.E. Ferguson, 1975: Measurements of stratospheric $CFCl_3$, CF_2Cl_2, and N_2O. *Geophys Res. Lett.*, **2**, 393–396.
20. Zerefos, C.S. and P.J. Crutzen, 1975: Stratospheric thickness variations over the northern hemisphere and their possible relation to solar activity. *J Geophys. Res.*, **80**, 5041–5043.
21. Crutzen, P.J., 1976: Upper limits on atmospheric ozone reductions following increased application of fixed nitrogen to the soil. *Geophys. Res. Lett.*, **3**. 169–172.
22. Crutzen, P.J., 1976: The possible importance of CSO for the sulfate layer of the stratosphere. *Geophys. Res. Lett.*, **3**, 73–76.
23. Crutzen, P.J. and D.H. Ehhalt, 1977: Effects of nitrogen fertilizers and combustion on the stratospheric ozone layer. Ambio, **6**, 1–3, 112–117.
24. Crutzen, P.J. and G.C. Reid, 1976: Comments on "Biotic extinctions by solar flares". *Nature*, **263**, 259.
25. Fehsenfeld, F.C., P.J. Crutzen, A.L. Schmeltekopf, C.J. Howard, D.L. Albritton, E.E. Ferguson, J.A. Davidson and H.I. Schiff, 1976: Ion chemistry of chlorine compounds in the troposphere and stratosphere. *J. Geophys. Res.*, **81**, 4454–4460.
26. Reid, G.C., I.S.A. Isaksen, T.E. Holzer and P.J. Crutzen, 1976: Influence of ancient solar proton events on the evolution of life. *Nature*, **259**, 177–179.
27. Crutzen, P.J. and J. Fishman, 1977: Average concentrations of OH in the troposphere, and the budgets of CH_4, CO, H_2 and CH_3CCl_3. *Geophys. Res. Lett.*, **4**, 321–324.
28. Fishman, J. and P.J. Crutzen, 1977: A numerical study of tropospheric photochemistry using a one-dimensional model. *J. Geophys. Res.*, **82**, 5897–5906.
29. Heath, D.F., A.J. Krueger and P.J. Crutzen, 1977: Solar proton event: influence on stratospheric ozone. *Science*, **197**, 886–889.
30. Hidalgo, H. and P.J. Crutzen, 1977: The tropospheric and stratospheric composition perturbed by NO_x emissions of high altitude aircraft. *J. Geophys. Res.*, **82**, 5833–5866.
31. Isaksen, I.S.A. and P.J. Crutzen, 1977: Uncertainties in calculated hydroxyl radical densities in the troposphere and stratosphere. *Geophysica Norvegica*, **31**, 4, 1–10.
32. Isaksen, I.S.A., K.H. Midtboe, J. Sunde and P.J. Crutzen, 1977: A simplified method to include molecular scattering and reflection in calculations of photon fluxes and photo-dissociation rates. *Geophysica Norvegica*, **31**, 11–26.
33. Schmeltekopf, A.L., D.L. Albritton, P.J. Crutzen, D. Goldan, W.J. Harrop, W.R. Henderson, J.R. McAfee, M. McFarland, H.I. Schiff, T.L. Thompson, D.J. Hofmann and N.T. Kjome, 1977: Stratospheric nitrous oxide altitude profiles at various latitudes. *J. Atmos. Sci.*, **34**, 729–736.
34. Crutzen, P.J. and C.J. Howard, 1978: The effect of the HO_2 + NO reaction rate constant on one-dimensional model calculations of stratospheric ozone perturbations. *Pure Appl. Geophys.*, **116**, 487–510.
35. Crutzen, P.J., I.S.A. Isaksen and J.R. McAfee, 1978: The impact of the chlorocarbon industry on the ozone layer. *J. Geophys. Res.*, **83**, 345–363.

36. Fishman, J. and P.J. Crutzen, 1978: The origin of ozone in the troposphere. *Nature*, **274**, 855–858.
37. Reid, G.C., J.R. McAfee and P.J. Crutzen, 1978: Effects of intense stratospheric ionization events. *Nature*, **257**, 489–492.
38. Zimmerman, P.R., R.B. Chatfield, J. Fishman, P.J. Crutzen and P.L. Hanst, 1978: Estimates on the production of CO and H_2 from the oxidation of hydrocarbon emissions from vegetation. *Geophs. Res. Lett.*, **5**, 679–682.
39. Crutzen, P.J., 1979: The role of NO and NO_2 in the chemistry of the troposphere and stratosphere. *Ann. Rev. Earth Planet. Sci.*, **7**, 443–472.
40. Crutzen, P.J., 1979: Chlorofluoromethanes: threats to the ozone layer. *Rev. Geophys. Space Phys.*, **17**, 1824–1832.
41. Crutzen, P.J., L.E. Heidt, J.P. Krasnec, W.H. Pollock and W. Seiler, 1979: Biomass burning as a source of atmospheric gases CO, H_2, N_2O, NO, CH_3Cl and COS. *Nature*, **282**, 253–256.
42. Dickerson, R.R., D.H. Stedman, W.L. Chameides, P.J. Crutzen and J. Fishman, 1979: Actinometric measurements and theoretical calculations of $J(O_3)$, the rate of photolysis of ozone to $O(^1D)$. *Geophys. Res. Lett.*, **6**, 833–836.
43. Fishman, J., V. Ramanathan, P.J. Crutzen and S.C. Liu, 1979: Tropospheric ozone and climate. *Nature*, **282**, 818–820.
44. Fishman, J., S. Solomon and P.J. Crutzen, 1979: Observational and theoretical evidence in support of a significant in situ photochemical source of tropospheric ozone. *Tellus*, **31**, 432–446.
45. Berg, W.W., P.J. Crutzen, F.E. Grahek, S.N. Gitlin and W.A. Sedlacek, 1980: First measurements of total chlorine and bromine in the lower stratosphere. *Geophys. Res. Lett.*, **7**, 937–940
46. Crutzen, P.J. and S. Solomon, 1980: Response of mesospheric ozone to particle precipitation. *Planet. Space Sci.*, **28**, 1147–1153.
47. Heidt, L.E., J.P. Krasnec, R.A. Lueb, W.H. Pollock, B.E. Henry and P.J. Crutzen, 1980: Latitudinal distributions of CO and CH_4 over the Pacific. *J. Geophys. Res.*, **85**, 7329–7336.
48. Seiler, W. and P.J. Crutzen, 1980: Estimates of gross and net fluxes of carbon between the biosphere and the atmosphere from biomass burning. *Climatic Change*, **2**, 207–247
49. Thomas, G.E., C.A. Barth, E.R. Hansen, C.W. Hord, G.M. Lawrence, G.H. Mount, G.J. Rottman, D.W. Rusch, A.I. Stewart, R.J. Thomas, J. London, P.L. Bailey, P.J. Crutzen, R.E. Dickinson, J.C. Gille, S.C. Liu, J.F. Noxon and C.B. Farmer, 1980: Scientific Objectives of the Solar Mesosphere Explorer Mission. *Pure Appl. Geophys.*, **118**, 591–615.
50. Rodhe, H., P. Crutzen and A. Vanderpol, 1981: Formation of sulfuric acid in the atmosphere during long range transport. *Tellus*, **33**, 132–141.
51. Rusch, D.W., J.C. Gérard, S. Solomon, P.J. Crutzen and G.C. Reis, 1981: The effects of particle precipitation events on the neutral and ion chemistry of the middle atmosphere–I. Odd Nitrogen. *Planet. Space Sci.*, **29**, 767–774.
52. Solomon, S. and P.J. Crutzen, 1981: Analysis of the August 1972 solar proton event including chlorine chemistry. *J. Geophys. Res.*, **86**, 1140–1146.
53. Solomon, S., D.W. Rusch, J.C. Gérard, G.C. Reid and P.J. Crutzen, 1981: The effect of particle precipitation events on the neutral and ion chemistry of the middle atmosphere–II. Odd Hydrogen. *Planet. Space Sci.*, **29**, 885–892.
54. Baulch, D.L., R.A. Cox, P.J. Crutzen, R.F. Hampson Jr., J.A. Kerr and J. Troe, 1982: Evaluated kinetic and photochemical data for atmospheric chemistry: Supplement I. *J. Phys. Chem. Ref. Data*, **11**, 327–496.
55. Crutzen, P.J., 1982: The global distribution of hydroxyl. In: *Atmospheric Chemistry*, ed. E.D. Goldberg, pp. 313–328. Dahlem Konferenzen 1982. Berlin, Heidelberg, New York: Springer-Verlag.
56. Crutzen, P.J. and J.W. Birks, 1982: The atmosphere after a nuclear war: Twilight at Noon. *Ambio*, **2&3**, 114–125.
57. Hahn, J. and P.J. Crutzen, 1982: The role of fixed nitrogen in atmospheric photochemistry. *Phil. Trans. R. Soc. Lond.*, **B 296**, 521–541.

58. Solomon, S., P.J. Crutzen and R.G. Roble, 1982: Photochemical coupling between the thermosphere and the lower atmosphere. 1. Odd nitrogen from 50–120 km. *J. Geophys. Res.,* **87,** 7206–7220.
59. Solomon, S., E.E. Ferguson, D.W. Fahey and P.J. Crutzen, 1982: On the chemistry of H$_2$O, H$_2$ and meteoritic ions in the meosphere and lower thermosphere. *Planet. Space Sci.,* **30,** 1117–1126.
60. Solomon, S., G.C. Reid, R.G. Roble and P.C. Crutzen, 1982: Photochemical coupling between the thermosphere and the lower atmosphere. 2. D-region ion chemistry and the winter anomaly. *J. Geophys. Res.,* **87,** 7221–7227.
61. Zimmerman, P.R., J.P. Greenberg, S.O. Wandiga and P.J. Crutzen, 1982. Termites: A Potentially Large Source of Atmospheric Methane, Carbon Dioxide, and Molecular Hydrogen. *Science,* **218,** 563–565.
62. Bolin, B., P.J. Crutzen, P.M. Vitousek, R.G. Woodmansee, E.D. Goldberg and R.B. Cook, 1983: Interactions of biochemical cycles, in: B. Bolin and R.B. Cook, Eds: *The Major Biochemical Cycles and Their Interactions,* SCOPE 21, pp. 1–40, Wiley, Chichester.
63. Crutzen, P.J., 1983: Atmospheric interactions–homogeneous gas reactions of C, N, and S containing compounds, in: B. Bolin and R.B. Cook, Eds.: *The Major Biochemical Cycles and Their Interactions,* SCOPE 21, pp. 67–114, Wiley, Chichester.
64. Crutzen, P.J. and L.T. Gidel, 1983: A two-dimensional photochemical model of the atmosphere. 2. The tropospheric budgets of the anthropogenic chlorocarbons, CO, CH$_4$, CH$_3$Cl and the effect of various NO$_x$ sources on tropospheric ozone. *J. Geophys. Res.,* **88,** 6641–6661.
65. Crutzen, P.J. and U. Schmailzl, 1983: Chemical budgets of the stratosphere. *Planet. Space. Sci.,* **31,** 1009–1032.
66. Frederick, J.E., R.B. Abrams and P.J. Crutzen, 1983: The Delta Band Dissociation of Nitric Oxide: A Potential Mechanism for Coupling Thermospheric Variations to the Mesosphere and Stratosphere. *J. Geophys. Res.,* **88,** 3829–3925.
67. Gidel, L.T., P.J. Crutzen and J. Fishman, 1983: A two-dimensional photochemical model of the atmosphere. 1: Chlorocarbon emissions and their effect on stratospheric ozone. *J. Geophys. Res.,* **88,** 6622–6640.
68. Chatfield, R.B. and P.J. Crutzen, 1984: Sulfur Dioxide in Remote Oceanic Air: Cloud Transport of Reactive Precursors. *J. Geophys. Res.,* **89** (D5), 7111–7132.
69. Crutzen, P.J., I.E. Galbally and C. Brühl, 1984: Atmospheric Effects from Postnuclear fires. *Climatic Change,* **6,** 323–364.
70. Crutzen, P.J., M.T. Coffey, A.C. Delany, J. Greenberg, P. Haagenson, L. Heidt, R. Heidt, L. Lueb, W.G. Mankin, W. Pollock, W. Seiler, A. Wartburg and P. Zimmerman, 1985: Observations of air composition in Brazil between the equator and 20°S during the dry season. *Acta Amazonica,* Manaus, 15(1–2): 77–119.
71. Crutzen, P.J., A.C. Delany, J. Greenberg, P. Haagenson, L. Heidt, R. Lueb, W. Pollock, W. Seiler, A. Wartburg and P. Zimmerman, 1985: Tropospheric Chemical Composition Measurements in Brazil During the Dry Season. *J. Atmos. Chem.,* **2,** 233–256.
72. Crutzen, P.J., D.M. Whelpdale, D. Kley and L.A. Barrie, 1985: The cycling of sulfur and nitrogen in the remote atmosphere, in: *The Biogeochemical Cycling of Sulfur and Nitrogen in the Remote Atmosphere* (J.N. Galloway, R.J. Charlson, M.O. Andreae and H. Rodhe, Eds.), NATO ASI Series C 158, Reidel, Dordrecht, Holland, 203–212.
73. Delany, A.C., P. Haagenson, S. Walters, A.F. Wartburg and P.J. Crutzen, 1985: Photochemically produced ozone in the emission of large scale tropical vegetation fires. *J. Geophys. Res.,* **90** (D1), 2425–2429.
74. Crutzen, P.J. and F. Arnold, 1986: nitric acid cloud formation in the cold Antarctic stratosphere: a major cause for the springtime "ozone hole". *Nature,* **324,** 651–655.
75. Crutzen, P.J., I. Aselmann and W. Seiler, 1986: Methane production by domestic animals, wild ruminants, other herbivorous fauna, and humans. *Tellus,* **38B,** 271–284.
76. Crutzen, P.J. and T.E. Graedel, 1986: The role of atmospheric chemistry in environment–development interactions, in *Sustainable Development of the Environment* (W.C. Clark and R.E. Munn, Eds.), Cambridge University Press, 213–251.
77. Bingemer, H.G. and P.J. Crutzen, 1987: The production of methane from solid wastes. *J. Geophys. Res.,* **92** (D2), 2181–2187.

78. Crutzen, P.J., 1987: Role of the tropics in the atmospheric chemistry, in: R. Dickinson, Ed: *Geophysiology of the Amazon*, Wiley, Chichester–New York, 107–131.

79. Crutzen, P.J., 1987: Acid rain at the K/T boundary, *Nature*, **330**, 108–109.

80. Barrie, L.A., J.W. Bottenheim, R.C. Schnell, P.J. Crutzen and R.A. Rasmussen, 1988: Ozone destruction and photochemical reactions at polar sunrise in the lower Arctic atmosphere. *Nature*, **334**, 138–141.

81. Brühl, C. and P.J. Crutzen, 1988: Scenarios of possible changes in atmospheric temperatures and ozone concentrations due to man's activities as estimated with a one-dimensional coupled photochmical climate model. *Climate Dynamics* **2**, 173–203.

82. Crutzen, P.J., 1988: Tropospheric ozone: an Overview pp. in: *Tropospheric Ozone*, I.S.A. Isaksen Ed., Reidel, Dordrecht, 3–32.

83. Crutzen, P.J., 1988: Variability in Atmospheric-Chemical Systems, in: *Scales and Global Change*, SCOPE 35, T. Rosswall, R.G. Woodmansee and P.G. Risser, Eds., Wiley, Chichester, 81–108.

84. Crutzen, P.J., C. Brühl, U. Schmailzl and F. Arnold, 1988: Nitric acid haze formation in the lower stratosphere: a major contribution factor to the development of the Antarctic "ozone hole", in *Aerosols and Climate* (M.P. McCormick and P.V. Hobbs, Editors), A. Deepak Publ., Hampton, Virginia, USA, pp. 287–304.

85. Hao, W.M., D. Scharffe, E. Sanhueza and P.J. Crutzen, 1988: Production of N_2O, CH_4, and CO_2 from Soils in the Tropical Savanna During the Dry Season. *J. Atmos. Chem.*, **7**, 93–105.

86. Horowitz, A., G. von Helden, W. Schneider, P.J. Crutzen and G.K. Moortgat, 1988: Ozone generation in the 214 nm photolysis of oxygen at 25°C. *J. Phys. Chem.*, **92**, 4956–4960.

87. Liu, S.C., R.A. Cox, P.J. Crutzen, D.H. Ehhalt, R. Guicherit, A. Hofzumahaus, D. Kley, S.A. Penkett, L.F. Phillips, D. Poppe and F.S. Rowland, 1988: Group Report: Oxidizing Capacity of the atmosphere, in: *The Changing Atmosphere*, F.S. Rowland and I.S.A. Isaksen, Eds., Wiley, Chichester, p. 219–232.

88. Wilson, S.R., P.J. Crutzen, G. Schuster, D.W.T. Griffith and G. Helas, 1988: Phosgene measurements in the upper troposphere and lower stratosphere. *Nature*, **334**, 689–691.

89. Aselmann, I. and P.J. Crutzen, 1989: Global distribution of natural freshwater wetlands and rice paddies, their net primary productivity, seasonality and possible methane emissions. *J. Atmos. Chem.*, **8**, 307–358.

90. Brühl, C. and P.J. Crutzen, 1989: On the disporportionate role of tropospheric ozone as a filter against solar UV-B radiation. *Geophys. Res. Lett.*, **16**, 703–706.

91. Crutzen, P.J. and C. Brühl, 1989: The impact of observed changes in atmospheric composition on global atmospheric chemistry and climate, in: *The Environmental Record in Glaciers and Ice Sheets*, Dahlem Konferenzen 1988, H. Oeschger and C.C. Langway, *Eds., Wiley*, Chichester, pp. 249–266.

92. Pearman, G.I., R.J. Charlson, T. Class, H.B. Clausen, P.J. Crutzen, T. Hughes, D.A. Peel, K.A. Rahn, J. Rudolph, U. Siegenthaler and D.S. Zardini, 1989: Group Report: What Anthropogenic Impacts are recorded in Glaciers? In: Dahlem Workshop Reports: *The Environmental Record in Glaciers and Ice Sheets*, Dahlem Konferenzen, H. Oeschger and C.C. Langway, Eds., Wiley, Chichester, pp. 269–286.

93. Robertson, R.P., M.O. Andreae, H.G. Bingemer, P.J. Crutzen, R.A. Delmas, J.H. Duizer, I. Fung, R.C. Harriss, M. Kanakidou, M. Keller, J.M. Melillo and G.A. Zavarzin, 1989: Group Report: Trace gas exchange and the chemical and physical climate: Critical interactions, In: Dahlem Workshop Reports: *Exchange of Trace Gases between Terrestrial Ecosystems and the Atmosphere*, Eds. M.O. Andreae and D.S. Schimel, Life Sciences Research Report 47, Wiley, Chichester, pp. 303–320.

94. Simon, F.G., J.P. Burrows, W. Schneider, G.K. Moortgat and P.J. Crutzen, 1989: Study of the reaction $ClO + CH_3O_2 \rightarrow$ Products at 300 K. *J Phys. Chem.*, **93**, 7807–7813.

95. Zimmermann, P.H., H. Feichter, H.K. Rath, P.J. Crutzen and W. Weiss, 1989: A global three-dimensional source-receptor model investigating [85]Kr. *Atmos. Environ.* **23**, 25–35.

96. Brühl, C. and P.J. Crutzen, 1990: Ozone and climate changes in the light of the Montreal Protocol, a model study. *Ambio*, **19**, 293–301.

97. Chatfield, R.B. and P.J. Crutzen, 1990: Are there interactions of iodine and sulfur species in marine air photochemistry? J. Geophys. Res., **95**, 22319–22341.

98. Crutzen, P.J. and M.O. Andreae, 1990: Biomass burning in the tropics: impact on atmospheric chemistry and biogeochemical cycles. *Science*, **250**, 1669–1678.

99. Feichter, J. and P.J. Crutzen, 1990: Parameterization of vertical tracer transport due to deep cumulus convection in a global transport model and its evaluation with [222] Radon measurements. *Tellus*, **42B**, 100–117.

100. Graedel, T.E. and P.J. Crutzen, 1990: Atmospheric trace constituents, in: *The Earth as Transformed by Human Action*, Eds. B.L. Turner II et al., Cambridge University Press, pp. 295–311.

101. Hao, W.M., M.H. Liu and P.J. Crutzen, 1990: Estimates of annual and regional releases of CO_2 and other trace gases to the atmosphere from fires in the tropics, based on the FAO statistics for the period 1975–1980, in: *Fire in the Tropical Biota, Ecological Studies*, **84**, J.G. Goldammer, Ed., Springer-Verlag, Berlin, 440–462 .

102. Lelieveld, J. and P.J. Crutzen, 1990: Influence of cloud and photochemical processes on tropospheric ozone. *Nature*, **343**, 227–233.

103. Lobert, J.M., D.H. Scharffe, W.M. Hao and P.J. Crutzen, 1990: Importance of biomass burning in the atmospheric budgets of nitrogen-containing gases. *Nature*, **346**, 552–554.

104. Sanhueza, E. W.M. Hao, D. Scharffe, L. Donoso and P.J. Crutzen, 1990: N_2O and NO emissions from soils of the northern part of the Guayana Shield, Venezuela. *J. Geophys. Res.*, **95** (D13) , 22481–22488.

105. Scharffe, D. W.M. Hao, L. Donoso, P.J. Crutzen and E. Sanhueza, 1990: Soil fluxes and atmospheric concentrations of CO and CH_4 in the northern part of the Guayana Shield, Venezuela. *J. Geophys. Res.*, **95** (D13), 22475–22480.

106. Crutzen, P.J. and P.H. Zimmermann, 1991: The changing photochemistry of the troposphere. *Tellus*, **43 A/B,** 136–151.

107. Hao, W.M., D. Scharffe, J.M. Lobert and P.J. Crutzen, 1991: Emissions of N_2O from the burning of biomass in an experimental system. *Geophys. Res. Lett.*, **18**, 999–1002.

108. Kanakidou, M., H.B. Singh, K.M. Valentin and P.J. Crutzen, 1991: A 2-D study of ethane and propane oxidation in the troposphere. *J. Geophys. Res.*, **96**, 15395–15413.

109. Kuhlbusch, A.T., J.M. Lobert, P.J. Crutzen and P. Warneck, 1991: Molecular nitrogen emissions from denitrification during biomass burning. *Nature*, **351**, 135–137.

110. Lelieveld, J. and P.J. Crutzen, 1991: The role of clouds in tropospheric photochemistry. *J. Atmos. Chem.*, **12**, 229–267.

111. Lobert, J.M., D.H. Scharffe, W.M. Hao, T.A. Kuhlbusch, R. Seuwen, P. Warneck and P.J. Crutzen, 1991: Experimental evaluation of biomass burning emissions: Nitrogen and carbon containing compounds, in: *Global Biomass Burning: Atmospheric, Climatic and Biosphere Implications*, Ed. J.S. Levine, MIT Press, Cambridge, MA, pp. 122–125.

112. Peter, Th., C. Brühl and P.J. Crutzen, 1991: Increase in the PSC-formation probability caused by high-flying aircraft. *Geophys. Res. Lett.*, **18**, 1465–1468.

113. Crutzen, P.J. and G.S. Golitsyn, 1992: Linkages between Global Warming, Ozone Depletion and Other Aspects of Global Environmental Change, in: *Confronting Climate Change*, I.M. Mintzer, Ed., pp. 15–32, Cambridge University Press.

114. Crutzen, P.J., R. Müller, Ch. Brühl and Th. Peter, 1992: On the potential importance of the gas phase reaction CH_3O_2 + ClO → ClOO + CH_3O and the hererogeneous reaction HOCl + HCl → H_2O + Cl_2 in "ozone hole" chemistry. *Geophys. Res. Lett.*, **19**, 1113–1116.

115. Kanakidou, M., P.J. Crutzen, P.H. Zimmermann and B. Bonsang, 1992: A 3-dimensional global study of the photochemistry of ethane and propane in the troposphere: Production and transport of organic nitrogen compounds, in: *Air Pollution Modeling and its Application IX*, H. van Dop and G. Kallos (Eds.), Plenum Press, New York, 415–426.

116. Langner, J., H. Rodhe, P.J. Crutzen and P. Zimmermann, 1992: Anthropogenic influence on the distribution of tropospheric sulphate aerosol. *Nature*, **359**, 712–715.

117. Lelieveld, J. and P.J. Crutzen, 1992: Indirect chemical effects of methane on climate warming. *Nature*, **355**, 339–342.

118. Luo, B.P., Th. Peter and P.J. Crutzen, 1992: Maximum supercooling of H_2SO_4 acid aerosol droplets. *Ber. Bunsenges. Phys. Chem.*, **96**, 334–338.

119. Singh, H.B. D. O'Hara, D. Herlth, J.D. Bradshaw, S.T. Sandholm, G.L. Gregory, G.W. Sachse, D.R. Blake, P.J. Crutzen and M. Kanakidou, 1992: Atmospheric measurements of PAN and other organic nitrates at high latitudes: possible sources and sinks. *J. Geophys. Res.*, **97**, 16511–16522.

120. Singh, H.B., D. Herlth, K. Zahnle, D. O'Hara, J. Bradshaw, S.T. Sandholm, R. Talbot, P.J. Crutzen and M. Kanakidou, 1992: Relationship of PAN to active and total odd nitrogen at northern high latitudes: possible influence of reservoir species on NO_x and O_3. *J. Geophys. Res.*, **97**, 16523–16530.

121. Berges, M.G.M., R.M. Hofmann, D. Scharffe and P.J. Crutzen, 1993: Measurement of nitrous oxide emissions from motor vehicles in tunnels. *J. Geophys. Res.*, **98**, 18527–18531.

122. Crutzen, P.J. and C. Brühl, 1993: A model study of atmospheric temperatures and the concentrations of ozone, hydroxyl, and some other photochemically active gases during the glacial, the preindustrial holocene and the present. *Geophys. Res. Lett.* **20**, 1047–1050.

123. Crutzen, P.J. and G.R. Carmichael, 1993: Modeling the influence of fires on atmospheric chemistry, in: *Fire in the Environment: The Ecological, Atmospheric, and Climatic Importance of Vegetation Fires*, (P.J. Crutzen and J.G. Goldammer, Eds.), op. cit., 90–105.

124. Dentener, F. and P.J. Crutzen, 1993: Reaction of N_2O_5 on tropospheric aerosols: Impact on the global distributions of NO_x, O_3 and OH. *J. Geophys. Res.*, **98**, 7149–7163.

125. Goldammer, J.G and P.J. Crutzen, 1993: Fire in the Environment: Scientific Rationale and Summary of Results of the Dahlem Workshop, in: *Fire in the Environment: The Ecological, Atmospheric and Climatic Importance*, (P.J. Crutzen and J.G. Goldammer, Eds.), op. cit., 1–14.

126. Kanakidou, M. and P.J. Crutzen, 1993: Scale problems in global tropospheric chemistry modeling: Comparison of results obtained with a three-dimensional model, adopting longitudinally uniform and varying emissions of NO_x and NMHC. *Chemosphere*, **26**, 787–801.

127. Kanakidou, M., F.J. Dentener and P.J. Crutzen, 1993: A global three-dimensional study of the degradation of HCFC's and HFC-134a in the troposphere. Proceedings of STEP-HALOCSIDE/AFEAS Workshop on *Kinetics and Mechanisms for the Reactions of Halogenated Organic Compounds in the Troposphere*, Dublin, Ireland, March 23–25, 1993, Campus Printing Unit, University College Dublin, 113–129.

128. Lelieveld, J., P.J. Crutzen and C. Brühl, 1993: Climate effects of atmospheric methane, *Chemosphere*, **26**, 739–768.

129. Müller, R. and P.J. Crutzen, 1993: A possible role of galactic cosmic rays in chlorine during polar night. *J. Geophys. Res.*, **98**, 20483–20490.

130. Peter, Th. and P.J. Crutzen, 1993: The role of stratospheric cloud particles in Polar ozone depletion. An overview. *J. Aerosol Sci.*, **24, Suppl. 1**, S119–S120.

131. Russell III, J.M., A.F. Tuck, L.L. Gordley, J.H. Park, S.R. Drayson, J.E. Harries, R.J. Cicerone and P.J. Crutzen, 1993: Haloe Antarctic observations in the spring of 1991, *Geophys., Res. Lett.*, **20**, 719–722.

132. Schupp, M., P. Bergamaschi, G.W. Harris and P.J. Crutzen, 1993: Development of a tunable diode laser absorption spectrometer for measurements of the $^{13}C/^{12}C$ ratio in methane. *Chemosphere*, **26**, 13–22.

133. Carslaw, K.S., B.P. Luo, S.L. Clegg, Th. Peter, P. Brimblecombe and P.J. Crutzen, 1994: Stratospheric aerosol growth and HNO_3 gas phase depletion from coupled HNO_3 and water uptake by liquid particles. *Geophys. Res. Lett.*, **21**, 2479–2482.

134. Chen, J.-P. and P.J. Crutzen, 1994: Solute effects on the evaporation of ice particles, *J.Geophys. Res.*, **99**, 18847–18859.

135. Cox, R.A., A.R. MacKenzie, R. Müller, Th. Peter and P.J. Crutzen, 1994: Activation of stratospheric chlorine by reactions in liquid sulphuric acid, *Geophys. Res. Lett.*, **21**, 1439–1442.

136. Crowley, J.N., F. Helleis, R. Müller, G.K. Moortgat and P.J. Crutzen, 1994: CH3OCl: UV/visible absorption cross sections, J values and atmospheric significance, J. Geophys. Res., **99**, 20683–20688.

137. Crutzen, P.J., 1994: Global Tropospheric Chemistry, Proceedings of the NATO Advanced Study Institute on *Low Temperature Chemistry of the Atmosphere*, Maratea, Italy, August 29–September 11, 1993, NATO ASI Series I, **21** (Eds. G.K. Moortgat et al.), Springer, Heidelberg, 465–498

138. Crutzen, P.J., 1994: Global budgets for non-CO_2 grennhouse gases, *Environmental Monitoring and Assessement*, **31**, 1–15.

139. Crutzen, P.J., J. Lelieveld and Ch. Brühl, 1994: Oxidation processes in the atmosphere and the role of human activities: Observations and model results, in: *Environmental Oxidants*, J.O. Nriagu and M.S. Simmons (Eds.), Vol. 28 in "Advances in Environmental Science and Technology", Wiley, Chichester, 63–93.

140. Dentener, F.J. and P.J. Crutzen, 1994: A three dimensional model of the global ammonia cycle, *J. Atmos. Chem.* **19**, 331–369.

141. Deshler, T., Th. Peter, R. Müller and P.J. Crutzen, 1994: The lifetime of leewave-induced ice particles in the Arctic stratosphere: I. Balloonborne observations. *Geophys. Res. Lett.*, **21**, 1327–1330.

142. Lelieveld, J. and P.J. Crutzen, 1994: Emissionen klimawirksamer Spurengase durch die Nutzung von Öl und Erdgas. *Energiewirtschaftliche Tagesfragen*, **7**, 435–440.

143. Lelieveld, J. and P.J. Crutzen, 1994: Role of deep cloud convection in the ozone budget of the troposphere, *Science*, **264**, 1759–1761.

144. Luo, B.P., Th. Peter and P.J. Crutzen, 1994: Freezing of stratospheric aerosol droplets. *Geophys. Res. Lett.*, **21**, 1447–1450.

145. Luo, B.P., S.L. Clegg, Th. Peter, R. Müller and P.J. Crutzen, 1994: HCl solubility and liquid diffusion in aqueous sulfuric acid under stratospheric conditions, *Geophys. Res. Lett.*, **21**, 49–52.

146. Müller, R., Th. Peter, P.J. Crutzen, H. Oelhaf, G. Adrian, Th. v. Clarman, A. Wegner, U. Schmidt and D. Lary, 1994: Chlorine chemistry and the potential for ozone depletion in the Arctic stratosphere in the winter of 1991/92. *Geophys. Res. Lett.*, **21**, 1427–1430.

147. Peter, Th. and P.J. Crutzen, 1994: Modelling the chemistry and micro-physics of the could stratosphere. Proceedings of the NATO Advanced Study Institute on *Low Temperature Chemistry of the Atmosphere*, Maratea, Italy, August 29–September 11, 1993, NATO ASI Series I, **21** (Eds. G.K. Moortgat et al.), Springer, Heidelberg, 499–530.

148. Peter, Th., P.J. Crutzen, R. Müller and T. Deshler, 1994: The lifetime of leewave-induced particles in the Artic stratosphere: II. Stabilization due to NAT-coating. *Geophys. Res. Lett.*, **21**, 1331–1334.

149. Sanhueza, E., L. Donoso, D. Scharffe and P.J. Crutzen, 1994: Carbon monoxide fluxes from natural, managed, or cultivated savannah grasslands, *J. Geophys. Res.*, **99**, 16421–16425.

150. Sassen, K., Th. Peter, B.P. Luo and P.J. Crutzen, 1994: Volcanic Bishop's ring: Evidence for a sulfuric acid tetrahydrate particle aerosole, *Appl. Optics.*,**33**, 4602–4606.

151. Singh, H.B., D. O'Hara, D. Herlth, W. Sachse, D.R. Blake, J.D. Bradshaw, M. Kanakidou and P.J. Crutzen, 1994: Acetone in the atmosphere: Distribution, sources and sinks. *J. Geophys. Res.*, **99**, 1805–1819.

152. Crutzen, P.J., 1995: On the role of CH_4 in Atmospheric chemistry: Sources, sinks and possible reductions in anthropogenic sources. *Ambio*, **24**, 52–55.

153. Crutzen, P.J., 1995: Ozone in the troposphere, in: *Composition, Chemistry, and Climate of the Atmosphere*. H.B. Sing (Ed.), Van Nostrand Reinhold Publ., New York, 349–393.

154. Crutzen, P.J., 1995: The role of methane in atmospheric chemistry and climate, in: *Ruminant Physiology: Digestion, Metabolism, Growth and Reproduction: Proceedings of the Eighth International Symposium on Ruminant Physiology*, Eds. W. v. Engelhardt, S. Leonhard-Marek, G. Breves and D. Giesecke, Ferdinand Enke Verlag, Stuttgart, 291–315.

155. Crutzen, P.J., J.-U. Grooß, C. Brühl, R. Müller, J.M. Russell III, 1995: A reevaluation of the ozone budget with HALOE UARS data: no evidence for the ozone defizit. *Science*, **268**, 705–708.

156. Finkbeiner, M., J.N. Crowley, O. Horie, R. Müller, G.K. Moortgat and P.J. Crutzen, 1995: Reaction between HO_2 and ClO: Product formation between 210 and 300 K. *J. Phys. Chem.*, **99**, 16274–16275.

157. Kanakidou, M., F.J. Dentener and P.J. Crutzen, 1995: A global three-dimensional study of the fate of HCFCs and HFC-134a in the troposphere. *J. Geophys. Res.*, **100**, 18781–18801.

158. Koop, T., U.M. Biermann, W. Raber, B.P. Luo, P.J. Crutzen and Th. Peter, 1995: Do stratospheric aerosol droplets freeze above the ice frost point?. *J. Geophys. Res.*, **22**, 917–920.

159. Kuhlbusch, T.A.J. and P.J. Crutzen, 1995: Toward a global estimate of black carbon in residues of vegetation fires representing a sink of atmospheric CO_2 and a source of O_2. *Global Biogeochem. Cycles*, **4**, 491–501.

160. Meilinger, S.K., T. Koop, B.P. Luo, T. Huthwelker, K.S. Carslaw, P.J. Crutzen and Th. Peter, 1995: Size-dependent stratospheric droplet composition in lee wave temperature fluctuations and their potential role in PSC freezing. *Geophys. Res. Lett.*, **22**, 3031–3034.

161. Rodhe, H. and P.J. Crutzen, 1995: Climate and CCN, *Nature*, **375**, 111.

162. Sander, R., J. Lelieveld and P.J. Crutzen, 1995: Modelling of the nighttime nitrogen and sulfur chemistry in size resolved droplets of an orographic cloud, *J. Atmos. Chem.*, **20**, 89–116.

163. Schade, G.W. and P.J. Crutzen, 1995: Emission of aliphatic amines from animal husbandry and their reactions: Potential source of N_2O and HCN. *J. Atmos. Chem.*, **22**, 319–346.

164. Singh, H.B., M. Kanakidou, P.J. Crutzen and D.J. Jacob, 1995: High concentrations and photochemical fate of oxygenated hydrocarbons in the global troposphere. *Nature*, **378**, 50–54.

165. Vömel, H., S.J. Oltmans, D. Kley and P.J. Crutzen, 1995: New evidence for the stratospheric dehydration mechanism in the aquatorial Pacific. *Geophys. Res. Lett.*, **22**, 3235–3238.

166. Wang, C. and P.J. Crutzen, 1995: Impact of a simulated severe local storm on the redistribution of sulfur dioxide. *J. Geophys. Res.*, **100**, 11357–11367.

167. Wang, C., P.J. Crutzen, V. Ramanathan, S.F. Williams, 1995: The role of a deep convective storm over the tropical Pacific Ocean in the redistribution of atmospheric chemical species. *J. Geophys. Res.*, **100**, 11509–11516.

168. Wayne, R.P., P. Biggs, J.P. Burrows, R.A. Cox, P.J. Crutzen, G.D. Hayman, M.E. Jenkin, G. Le Bras, G.K. Moortgat, U. Platt, G. Poulet and R.N. Schindler, 1995: Halogen oxides: Radicals, sources and reservoirs in the laboratory and in the atmosphere. *Atmos. Environ.*, **29**, 2675–2884 (special issue).

169. Crutzen, P.J., 1995: Introductory Lecture. Overview of tropospheric chemistry: Developments during the past quarter century and a look ahead. *Faraday Discuss. 100*, 1–21.

170. Sander, R. and P.J. Crutzen, 1996: Model study indicating halogen activation and ozone destruction in polluted air masses transported to the sea. *J. Geophys. Res. 101*, 9121, 1996

IN PRESS

SUBMITTED

171. Berges, M.G.M. and J. Crutzen, 1995 revised 1996: Estimates of global N2O emissions from cattle, pig and chicken manure, including a discussion of CH_4 emissions. *J. Atmos. Chem.*

172. Brühl, C., S.R. Drayson, J.M. Russell III, P.J. Crutzen, J.M. McInemey, P.N. Purcell, H. Claude, H. Gernandt, T.I. McGee, I.S. McDemid, M.R. Gunson, 1995: HALOE ozone channel validation. *J. Geophys. Res.*

173. Kuhlbusch, T.A.J., M.O. Andreae, H. Cachier, J.G. Goldammer, J.-P. Lacaux, R. Shea and P.J. Crutzen, 1995: Black carbon formation by savanna fires: Measurements and implications for the global carbon cycle. *J. Geophys. Res.*

174. Müller, R., P.J. Crutzen, J.U. Grooß, Ch. Brühl, J.M Russel and A.F. Tuck, 1994: Chlorine activation and ozone depletion in the Artic vortex during the winters of 1992 and 1993 observed by the halogen occulation experiment on the Upper Atmosphere Research Satellite. (to be submitted to *J. Geophys. Res.*)

175. Bergamaschi, P., C. Brühl, C.A.M. Brenninkmeijer, G. Saueressig, J.N. Crowley, J.U. Grooß, H. Fischer and P.J. Crutzen, 1996: Implications of the large carbon kinetic isotope effect in the reaction $CH_4 + Cl$ for the $^{13}C/^{12}C$ ratio of stratospheric CH_4. *Geophys. Res. Lett.*

176. Kley, D., H.G.J. Smit, H. Vömel, S. Oltmans, H. Grassl, V. Ramanathan and P.J. Crutzen, 1996: Extremely low upper tropospheric ozone observations in the convective regions of the Pacific. *Science.*

177. Shorter, J.H., J.B. McManus, C.E. Kolb, E.J. Allwine, B.K. Lamb, B.W. Mosher, R.C. Harriss, U. Parchatka, H. Fischer, G.W. Harris and P.J. Crutzen, 1996. *J. Atmos. Chem.*

178. Vogt, R. and P.J. Crutzen, 1996: A new mechanism for bromine and chlorine release from sea salt aerosol in the unpolluted boundary layer. *Nature.*

OTHER PUBLICATIONS

A1. Crutzen, P.J., 1969: Koldioxiden och klimatet (Carbon dioxide and climate), *Forskning och Framsteg,* **5,** 7–9.

A2. Crutzen, P.J., 1971: The photochemistry of the stratosphere with special attention given to the effects of NO_x emitted by supersonic aircraft, First Conference on CIAP, United States Department of Transportation, 880–88.

A3. Crutzen, P.J., 1971: On some photochemical and meteorological factors determining the distribution of ozone in the stratosphere: Effects on contamination by NO_x emitted from aircraft, Technical Report UDC 551.510.4, Institute of Meteorology, University of Stockholm.

A4. Crutzen, P.J., 1972: Liten risk för klimatändring (Small risk for climatic change; in Swedish). *Forskning och Framsteg,* **2,** 27.

A5. Crutzen, P.J., 1974: Artificial increases of the stratospheric nitrogen oxide content and possible consequences for the atmopsheric ozone, Technical Report UDC 551.510.4:546.2, Institute of Meteorology, University of Stockholm.

A6. Crutzen, P.J., 1974: Väderforskning med matematik (Weather research with mathematics; in Swedish). *Forskning och Framsteg,* **6,** 22–23, 26.

A7. Crutzen, P.J., 1975: Physical and chemical processes which control the production, destruction and distribution of ozone and some other chemically active minor constituents. *GARP Publications Series 16,* World Meterological Organization, Geneva, Switzerland.

A8. Crutzen, P.J., 1975: A two-dimensional photochemical model of the atmosphere below 55 km. In: *Estimates of natural and man-caused ozone perturbations due to NO_x. Proceedings of 4th CIAP Conference,* U.S. Department of Transportation, Cambridge, DOT-TSC-OST-75-38, T.M. Hard and A.J. Brodrick (eds.), 264–279.

A9. Crutzen, P.J., 1976: Ozonhöljet tunnas ut: Begränsa spray-gaserna (The ozone shield is thinning: limit the use of aerosol propellants; in Swedish). *Forskning och Framsteg,* **5,** 29–35.

A10. Crutzen, P.J., 1977: The Stratosphere-Meosphere. *Solar Output and Its Variations,* O.R. White (ed.), Colorado Associated University Press, Boulder, Colorado, 13–16.

A11. Crutzen, P.J., J. Fishman, L.T. Gidel and R.B. Chatfield, 1978: Numerical investigations of the photochemical and transport processes which affect halocarbons and ozone in the atmosphere. *Annual Summary of Research,* Dept. of Atmospheric Science, Colorado State Univ., Fort Collins, CO.

A12. Fishman, J. and P.J. Crutzen, 1978: The distribution of the hydroxyl radical in the troposphere. Atmos. Sci. Paper 284 (Dept. of Atmos. Sci., Colorado State University, Fort. Collins, CO).

A13. Crutzen, P.J. L.T. Gidel and J. Fishman, 1979: Numerical investigations of the photochemical and transport processes which affect ozone and other trace constituents in the atmosphere. *Annual Summary of Research*, Dept. of Atmospheric Science, Colorado State University, Fort Collins, CO.

A14. Crutzen, P.J., 1981: Atmospheric chemical processes of the oxides of nitrogen including nitrous oxide. In: *Denitrification, Nitrification and Atmospheric Nitrous Oxide*. Ed. C.C. Delwiche, John Wiley and Sons, New York 1981, 17–44.

A15. Birks, J.W. and P.J. Crutzen, 1983: Atmospheric effects of a nuclear war, *Chemistry in Britain*, **19**, 927–930.

A16. Galbally, I.E., P.J. Crutzen and H. Rodhe, 1983: Some changes in the Atmosphere over Australia that may occur due to a Nuclear War, pp. 161–185 in *"Australia and Nuclear War"*, (Ed. M.A. Denborough), Croom Helm Ltd., Canberrra, Australia, 270 pp.

A17. Brühl, C. and P.J. Crutzen, 1984: A radiative vonvective model to study the sensitivity of climate and chemical composition to a variety of human activities, Proceedings of a working party meeting, Brussels, 18th May 1984, Ed. A. Ghazi, CEC, pp. 84–94.

A18. Crutzen, P.J. and M.O. Andreae, 1985: Atmospheric Chemistry, in T.F. Malone and J.G. Roederer, Eds., *Global Change*, Cambridge University Press, Cambridge, 75–113.

A19. Crutzen, P.J. and I.E. Galbally, 1985: Atmospheric conditions after a nuclear war. In: *Chemical Events in the Atmosphere and Their Impact on the Environment*. (G.B. Marini-Bettolo, Editor), Pontificiae Academiae Scientiarum Scripta Varia, Città del Vaticano, pp. 457–502.

A20. Crutzen, P.J. and J. Hahn, 1985: Atmosphärische Auswirkungen eines Atomkrieges. *Physik in unserer Zeit*, **16**, 4–15.

A21. Klose, W., H. Butin. P.J. Crutzen, F. Führ, H. Greim, W. Haber, K. Hahlbrock, A. Hüttermann. W. Klein, W. Klug, H.U. Moosmayer, W. Obländer, B. Prinz, K.E. Rehfuess and O. Rentz, 1985: Forschungsbeirat Waldschäden/Luftverunreinigungen der Bundesregierung und der Länder, Zwischenbericht Dezember 1984, in *Bericht über den Stand der Erkenntnisse zur Ursache*

A22. Crutzen, P.J., 1986: Globale Aspekte der amtosphärischen Chemie: Natürliche und anthropogene Einflüsse, Vorträge Rheinisch-Westfälische Akademie der Wissenschaften, S. 41–72, Westdeutscher Verlag GmbH, Opladen.

A23. Klose, W., H. Butin and P.J. Crutzen, 1986: u.a. Forschungsbeirat Waldschäden/Luftverunreinigungen der Bundesregierung und der Länder, 2. Bericht, 229 S.

A24. Crutzen, P.J., 1987: Recent depletions of ozone with emphasis on the polar "ozone hole". *Källa*, **28**, (Stockholm; in Swedish).

A25. Crutzen, P.J., 1987: Climatic Effects of Nuclear War, Annex 2 in *Effects of Nuclear War on Health and Health Services*, Report A40/11 of the World Health Organization to the 40th World Health Assembly, 18 March 1987; WHO, Geneva.

A26. Crutzen, P.J., 1987: Ozonloch und Spurengase–Menschliche Einflüsse auf Klima und Chemie der Atmosphäre, Max-Planck-Gesellschaft, Jahrbuch 1987, München, S. 27–40.

A27. Darmstadter, J., L.W. Ayres, R.U. Ayres, W.C. Clark, R.P. Crosson, P.J. Crutzen, T.E. Graedel, R. McGill, J.F. Richards and J.A. Torr, 1987: Impacts of World Development on Selected Characteristics of the Atmosphere: An Integrative Approach, Oak Ridge National Laboratory, 2 Volumes, ORNL/Sub/86-22033/1/V2, Oak Ridge, Tennessee 37931, USA.

A28. Brühl, C. and P.J. Crutzen, 1989: The potential role of odd hydrogen in the ozone hole photochemistry, in *Our Changing Atmosphere* (P.J. Crutzen, J.-C. Gerard and R. Zander, Editors), Université de Liège, Institut d'Astrophysique, B-4200 Cointe-Ougree, Belgium, pp. 171–177.

A29. Crutzen, P.J., W.M. Hao, M.H. Liu, J.M. Lobert and D. Scharffe, 1989: Emissions of CO_2 and other trace gases to the atmosphere from fires in the tropics, in P.J. Crutzen, J.C. Gerard and R. Zander, Eds.: *Our Changing Atmosphere*, Proceedings of the 28th Liège International Astrophysical Colloqium, Université de Liège, Belgium, pp. 449–471.

A30. Graedel, T.E. and P.J. Crutzen, 1989: The Changing Atmosphere. *Scientific American,* pp. 58–68 (in deutsch: Veränderungen der Atmosphäre. *Spektrum der Wissenschaften,* pp. 58–68.

A31. Lelieveld, J., P.J. Crutzen and H. Rodhe, 1989: Zonal average cloud characteristics for global atmospheric chemistry modelling. Report CM-76, UDC 551.510.4, Glomac 89/1. International Meteorological Institute in Stockholm, University of Stockholm, 54 pp.

A32. Crutzen, P.J., 1990: Auswirkungen menschlicher Aktivitäten auf die Erdatmosphäre: Was zu forschen, was zu tun? *DLR-Nachrichten,* Heft **59,** S. 5–13.

A33. Crutzen, P.J., 1990: Comments on George Reid's "Quo Vadimus" contribution "Climate". In: Quo Vadimus. Geophysics for the next generation, Geophysical Monograph 60, IUGG Vol **10,** Eds. G.D. Garland and John R. Apel, *American Geophysical Union,* Washington, USA, p. 47.

A34. Crutzen, P.J., 1990: Global changes in tropospheric chemistry, *Proceedings of Summer School on Remote Sensing and the Earth's Environment,* Alpbach, Austria, 26 July–4 August 1989, pp. 105–113.

A35. Crutzen, P.J., C. Brühl, 1990: The potential role of HO_x and ClO_x interactions in the ozone hole photochemistry, in *Dynamics, Transport and Photochemistry in the Middle Atmosphere of the Southern Hemisphere,* Ed. A. O'Neil, Kluwer, Dor-drecht, 203–212.

A36. Crutzen, P.J. and C. Brühl, 1990: The atmospheric chemical effects of aircraft operations. In: *Air Traffic and the Environment–Background, Tendencies and Potential Global Atmospheric Effects. Proceedings of a DLR International Colloqium Bonn, Germany, November 15/16, 1990,* Ed. Schumann, Springer-Verlag, Heidelberg 1990, pp. 96–106.

A37. Horowitz, A., G. von Helden, W. Schneider, F.G. Simon, P.J. Crutzen and G.K. Moortgat, 1990: Oxygen photolysis at 214 nm and 25 °C, *Proceedings of the Quadrennial Ozone Symposium,* Göttingen 8–13 August 1988, Eds. R.D. Boikov and P. Fabian, Deepak Publ. Co., pp. 690–693.

A38. Brühl, Ch., P.J. Crutzen, E.F. Danielsen, H. Graßl, H.-D. Hollweg and D. Kley, 1991: Umweltverträglichkeitsstudie für das Raumtransportsystem SÄNGER, Teil 1 Unterstufe, Ed. Max-Planck-Institut für Meteorologie Hamburg, 142 pp.

A39. Crutzen, P.J., 1991: Methane's sinks and sources. *Nature,* **350,** pp. 380–382.

A40. Lelieveld, J. and P.J. Crutzen, 1991: Climate discussion and fossil fuels. *Oil Gas–European Magazine,* **4,** 11–15.

A41. Crutzen, P.J., 1992: Ozone depletion: Ultraviolet on the increase. *Nature,* **356,** 104–105.

A42. Crutzen, P.J., 1992: Menschliche Einflüsse auf das Klima und die Chemie der globalen Atmosphäre, in: *Stadtwerke der Zukunft–ASEW–Fachtagung Kassel, 1991,* Ed. ASEW, Köln, Ponte Press, Bochum, 7–27.

A43. Graedel T.E. and P.J. Crutzen, 1992: Ensemble assessments of atmospheric emissions and impacts, in: *Energy and the Environment in the 21st Century,* pp 1–24, Energy Laboratory, Massachusetts Institute of Technology, Cambridge.

A44. Sander, R., J. Lelieveld and P.J. Crutzen, 1992: Model calculations of the nighttime aqueous phase oxidation of S(IV) in an orographic cloud. Proceedings of Joint CEC/EUROTRAC Workshop and LACTOZ-HALIPP Working Group on *Chemical Mechanisms Describing Tropospheric Processes,* Leuven, Belgium, September 23–25, 1992, Air Pollution Research Report 45, Ed. J. Peeters, E. Guyot SA, Brussels, 285–290.

A45. Brühl, Ch., P.J. Crutzen, H. Graßl and D. Kley, 1993: The impact of the spacecraft system SÄnger on the composition of the middle atmosphere, in: *AIAA Fourth International Aerospace Planes Conference,* Orlando/Florida, 1–4 December 1992, American Institute of Aeronautics and Astronautics, Washington DC, 1–9.

A46. Crutzen, P.J., 1993: Die Beobachtung atmosphärisch-chemischer Veränderungen: Ursachen und Folgen für Umwelt und Klima. In: *Klima: Vorträge im Wintersemester 1992/93,* Sammelband der Vorträge des Studium Generale der Ruprecht-Karls-Universität Heidelberg (Ed.), Heidelberger Verlagsanstalt, 31–48.

A47. Grooß, J.U., Th. Peter, C. Brühl and P.J. Crutzen, 1994: The influence of high flying aircraft on polar heterogeneous chemistry, Proceedings of an International Scientific Colloquium on *Impact of Emissions from Aircraft and Spacecraft upon the Atmosphere*, Köln, Germany, April 18–20, 1994, DLR-Mitteilung 94-06, U. Schumann and D. Wurzel (Eds.), 229–234.

A48. Kanakidou, M., P.J. Crutzen and P.H. Zimmermann, 1994: Estimates of the changes in tropospheric chemistry which result from human activity and their dependence on NO_x emissions and model resolution, *Proceedings of the Quadrennial Ozone Symposium*, June 4–13, 1992, Charlottsville, Virginia, U.S., NASA Conference Publication 3266, 66–69

A49. Müller, R. and P.J. Crutzen, 1994: On the relevance of the methane oxidation cycle to "ozone hole chemistry", *Proceedings of the Quadrennial Ozone Symposium*, June 4–13, 1992, Charlottsville, Virginia, U.S., NASA Conference Publication 3266, 298–301.

A50. Peter, Th. and P.J. Crutzen, 1994: Das Ozonloch: Wie kam es dazu und was sollten wir daraus lernen? In: *Der Mensch im Strahlungsfeld der Sonne. Konstanz und Wandel in Natur und Gesellschaft.* Ed. C. Fröhlich, FORUM DAVOS, Wissenschaftliches Studienzentrum, Davos, 31–44.

A51. Steil, B., C. Brühl, P.J. Crutzen, M. Dameris, M. Ponater, R. Sausen, E. Roeckner, U. Schlese and G.J. Roelofs, 1994: A chemistry model for use in comprehensive climate models, Proceedings of an International Scientific Colloquium on *Impact of Emissions from Aircraft and Spacecraft upon the Atmosphere*, Köln, Germany, April 18–20, 1994, DLR-Mitteilung 94-06, U. Schumann and D. Wurzel (Eds.), 235–240.

A52. Andreae, M.O., W.R. Cofer III, P.J. Crutzen, P.V. Hobbs, J.M. Hollander, T. Kuhlbusch, R. Novakov, J.E. Penner, 1995: Climate impacts of carbonaceous and other non-sulfate aerosols: A proposed study. Lawrence Berkely Laboratory Document–PUB-5411.

A53. Crutzen, P.J., 1995: On the role of ozone in atmospheric chemistry. In: "The Chemistry of the Atmosphere. Oxidants and Oxidation in the Earth's Atmosphere", Proceedings of the 7th BOC Priestley Conference, Lewisburg, Pennsylvania, U.S.A., June 25–27, 1994, Ed. A.R. Bandy, The Royal Society of Chemistry, Cambridge, UK, 3–36.

IN PRESS

A54. Lelieveld, J., P.J. Crutzen, D. Jacob and A. Thompson, 1995: modeling of biomass burning influences on tropospheric ozone, SAFARI BOOK.

A55. Kanakidou, M., F.J. Dentener and P.J. Crutzen: Chlorodifluoromethane (HCFC-22) and its oxidation products in the troposhere. *Proceedings of the 3rd Conference on Environmental Science and Technology*, September 6–9, 1993, Lesvos, Greece.

Address: Max-Planck-Institute for Chemistry
(office)   Department of Atmospheric Chemistry
                   P.O. Box 3060
                   D-55020 Mainz/GERMANY
                   Telephone: +49-(0) 6131-305-458/9
                   Telefax: +49-(0) 6131-305-511 (or 436)
                   E-mail: air@mpch-mainz.mpg.d400.de
(privat)   Am Fort Gonsenheim 36
                   55122 Mainz
                   Telephone: +49-(0) 6131-381094

MARIO J. MOLINA

I was born in Mexico City on March 19, 1943; my parents were Roberto Molina Pasquel and Leonor Henriquez de Molina. My father was a lawyer; he had a private practice, but he also taught at the National University of Mexico (Universidad Nacional Autonoma de Mexico (UNAM)). In his later years, after I had left Mexico, he served as Mexican Ambassador to Ethiopia, Australia and the Philippines.

I attended elementary school and high school in Mexico City. I was already fascinated by science before entering high school; I still remember my excitement when I first glanced at paramecia and amoebae through a rather primitive toy microscope. I then converted a bathroom, seldom used by the family, into a laboratory and spent hours playing with chemistry sets. With the help of an aunt, Esther Molina, who was a chemist, I continued with more challenging experiments along the lines of those carried out by freshman chemistry students in college. Keeping with our family tradition of sending their children abroad for a couple of years, and aware of my interest in chemistry, I was sent to a boarding school in Switzerland when I was 11 years old, on the assumption that German was an important language for a prospective chemist to learn. I remember I was thrilled to go to Europe, but then I was disappointed in that my European schoolmates had no more interest in science than my Mexican friends. I had already decided at that time to become a research chemist; earlier, I had seriously contemplated the possibility of pursuing a career in music–I used to play the violin in those days. In 1960, I enrolled in the chemical engineering program at UNAM, as this was then the closest way to become a physical chemist, taking math-oriented courses not available to chemistry majors.

After finishing my undergraduate studies in Mexico, I decided to obtain a Ph.D. degree in physical chemistry. This was not an easy task; although my training in chemical engineering was good, it was weak in mathematics, physics, as well as in various areas of basic physical chemistry–subjects such as quantum mechanics were totally alien to me in those days. At first I went to Germany and enrolled at the University of Freiburg. After spending nearly two years doing research in kinetics of polymerizations, I realized that I wanted to have time to study various basic subjects in order to broaden my background and to explore other research areas. Thus, I decided to seek admission to a graduate program in the United States. While pondering my future plans, I spent several months in Paris, where I was able to study mathematics on my own and I also had a wonderful time discussing all sorts of topics, ranging from politics, philosophy, to the arts, etc., with many good friends. Subsequently, I returned to Mexico as an Assistant Professor at the UNAM and I set up the first graduate program in chemical engineering. Finally, in 1968 I left for the University of California at Berkeley to pursue my graduate studies in physical chemistry.

During my first year at Berkeley, I took courses in physics and mathematics, in addition to the required courses in physical chemistry. I then joined the research group of Professor George C. Pimentel, with the goal of studying molecular dynamics using chemical lasers, which were discovered in his group a few years earlier. It was also at that time that I met Luisa Tan, who was a fellow graduate student in Pimentel's group and who later became my wife and close scientific collaborator.

George Pimentel was also a pioneer in the development of matrix isolation techniques, which is widely used in the study of the molecular structure and bonding of transient species. He was an excellent teacher and a wonderful mentor; his warmth, enthusiasm, and encouragement provided me with inspiration to pursue important scientific questions.

My graduate work involved the investigation of the distribution of internal energy in the products of chemical and photochemical reactions; chemical lasers were well suited as tools for such studies. At the beginning I had little experience with the experimental techniques required for my research, such as handling vacuum lines, infrared optics, electronic instrumentation, etc. I learned much of this from my colleague and friend Francisco Tablas, who was a postdoctoral fellow at that time. Eventually I became confident enough to generate original results on my own: my earliest achievement consisted of explaining some features in the laser signals–that at first sight appeared to be noise–as "relaxation oscillations," predictable from the fundamental equations of laser emission.

My years at Berkeley have been some of the best of my life. I arrived there just after the era of the free-speech movement. I had the opportunity to explore many areas and to engage in exciting scientific research in an intellectually stimulating environment. It was also during this time that I had my first experience dealing with the impact of science and technology on society. I remember that I was dismayed by the fact that high-power chemical lasers were being developed elsewhere as weapons; I wanted to be involved with research that was useful to society, but not for potentially harmful purposes.

After completing my Ph.D. degree in 1972, I stayed for another year at Berkeley to continue research on chemical dynamics. Then, in the fall of 1973, I joined the group of Professor F. Sherwood (Sherry) Rowland as a postdoctoral fellow, moving to Irvine, California, with Luisa; we married in July of that year. Sherry had pioneered research on "hot atom" chemistry, investigating chemical properties of atoms with excess translational energy and produced by radioactive processes. Sherry offered me a list of research options: the one project that intrigued me the most consisted of finding out the environmental fate of certain very inert industrial chemicals–the chloro-fluorocarbons (CFCs)–which had been accumulating in the atmosphere, and which at that time were thought to have no significant effects on the environment. This project offered me the opportunity to learn a new field–atmospheric chemistry–about which I knew very little; trying to solve a challenging problem appeared to be an excellent way to plunge into a new research area. The CFCs are compounds similar to others that Sherry and I had investigated from the point of view of molecular dynamics; we were familiar with their chemical properties, but not with their atmospheric chemistry.

Three months after I arrived at Irvine, Sherry and I developed the "CFC-ozone depletion theory." At first the research did not seem to be particularly interesting–I carried out a systematic search for processes that might destroy the CFCs in the lower atmosphere, but nothing appeared to affect them. We knew, however, that they would eventually drift to sufficiently high altitudes to be destroyed by solar radiation. The question was not only what destroys them, but more importantly, what are the consequences. We realized that the chlorine atoms produced by the decomposition of the CFCs would catalytically destroy ozone. We became fully aware of the seriousness of the problem when we compared the industrial amounts of CFCs to the amounts of nitrogen oxides which control ozone levels; the role of these catalysts of natural origin had been established a few years earlier by Paul Crutzen. We were alarmed at the possibility that the continued release of CFCs into the atmosphere would cause a significant depletion of the Earth's stratospheric ozone layer. Sherry and I decided to exchange information with the atmospheric sciences community: we went to Berkeley to confer with Professor Harold Johnston, whose work on the impact of the release of nitrogen oxides from the proposed supersonic transport (SST) aircraft on the stratospheric ozone layer was well known to us. Johnston informed us that months earlier Ralph Cicerone and Richard Stolarski had arrived at similar conclusions concerning the catalytic properties of chlorine atoms in the stratosphere, in connection with the release of hydrogen chloride either from volcanic eruptions or from the ammonium perchlorate fuel planned for the space shuttle.

We published our findings in Nature, in a paper which appeared in the June 28, 1974 issue. The years following the publication of our paper were hectic, as we had decided to communicate the CFC - ozone issue not only to other scientists, but also to policy makers and to the news media; we realized this was the only way to insure that society would take some measures to alleviate the problem.

To me, Sherry Rowland has always been a wonderful mentor and colleague. I cherish my years of association with him and my friendship with him and his wife, Joan. While he was on sabbatical leave in Vienna during the first six months of 1974, we communicated via mail and telephone. There were many exchanges of mail during this short period of time, which illustrated the frantic pace of our research at that time while we continued to refine our ozone depletion theory. Soon after, Sherry and I published several more articles on the CFC-ozone issue; we presented our results at scientific meetings and we also testified at legislative hearings on potential controls on CFCs emissions.

In 1975, I was appointed as a member of the faculty at the University of California, Irvine. Although I continued to collaborate with Sherry, as an assistant professor I had to prove that I was capable of conducting original research on my own. I thus set up an independent program to investigate chemical and spectroscopic properties of compounds of atmospheric importance, focusing on those that are unstable and difficult to handle in the laboratory, such as hypochlorous acid, chlorine nitrite, chlorine nitrate, peroxynitric acid, etc. It was in those years that Luisa, my wife, began collaborating with me, providing invaluable help in carrying out those difficult experiments. We also started then to raise a family: our son, Felipe, was born in 1977. Initially, Luisa had a teaching and research position at Irvine; however, after Felipe was born, she decided to work only part-time so that she could devote more time to Felipe. Fortunately for me, she decided to join my research group. Throughout the years, she has been very supportive and understanding of my preoccupation with work and the intense nature of my research.

Although my years at Irvine were very productive, I missed not doing experiments myself because of the many responsibilities associated with a faculty position: teaching courses, supervising graduate students, meetings, etc. After spending seven years at Irvine as Assistant and then Associate Professor, I decided to move to a non-academic position. I joined the Molecular Physics and Chemistry Section at the Jet Propulsion Laboratory in 1982. I had a smaller group–only a few postdoctoral fellows–but I also had the luxury of conducting experiments with my own hands, which I enjoyed very much. Indeed, I spent many hours in the laboratory in those years, conducting measurements and developing techniques for the study of newly emerging problems. Around 1985, after becoming aware of the discovery by Joseph Farman and his co-workers of the seasonal depletion of ozone over Antarctica, my research group at JPL investigated the peculiar chemistry which is promoted by polar stratospheric clouds, some of which consist of ice crystals. We were able to show that chlorine-activation reactions take place very efficiently in the presence of ice under polar stratospheric conditions; thus, we provided a laboratory simulation of the chemical effects of clouds over the Antarctic. Also, in order to understand the rapid catalytic gas phase reactions that were taking place over the South Pole, Luisa and I carried out experiments with chlorine peroxide, a new compound which had not been reported previously in the literature and which turned out to be important in providing the explanation for the rapid loss of ozone in the polar stratosphere.

In 1989 I returned to academic life, moving to the Massachusetts Institute of Technology, where I have continued with research on global atmospheric chemistry issues. After taking off for a few years, Luisa rejoined my research group. Our son is now in college: besides science, he also has an interest in music; he has been playing piano for over 10 years.

Although I no longer spend much time in the laboratory, I very much enjoy working with my graduate and postdoctoral students, who provide me with invaluable intellectual stimulus. I have also benefited from teaching; as I try to explain my views to students with critical and open minds, I find myself continually being challenged to go back and rethink ideas. I now see teaching and research as complementary, mutually reinforcing activities.

When I first chose the project to investigate the fate of chlorofluoro-carbons in the atmosphere, it was simply out of scientific curiosity. I did not consider at that time the environmental consequences of what Sherry and I had set out to study. I am heartened and humbled that I was able to do something that not only contributed to our understanding of atmospheric chemistry, but also had a profound impact on the global environment.

One of the very rewarding aspects of my work has been the interaction with a superb group of colleagues and friends in the atmospheric sciences community. I truly value these friendships, many of which go back 20 years or more, and which I expect to continue for many more years to come. I feel that this Nobel Prize represents a recognition for the excellent work that has been done by my colleagues and friends in the atmospheric chemistry community on the stratospheric ozone depletion issue.

POLAR OZONE DEPLETION

Nobel Lecture, December 8, 1995

by

Mario J. Molina

Department of Earth, Atmospheric and Planetary Sciences, and Department of Chemistry, Massachusetts Institute of Technology, Cambridge, MA 02139, USA

> *In On the rigor of Science*
> *...In that Empire, the Art of Cartography achieved such Perfection that the map of a single Province occupied an entire City, and the map of the Empire an entire Province. With time, those unwieldy maps did not satisfy and the Cartographers raised a Map of the Empire, with the size of the Empire and which coincided with it on every point ...*
>
> Jorge Luis Borges

INTRODUCTION

The ozone layer acts as an atmospheric shield which protects life on Earth against harmful ultraviolet radiation coming from the sun. This shield is fragile: in the past two decades it has become very clear that it can be affected by human activities.

Roughly 90% of the Earth's ozone resides in the stratosphere, which is the atmospheric layer characterized by an inverted–that is, increasing–temperature profile that rises typically from ~210 K at its base at 10–15 km altitude, to roughly 275 K at 50 km altitude. The maximum concentration of ozone is several parts per million; it is continuously being produced in the upper stratosphere by the action of solar radiation on molecular oxygen, and continuously being destroyed by chemical processes involving free radicals. Work by Paul Crutzen in the early 1970's [1] established that nitrogen oxides of natural origin, present at parts per billion levels in the stratosphere, are responsible for most of this chemical destruction, which occurs by means of catalytic cycles. Harold Johnston pointed out in 1972 [2] that the large fleets of supersonic aircraft which were being considered at that time could have seriously affected the ozone layer through their emissions of nitrogen oxides.

In early 1974 F. Sherwood (Sherry) Rowland and I proposed that chlorofluorocarbons (CFCs) would decompose in the stratosphere, releasing chlorine atoms which would catalytically destroy ozone [3]. The CFCs are industrial compounds which have been used as refrigerants, solvents, propellants for spray cans, blowing agents for the manufacture of plastic foams, etc. The two important properties which make these compounds very useful are (1) they can be readily transformed from a liquid into a vapor under mild temperature and pressure conditions, as can be seen in Table 1, which displays vapor pressures and boiling points for some of the most common CFCs; and (2) they are chemically very inert, and hence are non-toxic, non-flammable, and do not decompose inside a spray can or a refrigerator.

Table 1. Physical Properties of Some Common CFCs

Compound	Formula	Vapor Pressure (atm)		Boiling Point
		260 K	300K	
CFC-11	$CFCl_3$	0.22	1.12	23.8°C
CFC-12	CF_2Cl_2	1.93	6.75	-29.8°C
CFC-113	$CFCl_2CClF_2$	0.08	0.47	47.7°C

Measurements reported by James Lovelock and coworkers [4] had indicated that the CFCs were accumulating throughout the Earth's atmosphere. In our 1974 paper we suggested that the CFCs will not be destroyed by the common cleansing mechanisms that remove most pollutants from the atmosphere, such as rain, or oxidation by hydroxyl radicals. Instead, the CFCs will be decomposed by short wavelength solar ultraviolet radiation, but only after drifting to the upper stratosphere–above much of the ozone layer–which is where they will first encounter such radiation. Upon absorption of solar radiation the CFC molecules will rapidly release their chlorine atoms, which will then participate in the following catalytic reactions [3,5]:

$$Cl + O_3 \rightarrow ClO + O_2 \qquad (1)$$
$$ClO + O \rightarrow Cl + O_2 \qquad (2)$$
$$\text{Net: } O + O_3 \rightarrow 2\,O_2 \qquad (3)$$

These free radical chain reactions are terminated by reactions forming temporary reservoirs, the three most important ones being hydrogen chloride, chlorine nitrate, and hypochlorous acid:

$$Cl + CH_4 \rightarrow HCl + CH_3 \qquad (4)$$
$$ClO + NO_2 + M \rightarrow ClONO_2 + M \qquad (5)$$
$$ClO + HO_2 \rightarrow HOCl + O_2 \qquad (6)$$

The free radicals are regenerated by reactions such as the following:

$$HCl + OH \rightarrow Cl + H_2O \qquad (7)$$
$$ClONO_2 + h\nu \rightarrow Cl + NO_3 \qquad (8)$$
$$HOCl + h\nu \rightarrow Cl + OH \qquad (9)$$

There are other reactions which affect the chlorine balance in the stratosphere; a schematic representation of the most important ones is shown in

Figure 1. The net effect of this chemistry is that the inorganic chlorine produced by the decomposition of the CFCs exists in free radical form–i.e., as
"active chlorine"–only a few percent of the time. We have published more
detailed reviews of stratospheric halogen chemistry; see, e.g., Abbatt and
Molina [6] and Shen *et al.* [7]

In the decade following the publication of our Nature paper, field observations corroborated many of the predictions based on model calculations
and on laboratory measurements of reaction rates. However, the effects on
ozone were unclear, because the natural ozone levels have relatively large
fluctuations.

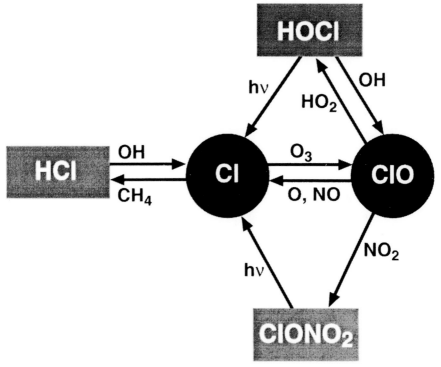

Fig.1: Schematic representation of chlorine chemistry in the stratosphere at mid and low latitudes.

POLAR OZONE LOSS

The rapid seasonal decline of stratospheric ozone over the South Pole–the
so called "Antarctic ozone hole"–is a startling phenomenon: a large fraction
of the total column ozone–more than a third–disappears in the spring
months over an area coinciding largely with the Antarctic continent (see
Figure 2). In recent years it has become more severe: in October 1992 and
1993 more than 99 % of the ozone disappeared at altitudes between about
14 and 19 km [8], where concentrations of this species are usually largest;
and in 1995, in mid-September, the ozone hole was the largest ever recorded
that early in the austral spring.

Significant ozone depletion over Antarctica started in the early 1980's, but it was not until 1985 that Farman *et al.* [9] announced their discovery of the ozone hole, suggesting furthermore a possible link with the growth of active chlorine in the stratosphere released by the decomposition of CFCs. Satellite data subsequently confirmed Farman *et al.*'s findings. This discovery was surprising, not only because of the magnitude of the depletion, but also because of its location. We had originally predicted that chlorine-initiated ozone loss would occur in regions of efficient ozone production, that is, predominantly in the upper stratosphere at middle and low latitudes; it is only there that the concentration of free oxygen atoms is large enough for catalytic reactions such as (1) and (2) to occur efficiently. Furthermore, the expectation was that at high latitudes active chlorine would be relatively less abun-

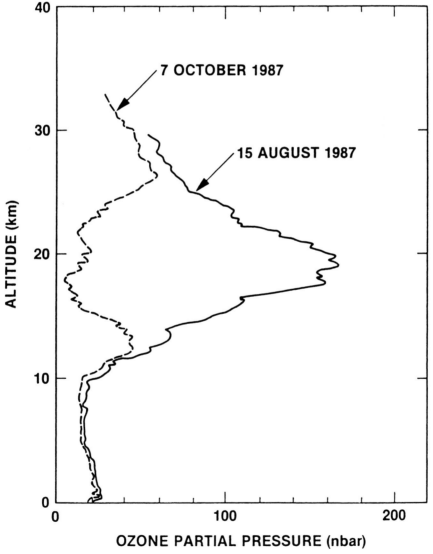

Fig. 2: Balloon measurements of ozone profiles over Halley Bay, Antarctica [30].

dant, because the reservoir species would be more stable–reactions such as
(7) and (8) slow down as a consequence of the decrease in temperature as
well as the decrease in solar radiation at the higher zenith angles, that is, the
lower sun elevation.

The initial question was, thus, whether the observed polar ozone loss was
of human origin, or else merely a periodic natural phenomenon, only never
before noticed. A prominent early theory put forth to explain this phenomenon had as a principal cause atmospheric dynamics driven by extreme
cold temperatures: the hypothesis was that upon first sunrise in the spring,
warming of the Antarctic stratosphere would lead to a net upward lifting of
ozone-poor air from the troposphere or lower stratosphere. If so, no ozone
destruction would be taking place; it would merely be redistributed periodi-
cally. Another theory suggested that the 11 year solar cycle was responsible
for the effect: chemical destruction of ozone involving catalytic NO_x cycles
would occur, with the catalysts being generated in the upper stratosphere by
high energy particles or by ultraviolet radiation following enhanced solar
activity. A third theory suggested that chlorine and bromine free radicals
were the catalysts for the chemical destruction of ozone, the source of these
radicals being compounds of industrial origin–CFCs and halons (these are
compounds containing bromine which are used as fire extinguishers). This
third theory is discussed next in more detail. It is based on our original CFC-
ozone depletion hypothesis, but it is modified and extended to take into
account the peculiar conditions prevailing over Antarctica: Figure 3 displays
schematically the most important reactions of the modified theory (compare to Figure 1, the original chemistry).

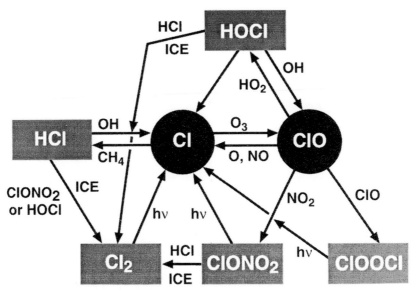

Fig. 3: Schematic representation of chlorine chemistry in the polar stratosphere.

GAS PHASE CHEMISTRY OF THE POLAR STRATOSPHERE

The large solar zenith angle characteristic of the polar stratosphere results in little, if any photodissociation of oxygen, and hence in practically no new ozone production. Similarly, ozone itself photolyzes only rather slowly in those regions, so that the concentration of free oxygen atoms remains very small–particularly in the lower stratosphere, where the ozone loss is largest. Consequently, catalytic cycles which require the presence of atomic oxygen are ineffective for ozone destruction, as mentioned above. Several cycles not involving oxygen atoms were proposed after the discovery of the ozone hole, such as the following one, based on the coupling of HO_x and ClO_x radicals [10]:

$$OH + O_3 \rightarrow HO_2 + O_2 \tag{9}$$
$$Cl + O_3 \rightarrow ClO + O_2 \tag{1}$$
$$ClO + HO_2 \rightarrow HOCl + O_2 \tag{10}$$
$$HOCl + h\nu \rightarrow OH + Cl \tag{11}$$

$$\text{Net: } 2\,O_3 + h\nu \rightarrow 3\,O_2 \tag{12}$$

Measurements of the abundance of HOCl in the Antarctic stratosphere indicate, however, that this cycle does not make a major contribution to ozone depletion. McElroy *et al.* [11] proposed another cycle involving bromine:

$$Cl + O_3 \rightarrow ClO + O_2 \tag{1}$$
$$Br + O_3 \rightarrow BrO + O_2 \tag{13}$$
$$ClO + BrO \rightarrow Cl + Br + O_2 \tag{14}$$

$$\text{Net: } 2\,O_3 \rightarrow 3\,O_2 \tag{12}$$

The reaction between ClO and BrO regenerates the free atoms in this cycle, but the reaction has actually three channels:

$$ClO + BrO \rightarrow Cl + Br + O_2 \tag{15}$$
$$ClO + BrO \rightarrow BrCl + O_2 \tag{16}$$
$$ClO + BrO \rightarrow Br + OClO \tag{17}$$

Laboratory investigations have shown that the first and third channels are approximately equally fast, whereas production of BrCl is relatively slow.

In 1987 we proposed a mechanism involving the self reaction of ClO to form chlorine peroxide [12], a compound which had not been previously characterized:

$$ClO + ClO + M \rightarrow Cl_2O_2 + M \tag{18}$$
$$Cl_2O_2 + h\nu \rightarrow 2\,Cl + O_2 \tag{19}$$
$$2\,(Cl + O_3 \rightarrow ClO + O_2) \tag{1}$$

$$\text{Net: } 2\,O_3 + h\nu \rightarrow 3O_2 \tag{12}$$

The reaction of ClO with itself has three bimolecular channels, as is the case for the ClO + BrO reaction:

$$ClO + ClO \rightarrow Cl_2 + O_2 \qquad (20)$$
$$ClO + ClO \rightarrow Cl + ClOO \qquad (21)$$
$$ClO + ClO \rightarrow Cl + OClO \qquad (22)$$

These bimolecular reactions are too slow to be of importance in the atmosphere. However, the termolecular reaction that leads to the formation of chlorine peroxide can occur efficiently in the lower stratosphere at high latitudes, as it is facilitated by the lower temperatures and higher pressures prevailing in those regions.

The structure of the product formed in the termolecular ClO self reaction has been shown by both theory [13] and experiment [14] to be indeed ClOOCl, rather than ClOClO (see Figure 4). Furthermore, photodissociation of ClOOCl yields predominantly Cl atoms, rather than ClO radicals, as we and others have shown in laboratory experiments [15, 16]. Formation of ClOO is also possible; however, even under polar stratospheric conditions ClOO rapidly decomposes to yield free chlorine atoms, since the Cl-OO bond strength is only about 5 kcal/mole.

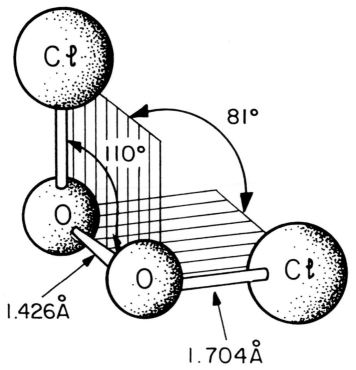

Fig. 4: Structure of chlorine peroxide [14].

POLAR STRATOSPHERIC CLOUDS

In contrast to the troposphere, the stratosphere is extremely dry and practically cloudless–the concentration of water vapor is measured in parts per million, and is, in fact, comparable to that of ozone. Normally, there is a thin layer of aerosol throughout the lower stratosphere–a haze–first described in 1961 by Junge and co-workers [17]. The aerosols droplets are mostly liquid and consist of supercooled aqueous sulfuric acid solutions–about 75% H_2SO_4 by weight at mid and low latitudes, with radius roughly 0.1 μm. The "background" number density falls typically in the 1–10 cm^{-3} range, but it may increase between one and two orders of magnitude following the injection of SO_2 directly into the stratosphere by major volcanic eruptions, such as El Chichon in 1982 and Mount Pinatubo in 1991. The SO_2 is oxidized in the stratosphere to form H_2SO_4 on a time scale of months, and the excess particles decay over a period of a few years. SO_2 emitted at the earth's surface by processes such as the combustion of coal is mostly oxidized and removed from the atmosphere (as acid rain) before reaching the stratosphere. The background aerosols arise mainly from COS, which is also emitted at the earth's surface but is sufficiently stable to reach the stratosphere, where it photooxidizes to yield SO_2 and subsequently H_2SO_4.

Polar stratospheric clouds (PSCs) form seasonally over the polar regions. Satellite-based instruments have been monitoring these clouds long before the discovery of the ozone hole by observing the attenuation of sun light, which is scattered at wavelengths characteristic of the size of the cloud particles. In the winter and spring months, sharp increases in the attenuation resulting from the presence of PSCs are evident, followed by a decrease in the summer after the cloud particles evaporate or sediment out.

The PSC particles form by condensation of vapors on the pre-existing sulfuric acid aerosols. Observations have also indicated the existence of different types of clouds, the two most common ones being often referred to as Type I and Type II PSCs. The later ones are present only when the temperatures fall below the frost point of water, which is typically about 185–187 K; the particles are roughly 10 micrometer in diameter and are composed principally of water-ice crystals. These Type II clouds are more prevalent over Antarctica, because of its lower temperatures compared to the Arctic. Type I PSCs are observed more frequently, forming at temperatures above the frost point of water but below 190–195 K, with particle diameters of roughly one micrometer. Thermodynamic considerations coupled with laboratory experiments indicate that Type I PSC particles consist most likely of nitric acid trihydrate (NAT), which is the most stable condensed phase under the conditions which lead to the formation of these clouds. However, more recent laboratory experiments indicate that the aerosol particles are not likely to freeze at temperatures above the water frost point [18]. Furthermore, investigations performed in the field [19] and in the laboratory [20, 21] indicate that the liquid sulfuric acid droplets should absorb significant amounts of nitric acid and of water vapors under polar stratospheric conditions, with a very rapid change in composition–and hence size–occurring over a narrow temperature range [22, 23], thus explaining the formation of Type I clouds below a certain temperature threshold. Hence, it appears that Type I PSCs are sometimes liquid–particularly over the Arctic–and sometimes solid.

CHEMISTRY ON CLOUD PARTICLES

Laboratory experiments, field observations and atmospheric modeling calculations have now established that chemical reactions occurring on PSC particles play a central role in polar ozone depletion. These reactions have two separate effects, referred to as chlorine activation and nitrogen deactivation: first, chlorine is transferred from the relatively inert reservoir compounds HCl and $ClONO_2$ into forms that can be readily photolyzed (mostly Cl_2); and second, nitrogen oxides are removed from the gas phase, through incorporation of nitric acid into the PSCs, thus preventing the formation of $ClONO_2$, a species that interferes with the catalytic chlorine cycles which destroy ozone. In fact, some PSC particles grow large enough to precipitate, permanently removing nitric acid from the stratosphere; this is the process labeled denitrification.

The following are the key chlorine activation reactions:

$$ClONO_2 + HCl \rightarrow Cl_2 + HNO_3 \tag{23}$$
$$ClONO_2 + H_2O \rightarrow HOCl + HNO_3 \tag{24}$$
$$HOCl + HCl \rightarrow Cl_2 + H_2O \tag{25}$$

The most important reaction is (23), the first one among these three. Our early laboratory experiments indicated that this reaction does not occur in the gas phase at significant rates [24a]. However, we noticed in those experiments that this and other reactions involving chlorine nitrate are very sensitive to surface effects; similar observations had been reported by Sherry Rowland's group [24b]. In 1986 Solomon *et al.* [10] suggested that reactions (23) and (24) would be promoted by PSCs, setting the stage for efficient ozone depletion over Antarctica. In 1987 we carried out a series of laboratory measurements showing that indeed reaction (23) is remarkably efficient, requiring only a few collisions of the reactant $ClONO_2$ with ice exposed to HCl vapor [25]. This peculiar chemical reactivity of ice was surprising; in contrast to liquid water–and the well-established chemistry of aqueous ions–reactions promoted by ice at temperatures down to 180 K were unprecedented. Our results were subsequently corroborated by several other groups (for a review, see Kolb *et al.* [26]).

MECHANISM FOR CHLORINE ACTIVATION REACTIONS

The net effect of reactions (24) and (25) occurring sequentially is reaction (23), suggesting that the mechanism for this last reaction involves two steps–reactions (24) and (25). Also, while the high efficiency of reaction (24) on ice is understandable–the reactant H_2O is already in condensed phase–both reactants for reaction (23) are originally in the gas phase; hence, it proceeds most likely through sequential, rather than simultaneous collisions of the reactants with the surface. This implies that at least one of the reactants–most likely HCl–has a high affinity for the condensed phase. Previous measurements had shown, though, that HCl is only sparingly soluble in ice; more recent investigations have indicated, however, that monolayer amounts of HCl are taken up by the ice surface under temperature and HCl partial pressure conditions similar to those prevailing in the polar stratosphere [26, 27].

Physical adsorption of HCl is expected to incorporate only negligible amounts of HCl on the ice surface [28], even allowing for a very strong hydrogen bond. These expectations led early on to the suggestion that reaction (23) should not occur on ice under stratospheric conditions of temperature and HCl partial pressure, and the results of laboratory measurements were considered to be artifacts resulting from grain boundary effects.

The high affinity of HCl for the ice surface can be explained, however, by assuming that the HCl solvates, forming hydrochloric acid, as is the case with liquid water. This process is exothermic by ~18 kcal/mole; in contrast, a hydrogen bond of HCl with ice is estimated to be only 5 kcal/mole. Solvation can occur because the surface layers of ice are disordered, with the water molecules there having much larger mobility than in the bulk crystal. This behavior leads to the formation of a quasi-liquid layer on the ice surface which can be experimentally observed at temperatures down to about 240 K. The presence of HCl strongly depresses the freezing point (a 9-molal HCl solution freezes at ~190 K), so that formation of a quasi-liquid HCl solution layer appears plausible under polar stratospheric conditions [28]. Furthermore, the large reaction probabilities for reactions (23) and (25) can be understood with the quasi-liquid layer model: the mechanism for these reactions most likely involves aqueous-like ions. In contrast, adsorbed molecular HCl is expected to behave rather similarly to gaseous HCl, and hence should react only very slowly with $ClONO_2$, as noted above.

Laboratory measurements of the uptake of HCl by NAT have also been conducted. For NAT, however, an additional parameter that needs to be taken into account is its H_2O vapor pressure: at any given temperature it can have a range of values, whereas for water-ice it has, of course, only one value. Observations show that when the H_2O vapor pressure of NAT approaches that of ice its surface takes up as much HCl as ice, whereas the amount decreases by more than two orders of magnitude as the H_2O vapor pressure drops. The reaction probabilities for reactions (23) and (25) behave accordingly: they have large values when the H_2O vapor pressure of NAT is within a factor of ~3 of that of ice [26, 29]. It appears that the uptake as well as the reaction probabilities are controlled by the availability of water at the surface.

The chlorine activation reactions (23–25) occur rather slowly on sulfuric acid aerosols at mid-latitudes, because HCl is only very sparingly soluble in relatively concentrated H_2SO_4 solutions, and because the water in those solutions is not available for reaction with $ClONO_2$. However, as the temperature drops towards high latitudes and the sulfuric acid solution becomes more dilute, the HCl becomes more soluble, and the efficiency of these reactions increases dramatically over a relatively narrow temperature range: under conditions of thermodynamic stability for type I PSCs, chlorine activation occurs rapidly on both liquid and solid particles [20].

MEASUREMENTS OF KEY SPECIES OVER ANTARCTICA

Beginning in 1986, a series of field experiments were designed to test the various hypothesis which had been put forth to explain the Antarctic ozone hole. For example, observations of upward flow in the polar stratosphere would have verified the "dynamics only" theory. The solar cycle theory would have been supported by observations of high NO_x levels in regions where ozone was being depleted; furthermore, according to this theory in years of low solar activity the polar stratosphere would recover to normal pre-ozone hole conditions. In contrast, expectations from the halogen theory were low levels of nitrogen oxides, and high levels of active chlorine.

The first observational campaign designed to test these theories–the National Ozone Experiment (NOZE)–was mounted in the winter of 1986. The NOZE scientists, based on McMurdo, Antarctica, and led by Susan Solomon, performed remote sensing measurements of key stratospheric species such as NO_2, O_3, and chlorine free radicals. Their findings provided the first evidence that the chemistry of the Antarctic stratosphere was highly perturbed, with the halogen chemistry mechanism appearing most likely. A second NOZE campaign, as well as the Airborne Antarctic Ozone Experiment (AAOE) were conducted in 1987. This expedition was based on Punta Arenas, Chile: the results from flights conducted over a six week period largely confirmed and extended the results from the NOZE experiment. The observations provided very clear evidence of a strong downward flow within the Antarctic polar vortex, which is the region of the stratosphere where the ozone hole develops: contrary to predictions based on the dynamics only theory, the levels of tracers such as the CFCs, CH_4 and N_2O were typical of air coming from higher altitudes–"aged" air which had been exposed to short-wavelength ultraviolet radiation. Furthermore, the concentration of NO_x was remarkably low, in contrast with expectations from the solar cycle theory.

The chemical CFC–halogen hypothesis received strong support from the results of the field experiments. The measurements conducted by James Anderson and co-workers [31] showed that the ClO concentrations in the polar vortex were greatly enhanced (see Figure 5), showing that more than half of the chlorine was present there as a free radical–chlorine that had been released in the stratosphere at lower latitudes by the decomposition of

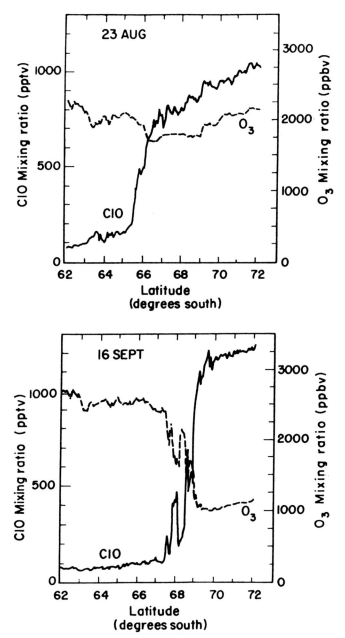

Fig. 5: Mixing ratios of ClO and ozone measured in 1987 during the AAOE expedition; adapted from [31].

the CFCs. These measurements, coupled with the observations of the rate of disappearance of ozone, indicated that the chlorine peroxide cycle that we had suggested (reactions 1, 18, and 19) was reponsible for more than two thirds of the ozone loss, with most of the rest of the loss resulting from bromine chemistry.

The unique instruments required for all these experiments–as well as the halogen hypothesis itself–were developed in a relatively short amount of time: the community was well prepared for these new developments, since it had vigorously continued for over a decade the pursuit of scientific research on the stratospheric ozone issue.

CONCLUSIONS

Much remains to be learned about stratospheric chemistry–and, in more general terms, about the physics and chemistry of the global atmosphere. On the other hand, the cause–effect relationship between human-produced chemicals and ozone depletion is rather well established now: the signals connected with the Antarctic ozone hole are very large indeed. The stratospheric ozone issue has shown us that mankind is quite capable of significantly affecting the atmosphere on a global scale: the most striking effects of CFCs, which are emitted mostly in the north, are seen as far away as possible from the sources, namely over the South Pole. This global problem has also shown us that different sectors of society can work together–the scientific community, industry, environmental organizations, government representatives and policy makers–to reach international agreements: the Montreal Protocol on Substances that Deplete the Ozone Layer has established a very important precedent for the solution of global environmental problems.

ACKNOWLEDGEMENTS

Many colleagues and collaborators have contributed to the work described above; while only some of them have been mentioned here, I would like to express my appreciation to all of them. I am grateful to my past and present students and associates for their dedication in carrying out research in my laboratory. I am especially indebted to my wife, Luisa, who has been my closest collaborator for many years. I also want to acknowledge the sponsors of our research, which have included the National Aeronautics and Space Administration and the National Science Foundation.

REFERENCES

1. Crutzen, P.J., *The Influence of Nitrogen Oxides on Atmosphere Ozone Content*, Q. J. R. Meteorol. Soc., **96**, 320–325 (1970).
2. Johnston, H.S., *Reduction of Stratospheric Ozone by Nitrogen Oxide Catalysts from Supersonic Transport Exhaust*, Science, **173**, 517–522 (1971).
3. Molina, M.J., and F.S. Rowland, *Stratospheric Sink for Chlorofluoromethanes: Chlorine Catalysed Destruction of Ozone*, Nature, **249**, 810–814 (1974).
4. Lovelock, J.E., R.J. Maggs, and R.J. Wade, *Halogenated Hydrocarbons in and over the Atlantic*, Nature, **241**, 194–196 (1973).
5. Stolarski, R.S., and R. Cicerone, *Stratospheric Chlorine: A Possible Sink for Ozone*, Can. J. Chem., **52**, 1610–1615 (1974).
6. Shen, T.-L., P.J. Wooldridge, and M.J. Molina, *Stratospheric Pollution and Ozone Depletion*, in "Composition, Chemistry, and Climate of the Atmosphere", ed. H.B. Singh, Van Nostrand Reinhold, New York (1995).

7. Abbatt, J.P.D. and M.J. Molina, *Status of Stratospheric Ozone Depletion*, Ann. Rev. Energy & Environ., **18**, 1–29 (1993).

8. World Meteorological Organization, *Scientific Assessment of Ozone Depletion: 1994.* WMO Global Ozone Research and Monitoring Project, Report No. 37 (1994).

9. Farman, J.C., B.G. Gardiner, and J.D. Shanklin, *Large Losses of Total Ozone in Antarctica Reveal Seasonal ClO_x/NO_x Interactions*, Nature, **315**, 207–210 (1985).

10. Solomon, S., R.R. Garcia, F.S. Rowland, and D.J. Wuebbles, *On the Depletion of Antarctic Ozone*, Nature, **321**, 755–758 (1986).

11. McElroy, M.B., R.J. Salawitch, S.C. Wofsy, and J.A. Logan, *Reduction of Antarctic Ozone due to Synergistic Interactions of Chlorine and Bromine*, Nature, **321**, 759–762 (1986).

12. Molina, L.T., and M.J. Molina, *Production of the Cl_2O_2 from the Self-Reaction of the ClO Radical*, J. Phys. Chem., **91**, 433–436 (1987).

13. McGrath, M.P., K.C. Clemitshaw, F.S. Rowland, and W.G. Hehre, *Structures, Relative Stabilities, and Vibrational Spectra of Isomers of Cl_2O_2: The Role of Chlorine Oxide Dimer in Antarctic Ozone Depleting Mechanism*, J. Phys. Chem., **94**, 6126–6132 (1990).

14. Birk, M., R.R. Friedl, E.A. Cohen, H.M. Pickett, and S.P. Sander, *The Rotational Spectrum and Structure of Chlorine Peroxide*, J. Chem. Phys., **91**, 6588–6597 (1989).

15. Molina, M.J., A.J. Colussi, L.T. Molina, R.N. Schindler, and T.-L. Tso, *Quantum Yield of Chlorine-Atom Formation in the Photodissociation of Chlorine Peroxide (ClOOCl) at 308 nm*, Chem. Phys. Lett., **173**, 310–315 (1990).

16. Cox, R.A. and G.D. Hayman, *The Stability and Photochemistry of Dimers of the ClO Radical and Implications for Antarctic Ozone Depletion*, Nature, **332**, 796–800 (1988).

17. Junge, C.E., C.W. Chagnon, and J.E. Manson, *Stratospheric Aerosols*, J, Meteorol., **18**, 81–108 (1961).

18. Koop, T., U.M. Biermann, W. Raber, B. Luo, P.J. Crutzen, and T. Peter, *Do Stratospheric Aerosol Droplets Freeze Above the Ice Frost Point?* Geophys. Res. Lett. **22**, 917–920 (1995).

19. Arnold, F., *Stratospheric Aerosol Increases and Ozone Destruction: Implications from Mass Spectrometer Measurements*, Ber. Bunsenges. Phys. Chem., **96**, 339–350 (1992).

20. Zhang, R., P.J. Wooldridge, and M.J. Molina, *Vapor Pressure Measurements for the $H_2SO_4/HNO_3/H_2O$ and $H_2SO_4/HCl/H_2O$ Liquid Systems, Incorporation of Stratospheric Acids into Background Sulfate Aerosols*, J. Phys. Chem., **97**, 8541–8548 (1993).

21. Molina, M.J., R. Zhang, P.J. Wooldridge, J.R. McMahon, J.E. Kim, H.Y. Chang, and K.D. Beyer, *Physical Chemistry of the $H_2SO_4/HNO_3/H_2O$ System: Implications for Polar Stratospheric Clouds*, Science, **261**, 1418–1423 (1993).

22. Beyer, K.D., S.W. Seago, H.Y. Chang, and M.J. Molina, *Composition and Freezing of Aqueous H_2SO_4/HNO_3 Solutions under Polar Stratospheric Conditions*, Geophys. Res. Lett., **21**, 871–874 (1994).

23. Carslaw, K.S., B.P. Luo, S.L. Clegg, Th. Peter, P. Brimblecombe, and P.J. Crutzen, *Stratospheric Aerosol Growth and HNO_3 Gas Phase Depletion from Coupled HNO_3 and Water Uptake by Liquid Particles*, Geophys. Res. Lett., **21**, 2479–2482 (1994).

24. (a) Molina, L.T., M.J. Molina, R.A. Stachnick, and R.D. Tom, *An Upper Limit to the Rate of the HCl + $ClONO_2$ Reaction*, J. Phys. Chem., **89**, 3779–3781 (1985).
    (b) Rowland, F.S., S. Sato, H. Khwaja, and S.M. Elliott, *The Hydrolysis of Chlorine Nitrate and its Possible Atmospheric Significance*, J. Phys. Chem., **90**, 1985–1988 (1986).

25. Molina, M.J., T.-L. Tso, L.T. Molina, and F.C.-Y, Wang, *Antarctic Stratospheric Chemistry of Chlorine Nitrate, Hydrogen Chloride and Ice: Release of Active Chlorine*, Science, **238**, 1253–1257 (1987).

26. Kolb, C.E., D.R. Worsnop, M.S. Zahniser, P. Davidovits, C.F. Keyser, M.-T. Leu, M.J. Molina, D.R. Hanson, A.R. Ravishankara, L.R. Williams, and M.A. Tolbert, *Laboratory Studies of Atmospheric Heterogeneous Chemistry*, in **Advances Series in Physical Chemistry**: "Progress and Problems in Atmospheric Chemistry", ed. J.R. Barker, World Scientific Publishing (1995).

27. Abbatt, J.P.D., K.D. Beyer, A.F. Fucaloro, J.R. McMahon, P.J. Wooldridge, R. Zhang, and M.J. Molina, *Interaction of HCl Vapor with Water-Ice: Implications for the Stratosphere*, J. Geophys. Res., **97**, 15819–15826 (1992).

28. Molina, M.J., *The Probable Role of Stratospheric 'Ice' Clouds: Heterogeneous Chemistry of the*

Ozone Hole, in "Chemistry of the Atmosphere: The Impact of Global Change", ed. J.G. Calvert, Blackwell Scientific Publications, Oxford, U.K. (1994).

29. Abbatt, J.P.D., and M.J. Molina, *Heterogeneous Interactions of $ClONO_2$ and HCl on Nitric Acid Trihydrate at 202 K*, J. Phys. Chem., **96**, 7674–7679 (1992).

30. Farman, J.C., *What Hope for the Ozone Layer Now?* New Scientist, **116**, No. 1586, 50–54 (1987).

31. Anderson, J.G., W.H. Brune, and M.H. Proffitt, *Ozone Destruction by Chlorine Radicals within the Antarctic Vortex: The Spatial and Temporal Evolution of $ClO-O_3$ Anticorrelation based on In Situ ER-2 Data*, J. Geophys. Res., **94**, 11,465–11,479 (1989).

F. Sherwood Rowland

F. SHERWOOD ROWLAND

I was born on June 28, 1927, the second of three sons, in the small central Ohio town of Delaware, the home of Ohio Wesleyan University. My father and mother had moved there the previous year when he took the position of Professor of Mathematics and Chairman of the Department at Ohio Wesleyan. All of my elementary and high school education was received in the Delaware public schools from an excellent set of teachers. The Delaware school system then believed in accelerated promotion, so that I entered first grade at age 5 and skipped the fourth grade entirely, with the result that I entered high school at 12 and graduated a few weeks before my sixteenth birthday. The college preparatory curriculum was strong on Latin, English, History, Science and Mathematics. The academic side of high school was easy for me, and I enjoyed it. In several summers of my early teens, the high school science teacher entrusted to me during his two week vacations the operation of the local volunteer weather station, an auxiliary part of the U.S. weather service–maximum and minimum temperatures and total precipitation. This was my first exposure to systematic experimentation and data collection.

Our home was filled with books, and all of us were avid readers. My reading at that time ran toward naval history, which was complemented with realistic scale-models and simulated naval battles using an elaborate mathematical system for rating each warship and the effects of combat on them. During my sophomore year in high school, my math teacher, who also coached tennis and basketball, encouraged me to take up tennis–which led me onto the varsity tennis team for my junior and senior years, and into a full decade of intense athletic competition. As a senior, I played on the varsity basketball team.

After graduation from high school in 1943, almost all of my male classmates immediately entered the military services. However, because I was still well under the compulsory draft age of 18, I enrolled at Ohio Wesleyan and attended the university year-round for the next two years. During these war years, only 30 or 40 civilian males were on campus, plus about 200 naval officer trainees and 1,000 women. With so few men available, I played on the University basketball and baseball teams, and wrote much of the sports page for the University newspaper.

My accelerated academic schedule made me eligible for my final year of university in June, 1945, as I approached my 18th birthday. However, with the fighting in the Pacific and the continuing military draft, I enlisted in a Navy program to train radar operators. The Pacific war ended while I was still in basic training near Chicago, and I served the next year in several midwestern Naval Separation Centers, as the 10,000,000 Americans who had preceded me into the military were returned to civilian life. A major amount of this Navy time was devoted to competitive athletics for the Navy base teams, and I emerged after 14 months as a non commissioned officer with a rating of Specialist (Athletics) 3rd class. My first real opportunity to see the rest of the United States came when I was transferred to San Pédro, California for discharge from the Navy.

I then hitchhiked 2000 miles back to Ohio, traveling through Yosemite and Yellowstone Park on the way.

This year away from the academic life convinced me that at age 19, there was little reason for me to seek a quick finish to my undergraduate education. I therefore arranged my schedule to take two more years rather than one to graduate, and continued to play basketball on the university team. My coursework at Ohio Wesleyan emphasized science within a liberal arts curriculum, with more or less equal amounts of chemistry, physics and mathematics, and majors in all three fields. As had been the case in high school, I really enjoyed the academic side of university life.

I do not honestly remember when the decision that I would go to graduate school was made. My father had studied for his Ph.D., and all of us took it for granted that I would, too. Furthermore, both my parents had firm convictions that the University of Chicago, which each had attended, was not just the best choice for graduate work, but the only choice. So I applied to the Department of Chemistry at the University of Chicago for Fall 1948, and was duly admitted. All service veterans were entitled to a certain number of months (27 in my case) of paid university education, and I had not used any of these credits during my undergraduate years at Ohio Wesleyan because faculty children did not pay tuition, and I lived at home. I therefore didn't apply for any of the teaching assistantships or academic fellowships, and was quite surprised after arriving in Chicago to find that many of my fellow students were being paid by the University to attend graduate school. In subsequent years, I was supported by an Atomic Energy Commission (A.E.C.) national fellowship.

At that time, the Chemistry Department of the University of Chicago had a policy of immediately assigning each new graduate student to a temporary faculty adviser prior to the choice of an individual research topic. My randomly assigned mentor was Willard F. Libby, who had just finished developing the Carbon-14 Dating technique for which he received the 1960 Nobel Prize. Bill Libby (although I never called him anything but "Professor Libby" until I was more than 40 years old) was a charismatic, brusque (on first meeting, "I see you made all A's in undergraduate school. We're here to find out if you are any damn good!") dynamo, with a very wide range of fertile ideas for scientific research. I settled automatically and happily into his research group, and became a radiochemist working on the chemistry of radioactive atoms. Almost everything I learned about how to be a research scientist came from listening to and observing Bill Libby.

The first nuclear reactor had been built by Enrico Fermi in 1942 under the football stands at the University of Chicago, and the post-war university had managed to capture many of the leading scientists from the Manhattan Project into the Physics and Chemistry departments. My impression at the time (and now in retrospect 45 years later) was that this was an unbelievably exciting time in the physical sciences at the University of Chicago. My physical chemistry course was taught by Harold Urey for two quarters and in the third quarter by Edward Teller; inorganic chemistry was given by Henry Taube; radiochemistry by Libby. I also attended courses on Nuclear Physics given by Maria Goeppert Mayer and by Fermi. (The chemistry student grapevine said, "Go to any lecture that Fermi gives on any subject"). Urey and Fermi already had been awarded Nobel prizes, and Libby, Mayer and Taube were to receive theirs in the future.

My thesis concerned the chemical state of cyclotron-produced radioactive bromine atoms. The nuclear process not only creates a radioactive atom, but breaks it loose from all of its chemical bonds. These highly energetic atoms exist only in very, very low concentration, but can subsequently be traced by their eventual radioactive decay. Bill Libby gave his graduate students an unusual amount of leeway in how they chose to use their time, and was a superb research superviser–supporting, encouraging, but never letting one forget that intensive critical thought, together with unrelenting hard work on experiments, underlay all progress in our research.

My interest in competitive athletics also continued unabated in graduate school. Because of the atypical structure of its undergraduate college system, the University of Chicago, unlike almost all other American universities, permitted graduate students to compete in intercollegiate athletics. During my first graduate year, I played both basketball and baseball for the University teams. I continued to play baseball for the University during the spring for two more years, and spent both of those summers playing semi-professional baseball for a Canadian team in Oshawa, Ontario. Each winter I also played for several basketball teams around the city of Chicago.

Without a doubt, however, the major extracurricular event of those four years at the University of Chicago was meeting and then marrying on June 7, 1952, Joan Lundberg, also a graduate of the University. We have now shared more than 43 years of married life–and shared is really the descriptive word. I finished my Ph. D. thesis in August of 1952, and we went off to Princeton University in September of that year for my new position of Instructor in the Chemistry Department. Our daughter Ingrid was born in Princeton in the summer of 1953, and our son Jeffrey in Huntington, Long Island, in the summer of 1955.

In each of the years from 1953–55, I spent the summer in the Chemistry Department of the Brookhaven National Laboratory. An early experiment there of putting a powdered mixture of the sugar glucose and lithium carbonate into the neutron flux of the Brookhaven nuclear reactor resulted in a one-step synthesis of radioactive tritium-labeled glucose, an article in *Science*, and a new sub-field of tritium "hot atom" chemistry. The A.E.C. also expressed considerable interest in this tracer chemistry, and offered support for continuation of the research.

In 1956, I moved to an Assistant Professorship at the University of Kansas, which had just completed a new chemistry building including special facilities for radiochemistry. Contract support from the A.E.C. was already approved, and in place when I arrived that summer. Several excellent graduate students interested in radiochemistry joined my research group that summer, and were shortly joined by others and by a series of postdoctoral research associates, including many from Europe and Japan. This research group was very productive for the next eight years, chiefly investigating the chemical reactions of energetic tritium atoms and I moved through the ranks to a full Professorship. Both Ingrid and Jeff grew up knowing the members of the group-meeting everyone at our regular home seminars, and from an early age occasionally visiting the laboratory. During these Kansas years, too, the everyday routine was that the entire family came home for lunch. Later on in California, Ingrid and Jeff each worked regularly (but unpaid) drafting slide and journal illustrations for the chemistry department, and thereby continuing to know the members of my research group.

The Irvine campus of the University of California was scheduled to open for students in September, 1965, and I went there in August, 1964, as Professor of Chemistry and the first Chairman of the Chemistry Department. The A.E.C. support turned out to be truly long-term, surviving this transfer, and then the transformations of the A.E.C. into the Energy Research and Development Administration and then into the Department of Energy. That basic contract finally terminated in 1994, by which time NASA was furnishing the major support for our continuing research.

"Hot atom" chemistry continued to play a major role in our research efforts at the University of California Irvine. However, I have deliberately followed a policy of trying to instill some freshness into our research efforts by every few years extending our work into some new, challenging aspect of chemistry–first, radioactive tracer photochemistry, using tritium and carbon-14; then chlorine and fluorine chemistry using the radioactive isotopes ^{38}Cl and ^{18}F.

When I decided in 1970 to retire from the Chemistry department chairmanship, I once again sought some new avenue of chemistry for our investigation. Because the state of the environment had become a significant topic for discussion both by the general public and within our family, I traveled to Salzburg, Austria, for an International Atomic Energy Agency meeting on the environmental applications of radioactivity. Afterward on the train to Vienna, I shared a compartment with an A.E.C. program officer also coming from the IAEA meeting. He learned in our conversation that I was personally interested in atmospheric science because of my early association and admiration for the ^{14}C work of Bill Libby, and further that my research had then been supported by the A.E.C. for the previous 14 years. I in turn learned that one of his A.E.C. responsibilities was the organization of a series of Chemistry-Meteorology Workshops, with the intention of encouraging more cross-fertilization between these two scientific fields.

In due course, I was invited to the second of these workshops in January, 1972, in Fort Lauderdale, Florida, where I heard a presentation about recent measurements by the English scientist, Jim Lovelock, of the atmospheric concentrations of a trace species, the man-made chlorofluorocarbon CCl_3F, on the cruise of the *Shackleton* to Antarctica. His shipboard observations showed its presence in both the northern and southern hemispheres, although in quite low concentration. One of the special advantages cited for this molecule was that it would be an excellent tracer for air mass movements because its chemical inertness would prevent its early removal from the atmosphere.

As a chemical kineticist and photochemist, I knew that such a molecule could not remain inert in the atmosphere forever, if only because solar photochemistry at high altitudes would break it down. However, many other possible chemical fates could be imagined, and I wondered whether any of these might occur. In early 1973, my regular yearly proposal was submitted to the A.E.C. and was duly approved and funded by them. In addition to the continuation of several radiochemistry experiments, I also included in the proposal a new direction–asking the question: what would eventually happen to the chlorofluorocarbon compounds in the atmosphere?

Later that year, Mario Molina, who had just completed his Ph. D. work as a laser chemist at the University of California Berkeley, joined my research group as a postdoctoral research associate. Offered his choice among several areas for our collaborative research, Mario chose the one furthest from his previous experience and from my own experience as well, and we began studying the atmospheric fate of the chlorofluorocarbon molecules.

Within three months, Mario and I realized that this was not just a scientific question, challenging and interesting to us, but a potentially grave environmental problem involving substantial depletion of the stratospheric ozone layer. A major part of both of our careers since has been spent on the continuing threads of this original problem.

Since 1973, the work of my research group has progressively involved more atmospheric chemistry and less radiochemistry until now our only important use of radioisotopes is directed toward problems associated with atmospheric chemistry. This research work has been conducted at the University of California Irvine by a strong, hard-working group of postdoctoral and graduate student research associates, together with some able technical specialists.

The chlorofluorocarbon-ozone problem became a highly visible public concern in late 1974, and brought with it many new scientific experiments, and also legislative hearings, extensive media coverage, and a much heavier travel schedule for me. This change came after both Ingrid and Jeff had moved away from home for their own university educations, leaving Joan free to accompany me in these travels. She has attended–and sat through with perceptive interest–countless scientific meetings since 1975. She quickly became quite conversant with the general scientific aspects of ozone depletion, and has been a knowledgeable and trusted confidante through all of the last two decades of ozone research. Ingrid and Jeff, too, have maintained close contact and support during those often controversial years.

In many ways, the understanding of atmospheric chemistry is still in an early stage. The necessary instrumental precision and sensitivity for dealing with chemical species in such low concentrations has only been progressively available over the last two decades, and of course the trace composition of the atmosphere is highly variable around the world. The research group has been heavily involved in a series of regional and global experiments, often since 1988 as participants in comprehensive aircraft-based atmospheric field research. Some of this research involves challenging and interesting scientific puzzles, and some can also be described as directed toward global environmental problems. As with the ozone depletion capability of the chlorofluorocarbons, one does not always know until well into the work whether it belongs to the second category as well as the first. We continue to find fascination in the chemistry of the atmosphere.

NOBEL LECTURE IN CHEMISTRY

December 8, 1995

by

F. SHERWOOD ROWLAND

Department of Chemistry, University of California, Irvine, California 92697-2025, USA

INTRODUCTION

There is a well-known mathematical exercise called the Konigsberg Bridge Problem in which the solution involves crossing each of the city's bridges once and only once. When my wife and I first visited Stockholm together in April 1974, we had the distinct impression that our host, Paul Crutzen, was trying to illustrate the corresponding solution for Stockholm by leading us across every single bridge, on foot. We were in Stockholm because I had sent Paul a preprint of our first scientific paper on the chlorofluorocarbon/stratospheric ozone problem [1], and he had invited me to present a lecture about this work to the Department of Meteorology of Stockholm University.

The starting point for that work was the discovery by Jim Lovelock that the molecule, CCl_3F, a substance for which no natural sources have been found, was present in the Earth's atmosphere in quantities roughly comparable to the total amount manufactured up to that date. Lovelock had earlier invented an extremely sensitive detection system employing electron capture (EC) by trace impurities, and attached it to the column of effluent gases from a gas chromatograph (GC), a device which separates a mixture of gases into its individual components. With this ECGC apparatus, Lovelock initially established that CCl_3F was always detectable in the atmosphere near his home in western Ireland, and then that it was also present in all of the air samples tested on the voyage of the R.V. Shackleton from England to Antarctica (Figure 1) [2]. The ECGC instrument is especially sensitive for CCl_3F, as well as for many other similar molecules in this chlorofluorocarbon class.

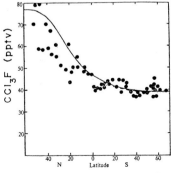

Figure (1). Mixing ratios of CCl_3F measured by James Lovelock [2] on board R. V. Shackleton in 1971.

The appearance in the atmosphere of a new, man-made molecule provided a scientific chemical challenge: Was enough known about the physico-chemical behavior under atmospheric conditions of molecules such as CCl_3F to allow prediction of its fate, once released into the environment? In 1973, I included in my yearly proposal to the U.S. Atomic Energy Commission, which had sponsored my research involving radioactive tracer species since 1956, a predictive study of the atmospheric chemistry of CCl_3F, in addition to the continuation of other studies already in progress. The A.E.C. agreed to permit this new venture, subject only to the requirement that I shift some of the funding already scheduled to be available for the studies on radio-activity.

When Mario Molina joined my research group as a postdoctoral research associate later in 1973, he elected the chlorofluorocarbon problem among several offered to him, and we began the scientific search for the ultimate fate of such molecules. At the time, neither of us had any significant experi-ence in treating chemical problems of the atmosphere, and each of us was now operating well away from our previous areas of expertise.

The search for any removal process which might affect CCl_3F began with the reactions which normally affect molecules released to the atmosphere at the surface of the Earth. Several well-established tropospheric sinks–chemi-cal or physical removal processes in the lower atmosphere–exist for most molecules released at ground level:

(1) Colored species such as the green molecular chlorine, Cl_2, absorb visi-ble solar radiation, and break apart, or photodissociate, into individual atoms as the consequence;

(2) Highly polar molecules, such as hydrogen chloride, HCl, dissolve in raindrops to form hydrochloric acid, and are removed when the drops actu-ally fall; and

(3) Almost all compounds containing carbon-hydrogen bonds, for exam-ple CH_3Cl, are oxidized in our oxygen-rich atmosphere, often by hydroxyl radical as in reaction (1).

$$CH_3Cl + HO \rightarrow H_2O + CH_2Cl \qquad (1)$$

However, CCl_3F and the other chlorofluorocarbons such as CCl_2F_2 and CCl_2FCClF_2* are transparent to visible solar radiation and those wavelengths of ultraviolet (UV) radiation which penetrate to the lower atmosphere, are basically insoluble in water, and do not react with HO, O_2, O_3, or other oxi-dizing agents in the lower atmosphere. When all of these usual decomposi-tion routes are closed, what happens to such survivor molecules?

---

* The chlorofluorocarbons (or CFCs) are often technically identified by a numerical formula which gives number of F atoms in the units digit, number of H atoms plus 1 in the tens digit, and number of C atoms minus 1 in the hundreds digit: $CCl_2FCClF_2 \rightarrow$ CFC-113; $CCl_3F \rightarrow$ CFC-11, dropping the 0 from 011; and $CCl_2F_2 \rightarrow$ CFC-12. During the past decade, the hydrogen-containing species have been distinguished from the fully-halogenated CFCs by subdivision into the two new categories of HCFCs (e.g. $CHClF_2 \rightarrow$ HCFC-22) and HFCs (e.g. $CH_2FCF_3 \rightarrow$ HFC-134A, with the A distinguishing this molecule from its isomer CHF_2CHF_2).

STRATOSPHERIC CHEMICAL PROCESSES

The radiation from the sun has wavelengths visible to humans from violet (400 nanometers, nm) to red (700 nm), plus invisible infrared (> 700 nm) and ultraviolet (< 400 nm) wavelengths. The energy of the radiation increases as the wavelengths shorten, and the absorption of highly energetic ultraviolet (UV) radiation usually causes the decomposition of simple atmospheric molecules.

All multi-atom compounds are capable of absorbing UV radiation if the wavelength is short enough, and almost all will decompose after absorbing the radiation. At the top of the atmosphere, the process is so rapid that a CFC molecule would last at most a few weeks if directly released there. However, CFC molecules in the lower atmosphere are protected against this very energetic UV radiation by O_2 and O_3 molecules at higher altitudes. In the upper atmosphere, the UV wavelengths below 242 nm can be absorbed by O_2 (Figure 2), splitting it into two oxygen atoms, as in equation (2). Each of

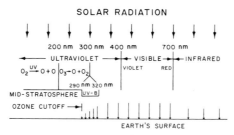

Figure (2) Solar spectrum, illustrating the absorption of ultraviolet radiation by O_2 and O_3, and the location of the UV-B range between 290 nm and 320 nm. Ultraviolet radiation < 290 nm, "the ozone cutoff", does not reach the Earth's surface.

these O atom products normally adds to another O_2 molecule in (3) to form ozone, O_3. Collision with some third molecule, M, is needed to stabilize the O_3 product. These ozone molecules in turn can absorb ultraviolet radiation in (4) and split off an O atom. Such absorption is especially strong for wavelengths shorter than 290 nm. Again, these O atoms usually reform ozone by reacting with O_2.

$$O_2 + UV \text{ light} \rightarrow O + O \qquad (2)$$
$$O + O_2 + M \rightarrow O_3 + M \qquad (3)$$
$$O_3 + UV \text{ light} \rightarrow O + O_2 \qquad (4)$$

Ozone is actually quite chemically reactive and sometimes intercepts the O atoms, as in (5). This set of four reactions involving O, O_2 and O_3 was already recognized by Chapman in 1930 [3].

$$O + O_3 \rightarrow O_2 + O_2 \qquad (5)$$

Through these processes, and others such as the free radical reactions of NO_x and ClO_x described below, a balance of ozone is maintained in the atmosphere by which about 3 parts in 10^7 of the entire atmosphere are present as O_3, versus almost 21% as O_2. About 90% of these ozone molecules are present at altitudes between 10 and 50 kilometers (km), i.e. in the stratosphere, where the mixing ratio of O_3 can rise as high as 1 part in 10^5.

The solar UV energy absorbed in (4) is converted into heat by processes such as energy transfer to M in (3), providing a heat source in the 30–50 km altitude range. This heat source creates the stratosphere, the region between 15 and 50 km in which the temperature increases with altitude. The ozone layer thus performs two important physical processes: it removes short wavelength UV radiation, and changes this energy into heat, both creating and maintaining the stratosphere.

Because both O_2 and O_3 can absorb short wavelength UV radiation, no solar radiation with wavelengths < 290 nm penetrates below the stratosphere. The CCl_3F molecule was known from laboratory studies to be able to absorb UV radiation at wavelengths < 220 nm, but to encounter such solar radiation in the atmosphere the molecule must first drift randomly through the atmosphere to altitudes higher than most of the O_2 and O_3 molecules–roughly to 25 or 30 km. More than 98% of the atmosphere and 80% of the ozone lies below 30 km altitude. In this rarefied air, the CFC molecules are exposed to very short wavelength UV radiation and decompose with the release of Cl atoms, as in (6) and (7). Because at any given time, only a very small fraction of CFC molecules are found at altitudes of 30 km or higher, the average molecule survives for many decades before it is decomposed by solar UV radiation.

$$CCl_3F + UV \text{ light} \rightarrow Cl + CCl_2F \qquad (6)$$
$$CCl_2F_2 + UV \text{ light} \rightarrow Cl + CClF_2 \qquad (7)$$

In 1974, we calculated the vertical profile to be expected for CCl_3F in the stratosphere, using several different sets of eddy diffusion coefficients, the parameter used to simulate vertical motions in these calculations [4]. The resulting vertical profiles for CCl_3F are all quite similar, as illustrated in Figure 3, because the decomposition rate for this molecule escalates rapidly with increasing altitude in the 20–30 km range. With each of these parameters, the estimated average lifetime in the atmosphere for CCl_3F was in the range from 40 to 55 years, and others extended the lifetime to 75 years. The range of lifetimes calculated for CCl_2F_2 was 75 to 150 years [1,4].

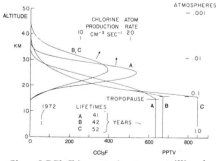

Figure (3) Vertical profiles of CCl_3F in pptv (parts per trillion by volume) at steady state, as calculated for three different diffusion models (A, B, C) and 1972 emission rate [4]. The peak photolytic destruction of CCl_3F occurs at an altitude of 28 kilometers.

The answer to our original scientific question is that the eventual fate of the CFC molecules is photodissociation in the mid-stratosphere with the release of atomic chlorine, but on a time scale of many decades. What is the fate of these chlorine atoms at an altitude of 30 km?

CHLORINE CHEMISTRY IN THE STRATOSPHERE

The major chemical components of the mid-stratosphere are well known, and their reaction rate constants with atomic Cl have been measured in the laboratory. By combining these data, the conclusion is readily reached that almost all chlorine atoms react with ozone by (8), forming another reactive molecule, ClO. This probability is about 1000 times more likely than reaction with methane, as in (9). The questions of the ultimate sinks, first for CCl_3F and then for Cl, have thus been answered, and the question now moves to a third molecule, the ClO product from (8): What happens to ClO at 30 km?

$$Cl + O_3 \rightarrow ClO + O_2 \tag{8}$$
$$Cl + CH_4 \rightarrow HCl + CH_3 \tag{9}$$

Two important answers appear: reaction with O atoms in (10) or with NO in (11). The combination of reactions (8) and (10) sums to the equivalent of (5), and constitutes a free radical catalytic chain reaction in which the Cl atom alternates among the chemical species Cl and ClO. The first step removes one O_3 molecule, while the second intercepts an O atom which could have become an O_3 by (3), but is instead also converted to O_2. The Cl atom, however, is only a catalyst, and remains to initiate the process once more.

$$ClO + O \rightarrow Cl + O_2 \tag{10}$$
$$ClO + NO \rightarrow Cl + NO_2 \tag{11}$$

This Cl/ClO cycle can be repeated hundreds or thousands of times, converting back to molecular O_2 in each pair of reactions one ozone molecule and one oxygen atom. When this catalytic efficiency of about 100,000 ozone molecules removed per chlorine atom is coupled with the yearly release to the atmosphere of about one million tons of CFC's, the original question chiefly of scientific interest has now been converted into a very significant global environmental problem–the depletion of stratospheric ozone by the chlorine contained in the chlorofluorocarbons.

The ClO_x chain of reactions (8) and (10) was discovered earlier in 1973 by Stolarski and Cicerone, who were interested in the possible natural release of HCl to the atmosphere from volcanoes, or of man-made chlorine in the exhaust of the rockets scheduled for propulsion of the space shuttle [5]. The ClO_x chain has a close analogy with the NO_x free radical catalytic chain of reactions (12) and (13), which also combine to convert O and O_3 into two molecules of O_2. This chain can be initiated in the stratosphere by the decomposition of the long-lived molecule nitrous oxide, N_2O [6]. Alternatively, the NO_x chain can also be triggered by the direct release of NO and NO_2 in the exhaust of high-flying aircraft [7, 8]. The potential effects on stratospheric ozone of NO_x from supersonic transport aircraft such as the Concorde and the proposed Boeing aircraft were discussed in detail during the period 1971 to 1974 under the Climatic Impact Assessment Program [9, 10].

$$NO + O_3 \rightarrow NO_2 + O_2 \tag{12}$$
$$NO_2 + O \rightarrow NO + O_2 \tag{13}$$

Reactions such as (9), (11) and (14) show that the distribution of chlorine is closely intertwined with the concentrations of other stratospheric species, including the nitrogen oxides, methane, and oxides of hydrogen. The most significant homogeneous gas phase interconnections among the active chain species (Cl, ClO) and the reservoir molecules (HCl, HOCl, $ClONO_2$) in the tropic and temperate zone stratosphere are shown in Figure 4. Because of the large-scale mixing processes which dominate throughout the troposphere and stratosphere, the total mixing ratio of Cl from any species released at the Earth's surface is essentially constant versus altitude. However, the fraction in each of the various chlorine-containing chemical forms varies significantly with altitude (Figure 5).

$$HO + HCl \rightarrow Cl + H_2O \tag{14}$$

TEMPERATE & TROPICAL STRATOSPHERE

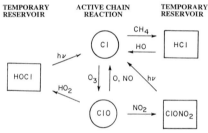

Figure (4) The most important stratospheric chemical transformations among inorganic chlorine species, except during the polar winter. Ozone depletion occurs only while chlorine is in the forms of Cl and ClO.

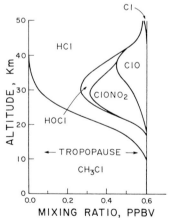

Figure (5) Schematic illustration of the distribution of chlorine among various chemical forms after release to the atmosphere in the form of methyl chloride (CH_3Cl). At steady state, the sum of the mixing ratios at all altitudes is constant, equal to the tropospheric mixing ratios of CH_3Cl alone.

Our original calculations about the behavior of the CFC species in the stratosphere were actually predictions of the vertical distribution because no
measurements were then available for any chlorinated species in the stratosphere, and certainly not for the CFCs. During 1975, two different research
groups sent evacuated containers equipped with pressure-sensitive valves up
on high altitude balloons, and recovered air samples from the stratosphere.
The measured mixing ratios for CCl_3F (Figure 6) were in excellent agreement with the vertical profiles calculated by us in the previous year. This fit
between theory and experiment demonstrates both that CFCs reach the stratosphere and that they are decomposed there by solar ultraviolet radiation at
the altitudes predicted earlier [11,12].

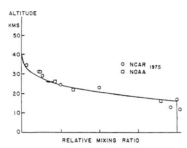

Figure (6) Mixing ratios for CCl_3F observed in 1975 by research groups from NCAR (o)
[12] and NOAA (□) [11] normalized to tropospheric value as 1.0. The solid line is the calculated value with diffusion coefficient C in Figure 3 [4].

GROWTH OF TROPOSPHERIC CFC CONCENTRATIONS

When upward transport of the CFCs into the stratosphere proved to move at
the same rate as for all other molecules, attention returned to the possibility
of tropospheric sinks. The rapid tropospheric removal processes (photolysis,
rainout, oxidation) were readily eliminated. Nevertheless, the calculated 50
to 100 year lifetimes for the various CFCs left room for concern about the
possibility of an accumulation of minor sinks, or even of an undiscovered tropospheric sink. While atmospheric and laboratory tests can be conducted
appropriate to each proposed individual sink, an even more comprehensive
approach is the measurement of the sum for all tropospheric removal processes, including those not yet specifically identified. This can be accom-
plished by measurement of the actual lifetime of the CFCs in the atmosphere itself, and requires accurate knowledge both of the amounts of a
particular CFC released to the atmosphere and of the amount still there.

An electron capture gas chromatographic trace, taken by the same technique originally developed by Lovelock, is shown in Figure 7 for an air sample collected in Tokyo in 1989. Tokyo is not unusual. Similar high concentrations are found in all major cities because most of the technological uses
of CFCs take place in urban areas. While this is an accurate assessment of the
halocarbon impurities found in Tokyo air on that date, it is not very useful

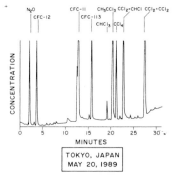

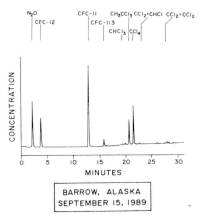

Figure (7) Analysis by electron capture gas chromatography of a Tokyo air sample collec-ted on May 20, 1989. Four of the peaks rise above the printed scale values. The detector has a different response for each compound, requiring calibration for each.

for a global assessment of total CFC concentrations because the measure-ment is not representative of the whole atmosphere.

The air volume associated with cities, and the amounts of CFCs in that air, represent minuscule fractions of the total atmosphere, and of its total CFC content. Instead, what is required are measurements at enough representa-tive remote locations to permit a global assay. A comparable electron cap-ture display from an air sample collected a few miles from Barrow, Alaska, (71°N Latitude) in 1989 is shown in Figure 8. Although the amounts are cer-tainly smaller, the three major CFC molecules are all clearly present near Barrow. Many of the other molecules found in Tokyo (Figure 7) have atmo-spheric lifetimes of one year or less, and are detected only in very low con-centration or not at all in northern Alaska.

Figure (8) Analysis by electron capture gas chromatography of an air sample collected in Barrow, Alaska, on Sept. 15, 1989. The detector has a different response for each com-pound, requiring calibration for each.

The west-east global mixing processes across longitudinal lines are suffi-ciently rapid that good estimates can be obtained for molecules as long-lived as the CFCs from a set of air samples collected within a short time period from remote locations across most of the north-south latitudinal spread. The

CCl₃F concentrations found at such a set of remote locations are illustrated in Figure 9 for the summer of 1979, and in Figure 10 for December 1987. Both of these figures illustrate that north-south mixing is quite rapid in the atmosphere, for the gradient from north-to-south is only about 10%, despite the release of about 95% of CCl₃F in the northern hemisphere, mostly between 30°N and 60°N latitudes. The average time for a molecule to mix from the northern to southern hemisphere, or vice versa, has been calculated from the release patterns in comparison with these measured concentrations and is approximately 15 months. The average CFC molecule therefore mixes back and forth between the northern and southern hemispheres 20 to 40 times or more before its ultimate decomposition in the stratosphere. Only the continued injection of CCl₃F into the northern hemisphere maintained the observed N/S concentration gradient from 1971 into the late 1980s.

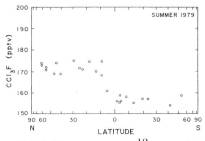

Figure (9) Mixing ratios of CCl₃F in parts per 10¹² versus latitude, as measured at remote sites in 1979.

Comparison of the data in Figures 9 and 10 with the Lovelock data demonstrates that the northern hemisphere concentration of CCl₃F increased from about 70 parts per trillion by volume (pptv = 10⁻¹²) in 1971 to 170 pptv in 1979 and 250 pptv in 1987. This steady increase in concentration with time is precisely what is expected for a molecule with an atmospheric lifetime in the 50 to 100 year range. In 1995, the best estimate for these CFC lifetimes made from a comparison of total atmospheric burden versus total release is 50 years for CCl₃F and 102 years for CCl₂F₂ [13]. Both estimates are well within the ranges of lifetimes calculated 20 years earlier [1,4]. This close agree-

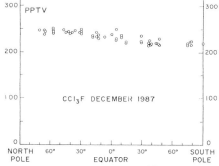

Figure (10) Mixing ratios of CCl₃F in parts per 10¹² versus latitude, as measured at remote sites in December 1987.

ment between measured lifetimes and those calculated on the assumption that no important tropospheric loss processes exist is a clear indication that this assumption is correct. The ultimate fate for the CFC compounds lies in their solar ultraviolet photodissociation in the mid-stratosphere.

A typical measurement with the current version of our analytical system for chlorocarbon molecules is shown in Figure 11 for an air sample collected through an external air intake on a C-130 Hercules aircraft 850 miles south of New Zealand on November 15, 1995. Even in this remote location, sixteen different low molecular weight compounds containing Cl or Br atoms are readily detected (Figure 11), together with others such as O=C=S and N_2O (not shown), the molecule whose atmospheric chemistry was considered by Crutzen 25 years ago [6].

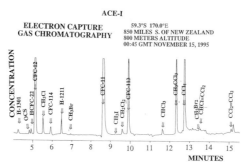

Figure (11) Electron capture gas chromatography of an air sample collected with an aircraft south of New Zealand on November 15, 1995.

SPRINGTIME LOSS OF OZONE IN THE ANTARCTIC

The greatest surprise in the CFC-ozone story was revealed in the spring of 1985, with the discovery by Joe Farman and his colleagues from the British Antarctic Survey of massive springtime losses of ozone over their station at Halley Bay, Antarctica (75.5°S Latitude) [14]. This station was established in preparation for the International Geophysical Year of 1957–1958, and was equipped with a Dobson ultraviolet spectrometer for the measurement of total ozone. The principle of this instrument, designed by Dobson in the 1920's, relies on the measurement at the surface of the Earth of the ratio of received solar radiation for two UV wavelengths. One of these is moderately absorbed by ozone, and one is almost unaffected by ozone, as illustrated in Figure 12. A typical wavelength pair is the C pair which uses 311.45 nm for the former, and 332.4 nm for the latter.

While the absorbing characteristics of most molecules vary continuously with wavelength, the ultraviolet region is often divided for convenience in description into three arbitrary wavelength regions: UV-A, 400 nm to 320 nm, most of which reaches the Earth's surface; UV-B, 320 nm to 290 nm, for which some reaches the surface; UV-C, < 290 nm, of which none reaches the surface. The various wavelength pairs used with a Dobson spectrometer usu-

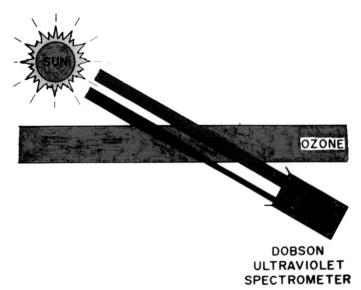

**DOBSON
ULTRAVIOLET
SPECTROMETER**

Figure (12) Standard technique for measurement of ozone with a Dobson ultraviolet spectrometer. The ratio of the two UV wavelengths is dependent upon the ozone content of the atmosphere and the solar zenith angle.

ally compare one UV-B wavelength against one from the UV-A range. Most concerns about UV effects on biology are directed toward the UV-B range because UV-C radiation doesn't get down to the levels where biological species flourish, and UV-A radiation at the surface is so intense that species without some mechanism for dealing with UV-A would long since have become extinct.

In the northern hemisphere, the highest total ozone concentrations are observed in the polar regions around the spring equinox (i.e. March/April), as shown in Figure 13. (The technical unit used to measure total ozone is the

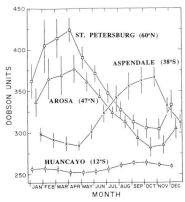

Figure (13) Seasonal variation of ozone concentrations as measured with a Dobson spectrometer for two northern hemispheric stations (St. Petersburg, Russia; Arosa, Switzerland), one tropical station (Huancayo, Peru), and one southern hemispheric station (Aspendale, Australia). The vertical bars indicate the natural variability of ozone measurements for that location and that calendar month. One Dobson Unit is one millimeter-atmosphere-centimeter and is approximately one part in 10^9 by volume.

milliatmosphere centimeter, but this has been almost universally superseded by its other name, the Dobson Unit, D.U. The Dobson unit represents approximately 1 part in 10^9 of the atmospheric molecules by volume, and the average global concentration is about 300 D.U.) The spring maximum values are higher at the higher latitudes, as at St. Petersburg, Russia, versus Arosa, Switzerland, while the highest of all are found in the north polar region. On the other hand, tropical locations have the lowest year-round levels of ozone, with little seasonal variation, as illustrated for Huancayo, Peru.

With no significant prior information about Antarctic meteorology in 1956, Dobson anticipated finding a spring equinoctial maximum over Halley Bay similar to that previously observed at Spitzbergen, Norway. However, instead of increasing steadily through the autumn and winter, as observed in the north, the southern polar ozone values remained essentially constant through the autumn and winter darkness, and into mid-spring. They then increased sharply to a peak in mid-November, as illustrated in Figure 14 for the summers of 1956–57, 1957–58 and 1958–59 [15]. Dobson recognized that this represented the discovery of a very strong Antarctic polar vortex, which prevented until its breakdown in mid-spring, the arrival over the polar regions of ozone-richer stratospheric air from the temperate zone.

This pattern of level ozone values into mid-spring was observed through-

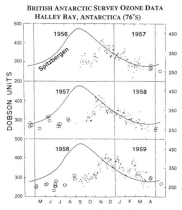

Figure (14) Daily ozone measurements at Halley Bay, Antarctica, during 1956–1959, in comparison with the pattern previously observed in the north polar region at Spitzbergen, Norway [15]. Dots indicate data using the sun as the UV source, and the less accurate circles with the moon as the source. Vertical scale: ozone in Dobson units.

out the 1960's and early 1970's. However, in the late 1970's, the average October ozone values over Halley Bay began to decrease, dropping below 200 D.U. in 1984 versus the 300 to 320 D.U. values of the 1960's (Figure 15). These data were reported in May 1985, together with the hypothesis that the ozone decrease was correlated with the increasing CFC concentrations in the atmosphere [14]. This loss of ozone begins soon after the end of the polar winter darkness, and proceeds very rapidly for the next six or seven weeks into mid-October.

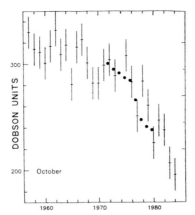

Figure (15) Average ozone concentrations in Dobson units for the month of October, 1957–1984, Halley Bay, Antarctica [14]. The black dots indicate on a relative scale the concentrations of CCl_3F versus time, plotted downward from zero at the top.

These observations of substantially less ozone over Halley Bay in October in the 1980's were quickly shown to be characteristic of the entire south polar region by the measurements of the TOMS (Total Ozone Mapping Spectrometer) on the Nimbus-7 satellite [16]. This instrument also measures ozone from the ratio of two UV wavelengths, utilizing solar UV backscattered from the troposphere through the stratosphere. More than 100,000 daily TOMS ozone measurements over the entire Southern hemisphere are expressed in Figure 16 through a color code for one early October day each in 1979, 1983, 1987 and 1992. In these TOMS displays, the lowest

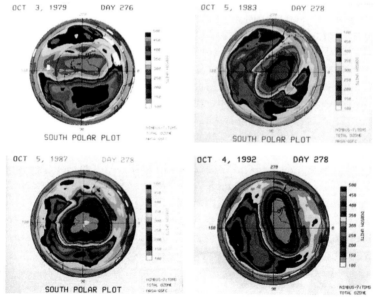

Figure (16) Southern hemispheric total ozone concentrations measured with the TOMS (Total Ozone Mapping Spectrometer) on the Nimbus 7 satellite: (a) October 3, 1979; (b) October 5, 1983; (c) October 5, 1987; (d) October 4, 1992. The color code for the measured numerical values is given to the right.

October ozone values fell rapidly from 250 D.U. to 175 D.U. and then 125 D.U. between 1979 and 1987. The ozone loss is not confined to the central area, as shown in Figure 16, with substantial reductions by 1983 in the areas originally covered by 450–500 D.U. in 1979. A totally different kind of ozone measurement, utilizing its chemical capability for oxidizing iodide ion to elemental iodine, was used on balloon sondes, demonstrating that most of this ozone loss occurred in the lower stratosphere (Figure 17). Again, the loss of ozone has been observed to be very rapid between late August and October.

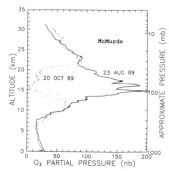

Figure (17) Partial pressures of ozone in nanobars versus altitude, as measured by balloon sondes from McMurdo, Antarctica. Solid line: August 23, 1989, in late winter at the end of the polar night. Dotted line: October 20, 1989, after two months of sunlight.

CHEMISTRY OF THE POLAR STRATOSPHERE

Numerous theories were suggested for this new observation of ozone loss, and can generally be classified into three groups: (a) natural change in the dynamics of the Antarctic stratosphere; (b) change in the natural chemistry, e.g. NO_x, of the stratosphere; or (c) chemical changes induced by mankind, especially through the introduction of artificial chlorine-containing compounds such as the CFCs. A decrease in the springtime stratospheric temperatures was observed to accompany the observations of lessened quantities of ozone, and was briefly considered as a possible cause for its destruction [17]. However, the lowered temperatures were soon established to be occurring after the ozone loss, and to be the consequence of reduced UV absorption by reduced amounts of ozone, not a cause of the ozone loss. The observed temperature decreases are in fact a slower warm-up from the wintertime temperature minimum.

Three polar expeditions in 1986 and 1987 provided the scientific basis for the conclusion that the ozone losses over Antarctica were indeed the consequence of chemical reactions driven by the much higher stratospheric chlorine concentrations of the mid-1980's versus those of 1950–1970. Two of these were ground-based expeditions to McMurdo, Antarctica, led by Susan Solomon. Two maxima were detected in 1986 in the vertical profile of the important free radical, ClO, by ground-based millimeter wave spectroscopy [18]. The larger maximum in ClO concentrations was found in the lower stratosphere, reaching ppbv (10^{-9}) concentrations, and coincident with the altitudes of major ozone loss. These results were then followed in 1987 by in situ measurements of ClO and O_3 within the polar vortex by instruments on the high-flying ER-2 aircraft [19]. These aircraft experiments were carried out from a base in Punta Arenas, Chile (53°S Latitude), and entered the polar vortex over the Antarctic peninsula. The data of Jim Anderson showed clearly (Figure 18) on the first successful flight on August 23 the ppbv levels of ClO over Antarctica. However, the simultaneous ER-2 data for ozone showed little or no change in concentration inside or outside the polar vortex. At this late-winter date, sunlight had only been available in the Antarctic stratosphere for a few days. In contrast, 24 days later on September 16, the ClO levels over Antarctica were again in the ppbv range, but this time approximately two-thirds of the ozone within the polar vortex had already disappeared. The edge of the Antarctic polar vortex was particularly ill-formed on September 16, and the high values of ClO and low values of O_3 tracked each other in perfect anti-coordination (Figure 18).

The chlorine chemistry of the lower Antarctic stratosphere has some substantial differences from that of the upper stratosphere in the temperate and tropical zones. Because the concentration of O atoms is very low in the polar lower stratosphere, reaction (10) is quite slow and the ClO_x chain of (8) + (10) is no longer effective. At these altitudes, the winter temperatures drop to the –80°C range, low enough for the formation there of polar stratospheric clouds (PSCs) which can contain both HNO_3 and H_2O [20–22]. Such low temperatures, and the accompanying clouds, are not generally found in the temperate zone stratosphere. However, the same air remains trapped in the strong Antarctic polar vortex throughout the total winter darkness, and the temperatures drop low enough for cloud formation.

These polar clouds furnish active surfaces on which chemical reactions can occur, including reactions such as (15) and (16) involving reservoir molecules HCl and $ClONO_2$ [23–26]. The products Cl_2 and HOCl are released from the PSC surface back into the gas phase, where they are photo-

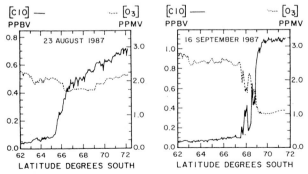

Figure (18) Chlorine oxide (–) and ozone (···) measurements with the ER-2 aircraft on flights from Punta Arenas, Chile, over Antarctica [19]. Left: August 23, 1987; Right: September 16, 1987.

lyzed when the sun returns from its months-long absence. The HNO_3 from such reactions remains in these clouds, tending to denitrify the air mass. When cold enough, sufficient water molecules can also add to the PSC particles to make them fall by gravitation, tending to dehydrate the air mass as well.

$$HCl + ClONO_2 \xrightarrow{PSC} Cl_2 + HNO_3 \tag{15}$$

$$H_2O + ClONO_2 \xrightarrow{PSC} HOCl + HNO_3 \tag{16}$$

An important aspect of the denitrification process is the very low residual concentration of NO and NO_2, with the absence of the latter severely curtailing the formation reaction for chlorine nitrate by (17). Without this removal process for ClO radicals, the ClO concentrations rise to mixing ratios in the ppbv range, and begin to react in significant numbers with other ClO radicals to form Cl_2O_2 by (18) [27]. The chlorine oxide dimer, ClOOCl, can then be destroyed by sunlight in (19), releasing one Cl atom each in (19) and (20). The sum of reactions [8] + [8] + [18] + [19] + [20] sums to [21] by which two O_3 molecules are transformed into three O_2 molecules. This alternate polar ClO_x chain reaction converts ozone into molecular O_2 without the need for the O atom step of reaction (10). Other reactions involving Br atoms from Halon molecules such as $CBrF_3$ and $CBrClF_2$ or methyl bromide (CH_3Br) are exceedingly efficient in removing ozone as well.

$$ClO + NO_2 + M \rightarrow ClONO_2 + M \tag{17}$$

$$ClO + ClO + M \rightarrow ClOOCl + M \tag{18}$$

$$ClOOCl + light \rightarrow Cl + ClOO \tag{19}$$

$$ClOO + M \rightarrow Cl + O_2 + M \tag{20}$$

$$O_3 + O_3 \rightarrow O_2 + O_2 + O_2 \tag{21}$$

OZONE LOSSES IN THE NORTHERN TEMPERATE ZONE

In 1985, the several existing statistical evaluations of the combined data from the Dobson stations in the northern hemisphere appeared to indicate that no significant loss in total ozone had occurred there [28]. The longest series of these ozone measurements without any significant breaks has been carried out since August 1931 in the Swiss alpine community of Arosa. In 1986, Neil Harris and I began a reexamination of the Arosa data series, treating it not as a single comprehensive set, but as a collection of twelve separate data sets, one for each calendar month. Interannual ozone comparisons show in Figure 13 substantial seasonal differences in the variability of the monthly averages. While each year individually shows the normal north temperate zone pattern of maximum ozone values around the spring equinox, and a minimum at the beginning of autumn, the year-to-year variability in average ozone concentration for each month is much greater for the January-to-April period than it is for July-to-October. When this 56-year data series from Arosa was broken into two periods–the 39 year period from 1931–1969, and the 17-year period from 1970–1986–the interesting result emerged that significantly less ozone was measured for several of the autumn–winter months after 1970 than in the previous four decades [29]. Figure 19 shows this Arosa winter-time ozone loss after the inclusion of data from 1987–88.

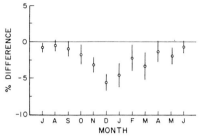

Figure (19) Changes in monthly average total ozone concentrations measured with a Dobson spectrometer at Arosa, Switzerland, for 1970–1988 versus 1931–1969 [Updated from 29]. Negative values represent smaller average concentrations after 1970.

We then extended this inquiry to two U.S. ozone stations, Caribou (Maine) and Bismarck (North Dakota), and found that these two also exhibited wintertime losses when the data for 1965–1975 were compared with 1976–1986 [29]. This comparison of consecutive 11-year periods corresponding minimizes any influence of variations in UV emission corresponding to the 11-year solar cycle on the evaluation of the long term trends. Subsequently, the WMO/NASA Ozone Trends Panel confirmed (Figure 20)

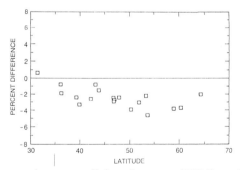

Figure (20) Changes in average "winter" season (DJFM) total ozone concentrations, 1976–1986 versus 1965–1975 for all Dobson stations north of 30° North Latitude with a complete 22 year (two complete solar cycles) record, 1965–1986 [30]. Negative values represent smaller average concentrations after 1975.

that all 18 of the ozone stations between 35°N and 60°N Latitudes with at least 22 years of data (i.e. also covering two complete 11-year solar cycles) had less wintertime ozone in 1976–1986 than in the earlier period [30]. An earlier National Academy of Sciences/National Research Council report had concluded that loss of ozone had already occurred at the altitudes between 35 and 40 km for which the ClOx chain reaction of (8) + (10) was predicted to be most effective [31]. However, the Ozone Trends Panel report was the first assessment document to assert that significant ozone loss had already been observed over the highly-populated areas of North America, Europe, the Soviet Union and Japan. The combination of very large springtime losses of ozone over Antarctica directly attributable to man-made chlorine compounds introduced into the stratosphere, and measurable ozone losses in the north temperate zone, brought a rapid change in the attitudes of CFC-producers and many of the users, and a complete phaseout of CFC production and uses became a widely accepted goal.

THE MONTREAL PROTOCOL FOR REGULATION OF CFCs

Regulations for the control of CFC emissions had first been introduced during the late 1970s for the use of CFC-11 and CFC-12 as propellant gases for aerosol sprays, but only in the United States, Canada, Sweden and Norway. In 1985, a United Nations meeting in Vienna, Austria, had agreed to a convention concerning protection of the stratospheric ozone layer. Then, in September 1987, a further agreement, known as the Montreal Protocol, called for actual limitations on the emissions of CFCs, with a 50% cutback in CFC production scheduled by the end of the century [32]. The terms of the Montreal Protocol called for periodic revisits to the controls, and a meeting in London in 1990 changed the restricted production into an outright phaseout of CFCs by the year 2000. A further meeting in Copenhagen in 1992 accelerated the phaseout to Jan. 1, 1996, for the major industrial countries, and included compounds such as methylchloroform (CH_3CCl_3), carbon tetrachloride (CCl_4), and the bromine-containing halons, i.e. H-1301$(CBrF_3)$, H-1211 $(CBrClF_2)$, H-2402 $(CBrF_2CBrF_2)$.

CHANGES IN ULTRAVIOLET FLUX AT EARTH'S SURFACE

Most of the measurements of total ozone, including the Dobson and TOMS data, actually depend upon the ratio of two wavelengths of solar ultraviolet radiation, as described earlier. A change in the amount of ozone along the path of the UV radiation is inferred from a change in the wavelength quality of the detected radiation. In these cases, the existence of a correlation between ozone and UV-B radiation under clear-sky conditions is a tautology. Less reported ozone means either more UV-B or less UV-A or both, and no mechanisms for significant changes in UV-A radiation have been put forward. However, because actual well-calibrated, precision absolute measurements of ultraviolet radiation over multi-decadal periods of time do not exist, no long-term comparison of trends in ozone concentrations and UV-B radiation has been possible.

In the past decade, instruments capable of such ultraviolet measurements have been deployed in various locations, especially in the south polar region, including three Antarctic stations–South Pole, McMurdo (78°S), and Palmer (64°S)–as well as Ushuaia, Argentina (55°S). Because the

Antarctic polar vortex, and its low ozone concentrations during recent Octobers, can be very elongated (See Figure 16, for Oct. 4, 1992), the possibility now exists for much higher UV-B exposures in the southern high latitudes than in previous times.

Comparison of the measured UV-B intensities for the same solar zenith angle in spring (i.e., depleted ozone) and autumn (relatively normal ozone) allow an experimental confirmation of the anticipated anti-correlation between UV-B and total ozone. The measured correlation in Figure 21 for the

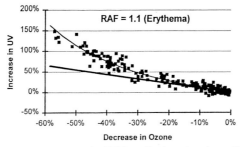

Figure (21) Percentage increases in UV-B radiation (erythemally weighted) versus percentage decreases in total ozone, at the South Pole, 1991–1992 [33].

South pole in 1991–1992 shows a 150% increase in UV-B for a 50% decrease in ozone content [33]. Furthermore, on October 26, 1993, when a particularly low ozone value existed over Palmer in the Antarctic peninsula, the UV-B intensity exceeded by 25% the highest intensity recorded in San Diego, California, for any day of 1993, the most intense there being June 21, 1993 [34]. The remarkable further observation was also recorded that the most intense weekly UV-B exposure in any single week at any of these stations was recorded at the South Pole. There, even though the sun is low on the horizon, the very low concentrations of ozone in the late Antarctic spring, the general absence of clouds, and the 24 hours per day of continuous sunlight combined to permit a higher weekly dose of UV-B than in any week in San Diego, California [34]. The often-heard statement that UV-B intensities in

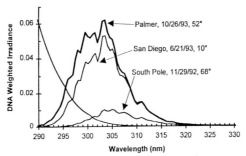

Figure (22) Maximum UV-B spectral intensities measured at three stations during 1992–1993, weighted by the measured absorption spectrum of DNA (shown at the left of the figure) [34]. The maximum intensity occurred at San Diego, California, on June 21, 1993, 10° Solar Zenith Angle; at Palmer in the Antarctic peninsula (64°S) on October 26, 1993, 52° SZA; and at the South Pole on November 29, 1992, 68° SZA. The relative integrated intensities are South Pole, 0.11; San Diego, 0.59; Palmer 0.74.

the Antarctic can never be very large is no longer true. The question of possible biological damage associated with UV-B radiation then requires the separate assessment for each biological species of whether such damage is the result of cumulative UV-B exposure over an entire season or of a single, extremely intense exposure; both circumstances exist elsewhere.

CHANGING O_3 AND CFC CONCENTRATIONS IN THE 1990S

Ozone concentrations have continued to fall during the early 1990's. The measured value at the South Pole was only 91 D.U. on Oct. 12, 1993, and the vertical profile data from the corresponding balloon sonde indicated that all ozone had been removed from a 5 kilometer altitude band between 14 and 19 kilometers. The TOMS satellite registered many new daily record low global ozone values in early 1992, and then for nearly 13 months from mid-April, 1992, a new daily low global ozone value was measured (Figure 23). This measurement series ended when the TOMS instrument finally failed on May 7, 1993. The ozone data interpretations were complicated by the eruption of Mount Pinatubo in the Philippines in June 1991. This volcano, while not providing any significant amount of hydrogen chloride to the stratosphere, did introduce particulate surfaces on which could occur reactions

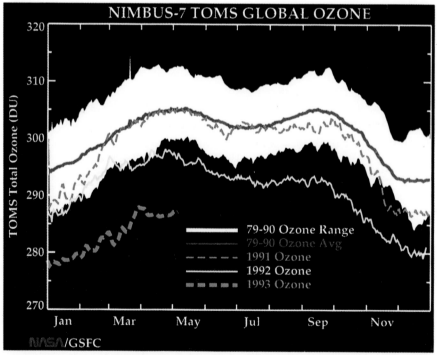

Figure (23) Global Ozone values as measured with the Total Ozone Mapping Spectrometer on the Nimbus 7 Satellite. The range of values recorded on each day from 1979–1990 is outlined in white, with the average for these years in red. The 1991, 1992, and 1993 values are shown in green, yellow and blue, respectively. The TOMS instrument failed on May 7, 1993.

similar to those taking place on the PSCs. The northern hemisphere Dobson stations continued to report declining ozone concentrations, with the losses now spreading into the summer. By 1995, the north temperate zone total ozone losses approached 10% during winter and spring and 5% in the summer and autumn [13].

In contrast, the measured tropospheric concentrations of the various halocarbon gases deviated in the 1990's from their earlier pattern of steady increases in response to the restrictions of the Montreal Protocol. The concentrations of the three major CFCs, which had risen almost linearly during the 1980's, markedly slowed this pace. By 1995, the concentrations of CCl_3F measured regularly at seven different stations varying in latitude from the polar Arctic to the South Pole had essentially leveled off, as shown in Figure 24. The CCl_2F_2 and CCl_2FCClF_2 concentrations were still rising, but at no more than half the rate of the 1980's. The sharpest changes were observed with the organic solvent methylchloroform, CH_3CCl_3, for which controls were added in revisions of the original Montreal Protocol. Methylchloroform has three hydrogen atoms which are susceptible to attack by HO radicals in the troposphere, in analogy with reaction (1), and therefore have an average lifetime of only about 5 years. With its rapidly diminishing emission rate under the limitations of the Protocol, the tropospheric concentrations of CH_3CCl_3 reached a maximum in 1992, and have declined significantly since (Figure 24).

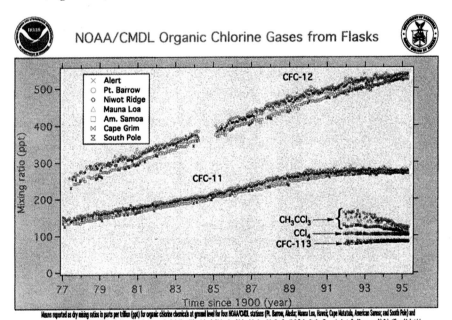

Figure (24) Measured Mixing Ratios in pptv, 1977–1995, for CCl_3F (CFC-11), CCl_2F_2 (CFC-12), CCl_2FCClF_2 (CFC-113), CH_3CCl_3 and CCl_4.

The situation at the end of 1995 is therefore a mixed one. The three most important CFC molecules have atmospheric lifetimes measured from many decades to a century or more, with the consequence that there will be significant quantities of them present in the atmosphere throughout the 21st century. The major springtime losses in the Antarctic will probably continue at least until mid-century. However, because the primary cause of these ozone losses are the man-made chlorine and bromine compounds limited by the Montreal Protocol, the amount of organochlorine in the troposphere will peak soon, if it hasn't already. When the weighted effects of bromine compounds are added in, the peak in tropospheric halocarbons will occur within this decade. Because of the delay time in rising upward, the stratospheric peak will follow within a few years. For the next few decades, the atmosphere will be loaded with chlorine and bromine compounds, so that a really large volcanic explosion with major increases in the particulate surface area of the stratosphere might cause transitory large ozone decreases. Another area for concern is the changing temperature structure of the stratosphere because of the steady increases in the greenhouse gases, especially carbon dioxide. The effect of more carbon dioxide is to decrease temperatures in the lower stratosphere, and this could lead to more extensive cloud formation, and more particulate surface area as well. However, with the limitations just expressed, the threat of extensive further stratospheric ozone depletion during the 21st century appears to be under control.

REFERENCES

1. M. J. Molina and F. S. Rowland, Stratospheric sink for chlorofluoromethanes: Chlorine atom-catalysed destruction of ozone, Nature, *249*, 810–812 (1974).
2. J. E. Lovelock, R. J. Maggs and R. J. Wade, Halogenated hydrocarbons in and over the Atlantic, Nature, *241*, 194–196 (1973).
3. S. Chapman, A theory of upper atmospheric ozone, Mem. Royal. Meteorol. Soc., *3*, 103–125 (1930)
4. F. S. Rowland and M. J. Molina, Chlorofluoromethanes in the environment, Rev. Geophys. Space Phys., *13*, 1–35 (1975).
5. R. S. Stolarski and R. J. Cicerone, Stratospheric chlorine: a possible sink for ozone, Can. J. Chem., *52*, 1610–1615 (1974).
6. P. J. Crutzen, The influence of nitrogen oxides on the atmospheric ozone content, Royal Meteorol. Soc. Quart. J., *96*, 320–325 (1970).
7. H. S. Johnston, Reduction of stratospheric ozone by nitrogen oxide catalysts from supersonic transport exhaust, Science, *173*, 517–522 (1971).
8. P. J. Crutzen, Ozone production rates in an oxygen-hydrogen-nitrogen oxide atmosphere, J. Geophys. Res., *76*, 7311–7327 (1971).
9. "Environmental Impact of Stratospheric Flight, Biological and Climatic Effects of Aircraft Emissions in the Stratosphere", Climatic Impact Committee, National Academy of Sciences, Washington, D. C., 1975.
10. "The Natural Stratosphere of 1974", CIAP Monograph 1, "A. J. Grobecker, editor. Final" Report, U. S. Dept. of Transportation, DOT-TST-75-51, Washington, D. C., 1975.
11. A. L. Schmeltekopf, P. D. Goldan, W. R. Henderson, W. J. Harrop, T. L. Thompson, F. C. Fehsenfeld, H. I. Schiff, P. J. Crutzen, I. S. A. Isaksen, and E. E. Ferguson, Measurements of stratospheric $CFCl_3$, CF_2Cl_2 and N_2O, Geophys. Res. Lett., *2*, 393–396 (1975).
12. L. E. Heidt, R. Lueb, W. Pollock and D. H. Ehhalt, Stratospheric profiles of CCl_3F

and CCl_2F_2, Geophys. Res. Lett., *2*, 445–447 (1975)

13. World Meteorological Organization, Global Ozone Research and Monitoring Project, Report No. 37, Scientific Assessment of Ozone Depletion: 1994, WMO/NASA, (1995).

14. J. C. Farman, B. G. Gardiner and J. D. Shanklin, Large losses of ozone in Antarctica reveal seasonal ClO_x/NO_x interaction, Nature, *315*, 207–210 (1985).

15. G. M. B. Dobson, Forty years' research on atmospheric ozone at Oxford: A history, Applied Optics, *7*, 401 (1968).

16. R. S. Stolarski, A. J. Krueger, M. R. Schoeberl, R. D. McPeters, P. A. Newman, and J. C. Alpert, "Nimbus 7 SBUV/TOMS measurements of the springtime Antarctic ozone decrease, Nature, *322*, 808–811 (1986)

17. (a) S. Chubachi and R. Kajawara, Total ozone variations at Syowa, Antarctica, Geophys. Res. Lett., *13*, 1197–1198 (1986); (b) Y. Sekiguchi, Antarctic ozone change correlated to the stratospheric temperature field, Geophys. Res. Lett., *13*, 1202–1205 (1986); (c) P. Newman and M. Schoeberl, October Antarctic temperature and total ozone trends from 1979–1985, Geophys. Res. Lett.,*13*, 1206–1209 (1986); (d) J. K. Angell, The close relationship between Antarctic total ozone depletion and cooling of the Antarctic lower stratosphere, Geophys. Res. Lett., *13*, 1240–1243 (1986).

18. R. L. deZafra, M. Jaramillo, A. Parrish, P. M. Solomon, B. Connor, and J. Barrett, High concentrations of chlorine monoxide at low altitudes in the Antarctic spring stratosphere, I, Diurnal variation, Nature, *328*, 408–411 (1987)

19. J. G. Anderson, W. H. Brune and M. H. Proffitt, Ozone destruction by chlorine radicals within the Antarctic vortex: the spatial and temporal evolution of ClO-O_3 anti-correlation based on in situ ER-2 data, J. Geophys. Res., *94*, 11465–11479 (1989).

20. M. P. McCormick, H. M. Steele, P. Hamill, W. P. Chu and T. J. Swissler, Polar stratospheric cloud sightings by SAM II, J. Atmos. Sci., *3*, 1387–1397 (1982).

21. P. Crutzen and F. Arnold, Nitric acid cloud formation in the cold Antarctic stratosphere: A major cause for the springtime "ozone hole", Nature, *324*, 651–655 (1986).

22. O. B. Toon, P. Hamill, R. P. Turco and J. Pinto, Condensation of HNO_3 and HCl in the winter polar stratosphere, Geophys. Res. Lett., *13*, 1284–1287 (1986).

23. (a) H. Sato and F. S. Rowland, Paper presented at the International Meeting on Current Issues in our Understanding of the Stratosphere and the Future of the Ozone Layer, Feldafing, West Germany, (1984); (b) F. S. Rowland, H. Sato, H. Khwaja and S. M. Elliott, The hydrolysis of chlorine nitrate and its possible atmospheric significance, J. Phys. Chem., *90*, 1085–1088 (1986).

24. S. Solomon, R. R. Garcia, F. S. Rowland and D. J. Wuebbles, On the depletion of Antarctic ozone, Nature, *321*, 755–758 (1986).

25. M. J. Molina, T.-L. Tso, L. T. Molina and F. C.-Y. Wang, Antarctic stratospheric chemistry of chlorine nitrate, hydrogen chloride and ice. Release of active chlorine, Science, *238*, 1253–1257 (1987).

26. M. A. Tolbert, M. J. Rossi, R. Malhotra and D. M. Golden, Reaction of chlorine nitrate with hydrogen chloride and water at Antarctic stratospheric temperatures, Science, *238*, 1258–1260 (1987).

27. L. T. Molina and M. J. Molina, Production of Cl_2O_2 from the self-reaction of the ClO radical, J. Phys. Chem., *91*, 433–436 (1987)

28. "Atmospheric Ozone 1985", WMO Report No. 16, 3 volumes, 1986.

29. N. R. P. Harris and F. S. Rowland, Trends in total ozone at Arosa, EOS, *67*, 875 (1986).

30. (a) Chapter Four, "Trends in Total Column Ozone Measurements", F. S. Rowland, Chair, in "Report of the International Ozone Trends Panel", WMO Report No. 18, Volume 1, 1990; (b) F. S. Rowland, N. R. P. Harris, R. D. Bojkov, and P. Bloomfield, "Statistical Error Analyses of Ozone Trends–Winter Depletion in the Northern Hemisphere", pp. 71–75, Ozone in the Atmosphere, R. D. Bojkov and P. Fabian, editors, Deepak Publishing, Hampton, Virginia, 1989.

31. "Causes and Effects of Stratospheric Ozone Reduction: An Update," Committee on

the Chemistry and Physics of Ozone Depletion, National Research Council, National Academy Press, Washington, D. C., 1982.

32. R. E. Benedick, "Ozone Diplomacy", Harvard University Press, Cambridge, Massachusetts, 1991.

33. C. R. Booth and S. Madronich, "Radiation amplification factors: improved formula accounts for large increases in ultraviolet radiation associated with Antarctic ozone depletion", Antarctic Research Series, *62*, 39–42 (1994), C. S. Weiler and P. A. Penhale, editors, American Geophysical Union, Washington, D. C., 1994.

34. C. R. Booth, T. B. Lucas, T. Mestechkina, and J. Tusson, Antarctic Journal of U.S., *39*, 256–259 (1994).